Data Communications Today
Networks, The Internet, and The Enterprise

Second Edition

Stan Gelber

To join a Prentice Hall Internet mailing list, point to:
http://www.prenhall.com/mail_lists/

Prentice Hall PTR
Upper Saddle River, New Jersey 07458
http://www.prenhall.com

Acquisitions editor: Mike Meehan
Cover designer: Kiwi Design
Cover design director: Jayne Conte
Manufacturing manager: Alexis R. Heydt
Marketing manager: Stephen Solomon
Compositor/Production services: Pine Tree Composition, Inc.

Prentice-Hall, Inc.
A Simon & Schuster Company
Upper Saddle River, New Jersey 07458

Printed in the United States of America
10 9 8 7 6 5 4 3 2 1

ISBN: 0-13-755950-X

Prentice-Hall International (UK) Limited, *London*
Prentice-Hall of Australia Pty. Limited, *Sydney*
Prentice-Hall Canada Inc., *Toronto*
Prentice-Hall Hispanoamericana, S.A., *Mexico*
Prentice-Hall of India Private Limited, *New Delhi*
Prentice-Hall of Japan, Inc., *Tokyo*
Simon & Schuster Asia Pte. Ltd., *Singapore*
Editora Prentice-Hall do Brasil, Ltda., *Rio de Janeiro*

Contents

APPENDICES

Preface

As an information systems professional, I spend a good part of my life in front of a computer working in a communications-oriented environment. I originally thought that writing a book on data communications would be logical and relatively straight forward. I was surprised at how difficult it was to take complex concepts and make them simple and interesting enough to keep one's attention. At the same time, I did not want to overwhelm you with raw data but rather to provide only the necessary information.

I also discovered that writing a book, as opposed to teaching a class, presented a different set of challenges. Virtually every week something new is being announced making it quite difficult to produce a book that is timely and up-to-date. This is primarily because of the rapid development and evolution of hardware, software and services designed for the data communications environment. Developing a training course is somewhat easier than writing a book, because the material is constantly updated and is primarily used as a supplement to lectures and demonstrations. To address this problem, I decided to concentrate on the basics, and on products and services that were associated with established standards.

The first edition of this book was the product of a course of data communications that I developed several years ago for use by training departments at organizations such as the Rolm Corporation, Xerox International, MCI, GTE, AT&T, Mitel and Bank of America. The course was designed to provide information to personnel requiring a basic understanding of data communications in order to support their corporate missions. Since that initial effort, much has changed in the world of data communications. The changes are primarily due to the millions of people who use the Internet and online services, and their wish to have a better understanding of how it all works.

In the first edition of this book I stated that I had reviewed almost every book I could get my hands on regarding the subject of data communications. Then as well as now, I continue to look for introductory books that I can recommend to my students for supplemental reading which leads me to the intended audience for this book.

Data Communications Today is intended for those who desire a basic but comprehensive understanding of the primary components which make up data communications networks as well as those who wish to build a foundation for understanding the more technical aspects of the subject. This book is divided into six parts, and is intended to present an overview of the basics of data communications rather than the technical complexities.

I wish to note that a case study has been included at the end of each chapter and is designed to provide an example of the construction and implementation of a data communications network as a point of reference. Although it appears that I have focused on the IBM world which will raise questions from informed readers who may think I have not addressed the world outside of IBM, I wish to note the basic concepts apply to all aspects of data communications, regardless of manufacturer.

While in some cases several non-IBM manufacturers do in fact use different methodologies and equipment for performing data communications functions, basic data communications concepts do not vary between manufacturers. The information contained herein is just as applicable to the student attempting to learn data communications, as well as to the PC user wanting to implement a communications capability.

SCOPE OF THE BOOK

Part I covers the foundation for data communications, and the evolution of communications into data communications.

Part II explains the role of systems programs, communications programs and applications programs in data communications. It describes how the various pieces of software interact to provide communications to the end user.

Part III presents information on how data is converted into analog and digital transmission signals and then transported via various types of communications facilities to its destinations.

Part IV introduces the various types of communications hardware, transmission media and hardware and logical interfaces used in a data communications network.

Part V begins a discourse on the standards, methods and procedures for implementing a data communications network. It presents important standards, architectural approaches and the various protocols in use.

Part VI introduces local area networking and covers the ins and outs of managing a network, and diagnosis and repair of a problematic network. The book concludes with a forecast of future trends in data communications.

Acknowledgments

Once again I would like to thank Jean Pierog who has taken pity on me and believes that she has saved me from myself by becoming my wife. Without her constant encouragement, her editing skills, and her perseverance in keeping me in good health despite the many bad habits I refuse to give up, another book has been produced and my plans to be a beach bum have been put on hold.

I would also like to thank the many business associates and friends who had faith in me by supporting my efforts to develop this book regardless of the fact that I still do not consider myself a writer. Without their constant feedback as well as their financial support for the courses, training programs, and seminars that provide a livelihood for me, I would not have had the time or energy to complete this undertaking.

Finally, I would like to acknowledge the wisdom of my mentor Dick Cowden who took on the difficult role of tutoring me in many aspects of basic electronic theory and practical application. Dick is truly one of the real pioneers of data processing and computer hardware who along with other notables such as Al Shugart and David Wheeler, introduced the world to disk drive and mass storage devices.

PART I

INTRODUCTION, OVERVIEW, AND A HISTORY OF DATA COMMUNICATIONS

CHAPTER 1

Introduction

A BRIEF HISTORY OF COMPUTERS AND COMMUNICATIONS

Simply put, data communications is the movement of information from one point to another across a communications network. The function of data communications is to transfer data from location A to location N efficiently, reliably and as inexpensively as possible. This function requires software, hardware, transmission links, and services.

The evolution of data communications technology is rapid and dynamic. Likewise, the ongoing development of the digital computer has enabled us to progress technologically at an accelerated rate. New discoveries, new technologies and the refinement of existing technologies create an incredibly exciting environment for rapid growth. It is sometimes easy to forget that 50 years ago, practical use of computers and data communications did not exist.

In the 18th century, a form of mechanical data processing existed for a number of activities, including cloth weaving and census taking. Today, while we view information processing as a computerized function using program logic, all of the necessary mechanical computer logic functions were present in those early hardware devices, except for the ability to communicate the data from one device to another.

We use the term "hardware" to describe the equipment required to process raw data into information or transfer the data across a communications link. (Figure 1.1 illustrates typical data processing hardware.) The term "software" refers to the processes and various logical instructions required to direct the hardware in specific tasks. (Figure 1.2 illustrates an example of operating systems, communications, and support software.)

To better understand the development of data communications hardware and software, it is important to emphasize that the evolution of the computer, like the evolution of data communications, was not the result of work by a single person or organization. Many teams of scientists working in many locations explored different, or in some cases the

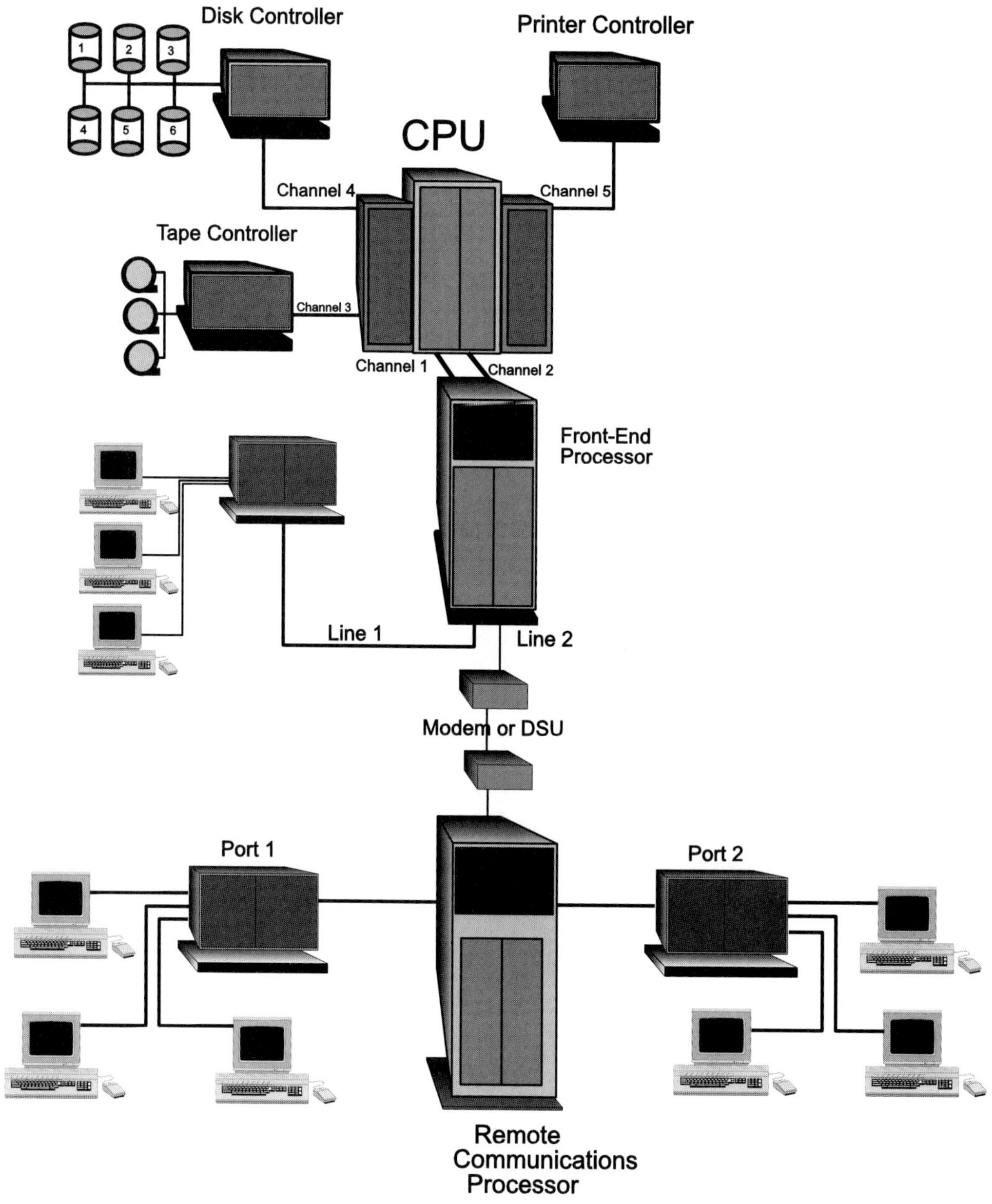

Figure 1.1 Example of a Remote and Local Communications Environment

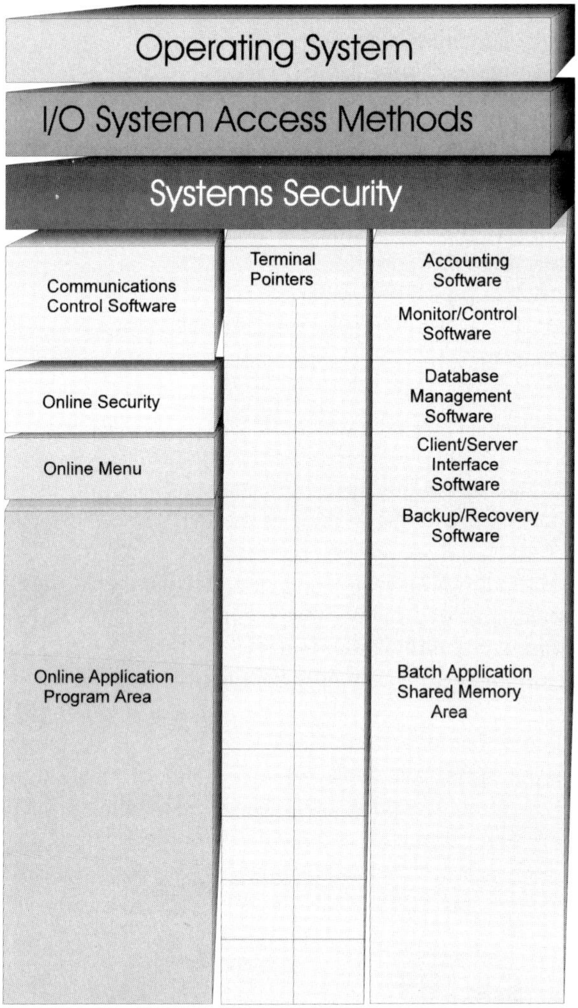

Figure 1.2 Illustration of Mapped Memory Showing Communications Tasks

same, ideas. Unfortunately, the development of the computer was not based on the concept that computers would some day be required to communicate with one another. This lack of foresight created a situation where computers and communications evolved in separate directions that were, for the most part, incompatible.

There is some question as to exactly when the first electronic computer made its appearance. Some attribute it to a German engineer named Konrad Zuis, who demonstrated a machine in the late 1930s as a product of the German war effort. Fortunately for the rest of the world, Hitler deemed it more important to build fighter planes than computers and the project ended before it could be employed by Hitler's war machine. Today, most people believe that the start of modern day computing began February 14, 1946 with the development of the ENIAC computer at the University of Pennsylvania. This development team was led by J. Presper Eckert and John Mauchley. The ENIAC was the first large scale electronic computer and had the capability of performing 357 multiplications per second.

In 1949, the first computer to use all electronic diode-based logic and capable of executing a stored program was developed by the U.S. National Bureau of Standards (NBS). A number of organizations continued development, and the Universal Automatic Computer (Univac 1) was completed in 1951 by the Remington Rand Corporation. Installed at the U.S. Census Bureau, this machine was considered the first commercially produced American computer.

By 1950, Remington Rand and the IBM Corporation were the two major companies seriously developing commercial computers. One of the most important early developments for data communications was IBM's introduction of the magnetic tape storage device in 1953. It represented a major advance in computer processing because it could store 100 characters on an inch of tape and move at 75 inches per second. This capability considerably increased the speed of processing data into information using the digital computer. This was driven in part by the growing need in the late 1950s, to move data from one location to another.

By the late 1950s, the need to move data from one location to another was growing. Fireman's Fund Insurance Company, among many other companies, used a variation of the IBM tape drive in combination with a Bell dataset (a device that connected to both the telephone network and the tape drive) to send computer data between the East and West coasts of the United States.

With the advent of this new technology, computer data was converted into a form suitable for transmission across the public telephone network. This information was received by a tape drive/dataset at the remote location using the same signaling principals as those employed in the telegraph system. While these early data communications activities were slow and analog in form, they created tremendous opportunities for business to expand outward to distant locations.

ANALOG AND DIGITAL COMMUNICATIONS

Communications can be classified into two distinct categories, *digital communications* and *analog communications*. Digital communications are generated by a microprocessor-controlled device or computer and consist of signals containing two different electrical voltages. Analog communications—such as voice, television, and fax—use continuous variable electrical signals to transport the information. (Figure 1.3 illustrates an analog and digital waveform as seen on an oscilloscope.) With the introduction of new digital technologies as well as a worldwide Integrated Services Digital Network (ISDN), these two classifications are becoming less distinct.

Although we still use nondigital transmission methods extensively, indications are that in the not-too-distant future, all forms of electronic communications will be accomplished via the digital route. This is evidenced by the rapid introduction of digital tele-

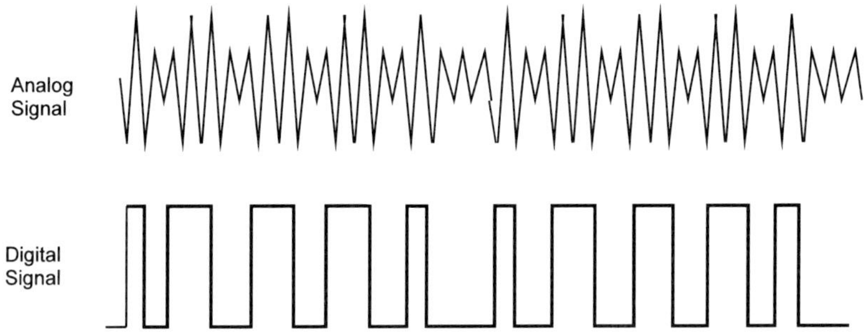

Figure 1.3 Examples of a Digital and an Analog Signal as Shown on an Oscilloscope

phone and digital television. In these cases, analog real-time communications sources are converted via digital (data) processing into a real-time digital signal (see Figure 1.4). This signal is then transmitted across a communications link and reconverted into an analog signal, once again, at the receiving side.

The benefits of digital communications, whether it is data generated by a computer, digitized voice or video, include accurate reproduction of the source material, elimination of noise that creates distortion and rapid transmission of the signal. Figure 1.5 shows an example of wide area network nodes connected together using digital links.

THE APPLICATION OF DATA COMMUNICATIONS

Data communications has become a vital function in today's world. A typical application of data communications can be demonstrated by illustrating a corporation with two distal computer centers (see Figure 1.6). These facilities are located 100 miles apart. One center functions as the corporate computer center and the other is a support computer center, located at a manufacturing facility.

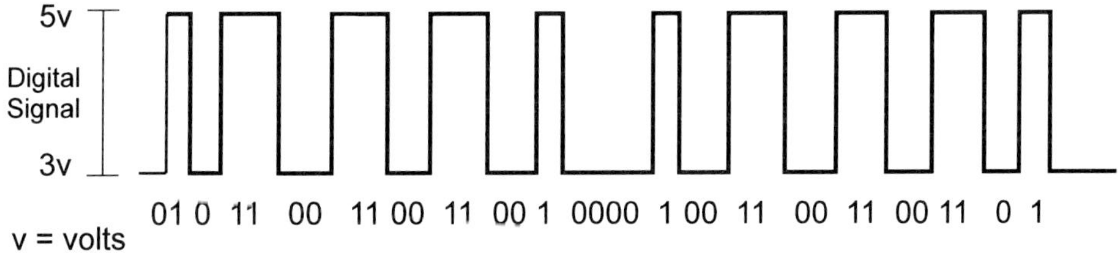

Figure 1.4 Digital Signal as Shown on an Oscilloscope

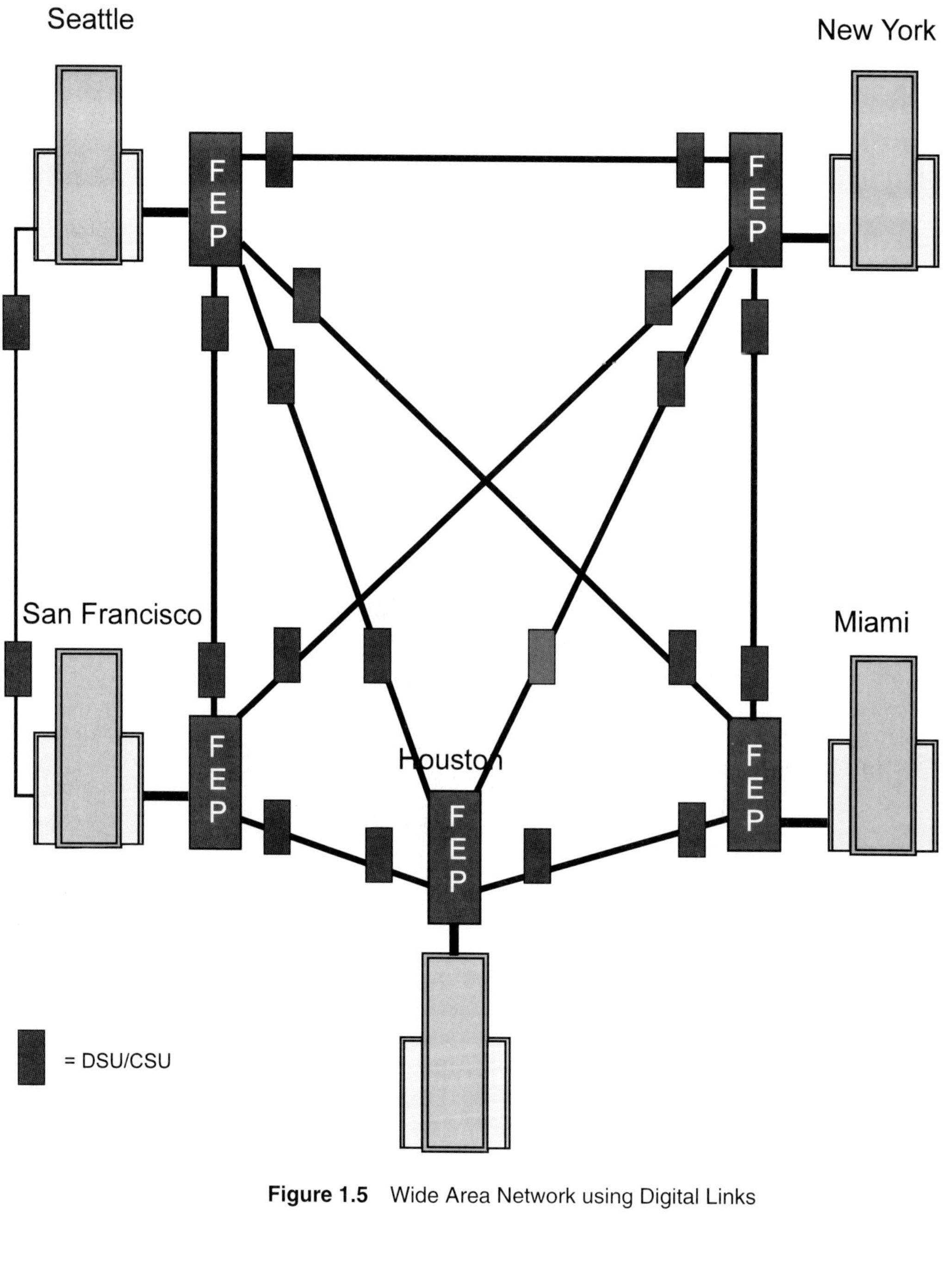

Figure 1.5 Wide Area Network using Digital Links

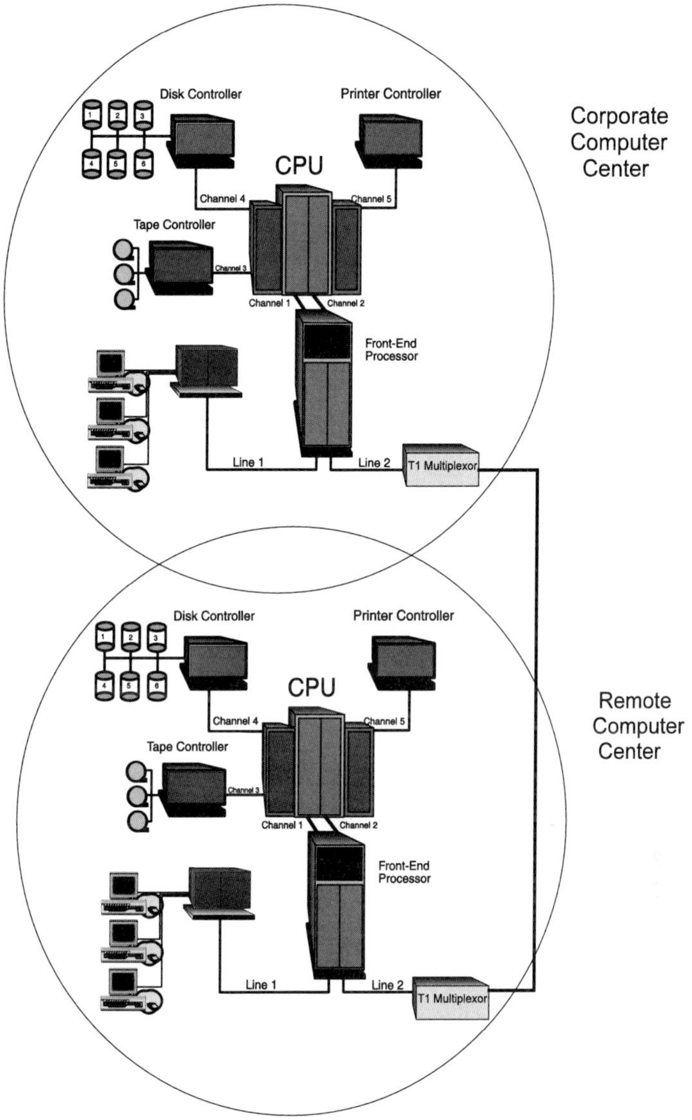

Figure 1.6 Distant Computer Centers Linked Together

At the corporate computer center, user terminals are directly connected (wired) to the computer and are used for a variety of activities that include data entry, sales and manufacturing inquiry and program development. Because terminals are directly connected by wire to the computer, they are limited to a distance of 5000 feet, but they have very fast access times. These local or directly attached terminals allow user departments to input their own data directly into the computer system. The computer supports remote access (the ability to talk to another computer or terminal via telephone links) and may be used for backup and communications activities, except that these functions have not yet been implemented.

Input data and update information are assembled manually and entered for processing, via an online terminal, onto tape or disk by each department's data entry section. Depending on the processing requirements, the computer will either store the data on disk or tape for future reference and processing, or process it immediately.

The remote computer center is configured similarly to the corporate center configuration; it also has user terminals that are attached via direct cabling throughout the factory. These terminals are used by manufacturing and factory support personnel to assemble, track, manage and ship products.

After analyzing the corporate computer environment and the organization's needs, the company determines its requirement for the electronic movement of data and information between the two data centers. In the past, this has been accomplished using courier services. However, the volume of data that must be shared and the necessity for each center to have it immediately available for user access mandates that a data communications network be implemented.

NETWORK IMPLEMENTATION

To establish a data communications network between two or more computers or digital devices, you must implement several interrelated elements. Once you have justified the need for a connection between the two locations, you will need to establish a communications link between them. This data link will allow the two computer centers to share resources and data. Typically, the connection is accomplished using long-established telephone company or other carrier facilities.

The communications link not only provides data and voice communications between the two computer center locations, it also provides backup facilities for each of the computer centers should a disaster strike one of them. The implementation of this link alone does not accomplish the data communications task. The link itself is nothing more than the medium connecting the computers together.

Additional hardware may be required at both ends of the link to support the data communications activities. In some cases, computer manufacturers include communications facilities as part of their hardware and software. However, customers usually need to purchase the required hardware and software at extra cost. The required hardware may include the following components.

Communications Processors. Depending on the number of terminals requiring access to the processing power of the host computer and the types of applications required by users, a communications processor may be mandatory. The communications processor is designed to provide software and hardware management of the terminal and communications network for the host computer. Figure 1.7 shows a typical communications processor configuration.

Multiplexers and Concentrators. Multiplexers and concentrators are designed to take advantage of the capabilities of the media by allowing multiple devices to share the available capacity. Depending on the number of remotely located terminals, a multiplexer may provide cost benefits when used.

Modems, Digital Service Units, and Channel Service Units. Modems, digital service units (DSU), and channel service units (CSU) convert digital signals used by the computer into signals that can be transported across communications links such as telephone lines. They are required for data communications between host computers and terminals in a remote environment.

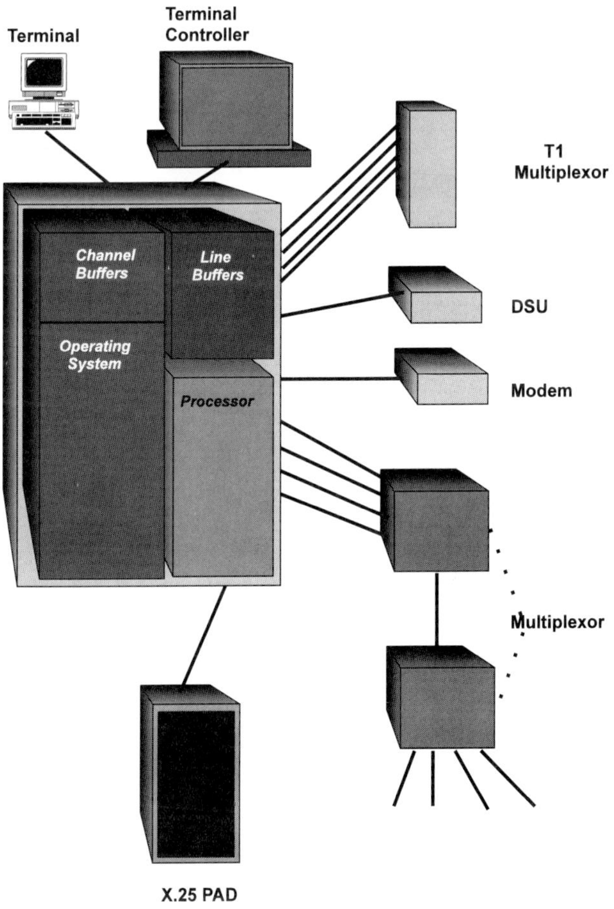

Figure 1.7 Link Connectivity

Network Control Monitors. Network control monitors may be required in large networks to monitor and advise online, terminal and link conditions across the network. After the required hardware is in place, software for both application processing and communications processing can be identified for specific hardware platforms. Many hardware manufacturers provide the necessary communications components in the basic computer hardware. Some also will provide the software, but this is the exception rather than the rule. As an example, IBM provides extensive communications capabilities, but they all cost extra and in many cases can only be leased rather than purchased. Figure 1.8 illustrates layered communications software resident in computer system memory.

Types of software required to support data communications functions are described in the following sections.

Communications Systems Software. This type of software may be required and is usually embedded in the operating system or becomes a systems task under control

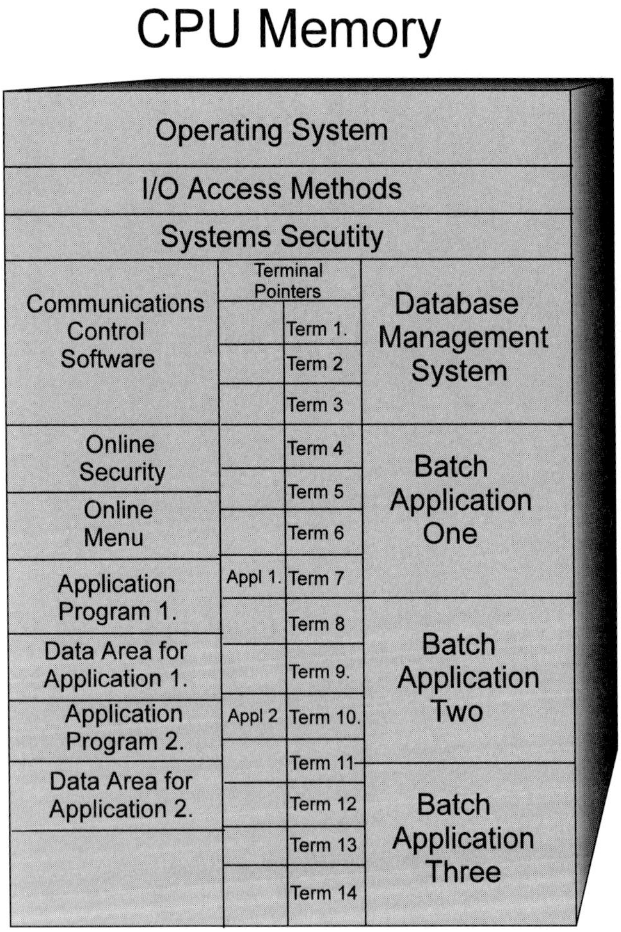

Figure 1.8 CPU Memory Map

of the operating system. Its role is to assist the operating system in managing local and remote terminal and network access.

Communications Control Software. Communications control software manages the online application programs, resolves logical transmission errors and ensures that network access conflicts are resolved in an equitable manner. This type of software assists the application program in providing terminal processing.

Online Application Software. Online applications are required to perform the necessary data processing as well as provide the logical interaction between the terminal user and the data. A typical online application such as loan administration provides access, processing and display formatting of loan information.

Network Control Software. Network control software may be required for managing network and internetwork access and communications. This type of software is required to ensure compatibility and management of communications activities between devices within the network as well as outside the network.

SHARING DATA USING DATA COMMUNICATIONS RESOURCES

The rationale for implementing computer communications is the ability to access vital information immediately at both ends of the link for decision-making activities. From a technological perspective, the ability to communicate information provides additional storage and access availability because important data can be distributed and quickly accessed from remote points. Data communications also leads to significant cost and time reductions associated with shipping reports and data from one location to another. Finally, productivity is markedly enhanced.

The ability to collect and disseminate information across long distances and to provide users with immediate access to data opens up a world of opportunities. The sharing of data and resources across vast distances has allowed mankind to operate in outer space, map gene structures, access airline schedules, monitor electrical power generation and perform countless other activities.

The ability to distribute data across regional areas and make it instantly available allows timely decisions based on accurate information. Technological advances, however, are not without cost and short-term disadvantages. Before implementing a data communications network, be sure that long-term benefits will be realized and expenses can be controlled.

ANALYZING REQUIREMENTS

The first step in selecting and implementing a data communications network is to analyze the company's organizational and information environments. Careful analysis should determine needs, costs, and resources required by the corporation. The following considerations are vital to the selection of an effective, efficient data communications system.

Data Traffic Requirements. An analysis of data traffic is required to justify the costs of implementing a data communications network.

Evaluation of Short- and Long-Term Goals. The team that is responsible for creating the data communications network should evaluate the organization's short- and long-term growth objectives as well as its financial status.

Cost Justification for Network Selection. Consider whether the expenses for the research necessary to determine the most appropriate data communications system are justified. Many covert expenses can drive up the cost of installing such a system. The major expenses involved in the selection, implementation and maintenance of a communications network are planning and ongoing costs for support and line charges.

Selection of Data Communications Equipment. Thousands of companies provide data communications equipment for the end user. The company planners must exercise care in the evaluation and selection of this equipment. Questions any planner should consider include:

- What are the life cycles of the equipment needed?
- Will requirements be met as they relate to capacity and reliability?
- Is the desired product (or products) available, or will it be necessary to accept additional capabilities that are not needed?
- Will the equipment be upgradable or will we have to purchase new equipment as the company's needs expand?
- What are the odds that the vendor will continue to be in business and provide support for the purchased equipment?
- How reliable is the equipment itself, and how reliable and efficient is the vendor in terms of repair turnaround time?

Technical Resources Needed to Support the Network. This is an area frequently overlooked by planners. As a planner, key questions you should ask include:

- Does the vendor have the technical expertise to support the installed equipment?
- What is the availability of expert support?
- Will the vendor supply or will we be able to hire technicians for on-site support of the vendor's products (because many products are complex and skilled technicians may be in short supply)?
- Does the vendor provide educational support for its products, and what are the costs for these services?

Selection of Application/System Software. A tremendous number of diverse online application programs for the user community exists, so how do you determine what is best for the company? Consider these questions:

- What have other companies experienced with a particular brand of program?
- Who supports the code, and what happens if the vendor goes out of business?
- Does the organization create its own applications, and will you be able to find programmers with the expertise to write and maintain the program code?
- What are the costs associated with purchasing the required code as opposed to developing the programs in-house?

Systems software can be difficult to install and maintain, and it can require specialized knowledge to keep in an operational state. Will the vendor provide technical support and education? Will it be responsible for providing corrections when problems occur? Many computer companies lease their software rather than sell it because they can better maintain it and it will generate continuous revenue. What are your recurring costs for its use? Are you able to acquire the expertise to develop the system software in-house using the programming staff? What will the net costs be to the company?

Technical Resources to Support Application/System Software. Most companies installing data communications products have technical resources already in place to support the computer environment, but will those resources be strained to support the new capabilities? Are technical resources available on the market or will you have to train additional employees to support the communications activities? What will the overall costs be for new hires, training, and vendor support?

Types of Communications Links and Services. One of the most difficult decisions in data communications is the type of link and service required by the organization to link facilities together. Communications links can range from satellite services to private microwave to dial-up telephone links. The type of link needed is dictated primarily by distance, access right of way, speed required and the necessary capacity. It is also important to acquire additional capacity to enable the system to grow.

Overhead Associated with Link Charges. One of the most important considerations is the ongoing monthly expense for communications lines. This represents a large expense and enlightened planning is imperative. Suppliers of communications services typically insist on long-term leases, but rates constantly change and you should ensure that the organization is not locked into a contract that could be economically unfavorable in the future.

CASE STUDY

The implementation of a data communications network is an evolutionary process that begins with the task of understanding the needs of the organization. Proper planning and the development of a step by step methodology with corresponding review processes are essential to a successful implementation. The case study presented in this book is about a fictitious company, the Bank of Scotts Valley, however, the solutions are applicable to similar real situations. Figure 1.9 illustrates the Bank of Scotts Valley operational area.

A small, newly created local bank is in the process of defining the requirements necessary to upgrade from a manual processing environment to a data processing/data communications one. The bank wishes to computerize and have online access to checking account information, savings account information, loan administration, stock transfers, mortgages and planning services. It also plans to offer Automatic Teller Machine (ATM) services to its customers and would be interested in participating in an international ATM network at some point in the future.

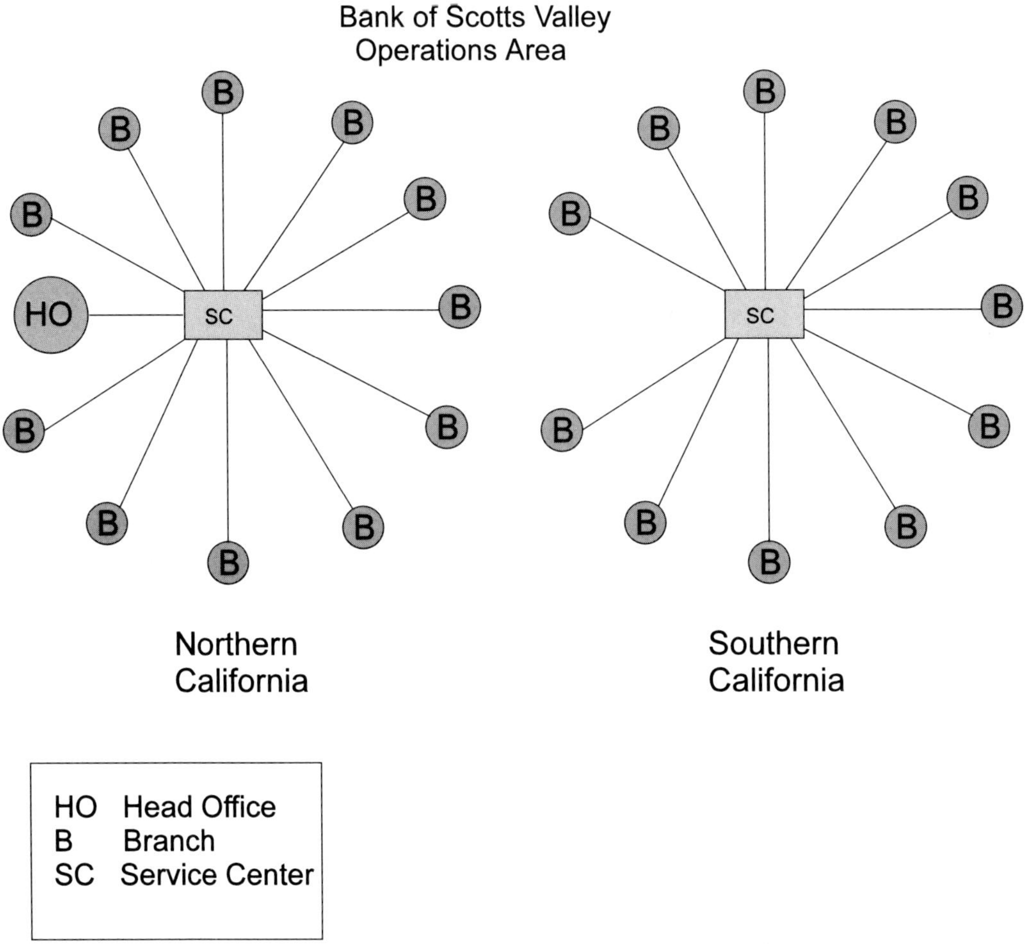

Figure 1.9 Bank of Scotts Valley Network Example

The bank's computer network will initially need to support a minimum of 30 branch offices and 10 business centers. Each of the branches will require eight to 15 teller workstations, two administrative workstations and up to three ATMs.

Careful planning of the network will be essential to ensure a stable, manageable, and reliable environment for the development of informational services. Key requirements include:

- Backup capability should the primary sites or lines become inoperable
- Connection capabilities for communication with other networks for the exchange of information
- Efficiencies of the network that would ensure acceptable levels of performance
- Cost-effectiveness for both short- and long-term time frames.

The bank's management has decided to implement a multiple-site computer environment. This will protect the bank's business activities should a physical disaster such as an earthquake occur. Throughout this book, I will describe the communications software required at the bank computer centers, along with the hardware needed to support the network. The types of communications services needed at both the branch level as well as the ATM level will be detailed, along with the method and facilities required to support them.

In addition, I will examine the required internal and vendor maintenance resources. The case study will conclude with a look at future services that might be implemented along with the type of organization needed to support those services.

REVIEW QUESTIONS

1. Why has the data communications industry grown at such an accelerated rate?
2. What does the term " data communications hardware" refer to?
3. What are the advantages of implementing a data communications network?
4. What are the disadvantages associated with data communications?
5. Does backup need to be considered in establishing communications between different locations?
6. What types of software are typically used for data communications?
7. What are the two transmission methods used today for data communications?
8. Is cost the only factor that needs to be considered in the implementation of a communications network?
9. What is the basic responsibility of a remote computer center?
10. What considerations should be evaluated when planning a data communications network?

CHAPTER 2

Evolution and Basic Communications Planning

EVOLUTION OF COMPUTERIZED COMMUNICATIONS

Although we have a tendency to view communications today from an electronic digital perspective, communication was a necessity of life a long time before the introduction of data communications and computerized processing. Communication was accomplished using a multitude of methods, including the written word, smoke signals, sound, reflections of sun rays using a shiny object, and the spoken word.

As mankind evolved and entered the mechanical and electronic eras, the need for much faster and more reliable communications became apparent. This chapter discusses some of the methods that have been developed.

The beginning of modern day digital communications probably can be attributed to Samuel Morse who, on May 24, 1844, established the first digital communications link between Washington D.C. and Baltimore, MD. This communications link was what we commonly call the telegraph system using Morse's code set. This very simple system consisted of a pair of wires attached to a power source with contacts at both ends.

Morse devised a code that associated letters and numbers with audible clicks resulting from the opening and closing of electrical contact keys located at both ends of the link. This code consisted of a number of clicks sent in a specific period of time. As an example, the letter "A" was represented by a click referred to as a dot, a pause and another click, referred to as a dash. Morse's code set consisted of a unique combination of dots and dashes that represented the letters of the English language, special characters (?!,.&*) and numbers. Figure 2.1 is an example of the Morse Telegraph System.

The principle reason for the development of the telegraph system was the expansion of the railroads westward and the thought that this new telegraph system could play a vital role in this expansion. It is important to recognize that the growth of the telegraph system paralleled that of the railroad and that the development of commerce required rapid and

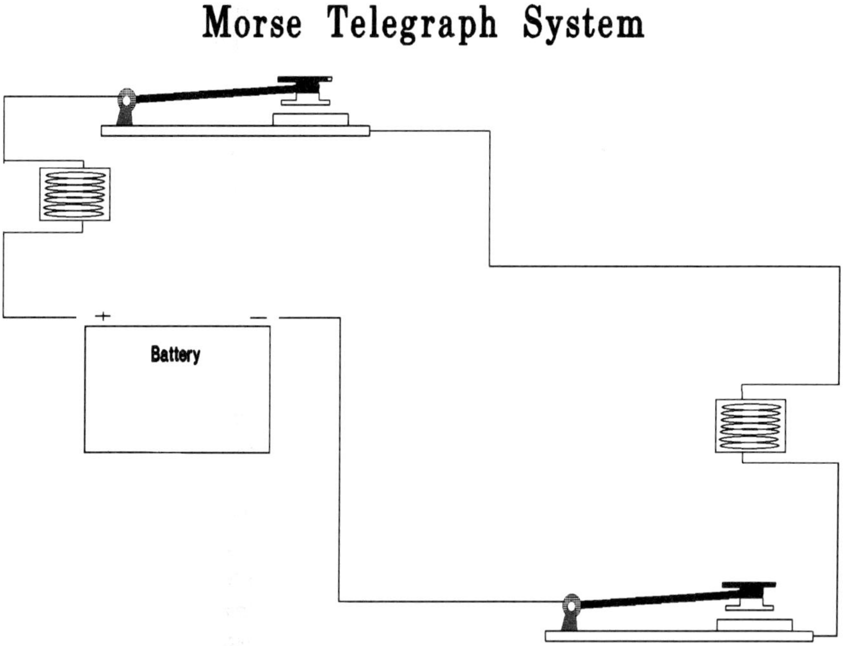

Figure 2.1 Example of the Morse Telegraph System

reliable communications. Morse's telegraph system played an instrumental role in communicating information such as train schedules and equipment availability. Figure 2.2 illustrates an example of the codeset developed by Morse.

It quickly became evident that the telegraph system had a major shortcoming, because its speed was based on an operator's ability to interpret the incoming code and translate it into an understandable format. No specific time intervals existed between the characters sent, so messages were often misinterpreted. The introduction of constant-length code devised by the Frenchman Emil Baudot in 1874 solved the problem. Figure 2.3 illustrates Baudot code. Emil Baudot's name is still used to identify the communications code set for telex service.

Baudot devised a five-element, constant-length code (each character had five components to it). This code was combined with a method to indicate the start and end of the character, which was introduced by Howard Krum in 1910. The resulting communications procedure was an important step toward the automation of the telegraph system and a major step toward today's communications code sets and high-speed transmission methodologies.

The combination of Baudot's code set and Krum's start-and-stop technology paved the way for machines that could read holes punched in paper tape, convert them to electrical signals, transmit the electrical signals from point A to point B and create a new paper

A	•—	N	—•	1	•————
B	—•••	O	———	2	••———
C	—•—•	P	•——•	3	•••——
D	—••	Q	——•—	4	••••—
E	•	R	•—•	5	•••••
F	••—•	S	•••	6	—••••
G	——•	T	—	7	——•••
H	••••	U	••—	8	———••
I	••	V	•••—	9	————•
J	•———	W	•——	0	—————
K	—•—	X	—••—	.	•—•—•—
L	•—••	Y	—•——	,	——••——
M	——	Z	——••	?	••——••

Morse Code Set

Figure 2.2 Morse Code Character Set

tape as an exact copy at the receiving end. Figure 2.3 depicts the Baudot Telegraph Code Set. To interpret the tape, another machine was developed that combined the elements of a typewriter with the components of a paper tape reader. This device could read the paper tape and produce an interpreted printed message.

In 1910, the U.S. Postal service started teleprinter service using this technique. The code developed by Baudot is based on Boolean logic; it consists of five binary numbers that, when combined, represent specific characters. These positional values are referred to as bit positions.

A specific character has a corresponding binary value. This code set was employed in a machine that used a paper tape having five columns, each of which represented a binary digit (positional bit). The presence or absence of holes in a particular horizontal row across the five columns represented the binary value for a character.

Modern data communications uses an expanded variation of Baudot's code due to the limitations placed on the number of combinations supported by only five positions. Even though the five-bit code covered all the letters in the alphabet as well as the basic numbers, no room was left over to expand the code set. Inclusion of start-and-stop indicators and other characters needed to communicate the status of the data being sent (such as sequence numbers, length of the data, etc.) were needed for a more extensive communications service. Figure 2.4 is an example of the Baudot Telegraph Engine

Today, communications code sets have seven, eight and nine binary positions. Two of the most common communications code sets in use today are the American Standard

Character Case Lower	Upper	Bit Pattern 5 4 3 2 1	Character Case Lower	Upper	Bit Pattern 5 4 3 2 1
A	-	0 0 0 1 1	Q	1	1 0 1 1 1
B	?	1 1 0 0 1	R	4	0 1 0 1 0
C	:	0 1 1 1 0	S	'	0 0 1 0 1
D	$	0 1 0 0 1	T	5	1 0 0 0 0
E	3	0 0 0 0 1	U	7	0 0 1 1 1
F	!	0 1 1 0 1	V	;	1 1 1 1 0
G	&	1 1 0 1 0	W	2	1 0 0 1 1
H	#	1 0 1 0 0	X	/	1 1 1 0 1
I	8	0 0 1 1 0	Y	6	1 0 1 0 1
J	Bell	0 1 0 1 1	Z	"	1 0 0 0 1
K	(	0 1 1 1 1	Letters (Shift)		1 1 1 1 1
L	)	1 0 0 1 0	Figures (Shift)		1 1 0 1 1
M	.	1 1 1 0 0	Space (SP) =		0 0 1 0 0
N	,	0 1 1 0 0	Carrage Return <		0 1 0 0 0
O	9	1 1 0 0 0	Line Feed =		0 0 0 1 0
P	0	1 0 1 1 0	Blank		0 0 0 0 0

Figure 2.3 Baudot Telegraph Code

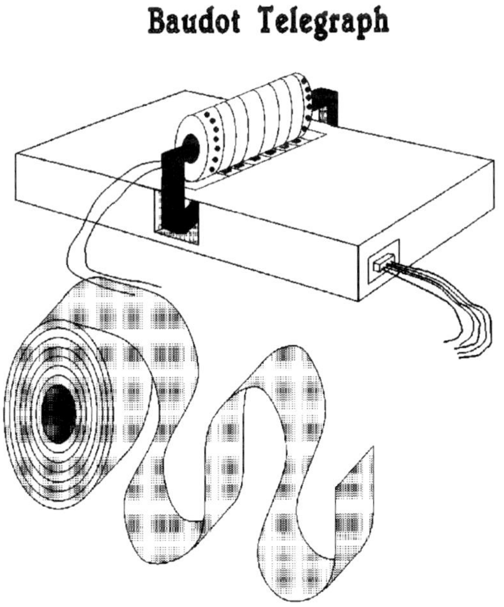

Figure 2.4 Baudot Telegraph Machine

Eight Bit ASCII Code Set

B8 B7 B6 B5

LOW High		0000	0001	0010	0011	0100	0101	0110	0111	1000	1001	1010	1011	1100	1101	1110	1111
B4 B3 B2 B1	HEX	0	1	2	3	4	5	6	7	8	9	A	B	C	D	E	F
0000	0	NUL	DLE	SP	0	@	P	'	p								
0001	1	SOH	DC1	!	1	A	Q	a	q								
0010	2	STX	DC2	"	2	B	R	b	r								
0011	3	ETX	DC3	#	3	C	S	c	s								
0100	4	EOT	DC4	$	4	D	T	d	t								
0101	5	ENQ	NAK	%	5	E	U	e	u								
0110	6	ACK	SYN	&	6	F	V	f	v								
0111	7	BEL	ETR	'	7	G	W	g	w								
1000	8	BS	CAN	(	8	H	X	h	x								
1001	9	HT	EM	)	9	I	Y	i	y								
1010	A	LF	SUB	*	:	J	Z	j	z								
1011	B	VT	ESC	+	;	K	[	k	{								
1100	C	FF	FS	.	<	L	\	l	|								
1101	D	CR	GS	-	=	M	]	m	}								
1110	E	SO	RS		>	N	^	n	~								
1111	F	SI	US	/	?	O	–	o	DEL								

Figure 2.5 ASCII Code Set

Code for Information Interchange (ASCII) and the Extended Binary Coded Decimal Interchange Code (EBCDIC). ASCII is a seven- or eight-bit code depending on which version is being used, standard or extended. It is an extension of a code set known as the six-bit trans code that was used between 1910 and the early 1950s. The EBCDIC is an eight-bit code set. IBM developed both of these code sets. Figure 2.5 shows an ASCII code chart. Figure 2.6 shows an EBCDIC chart.

HARDWARE REQUIREMENTS

Originally, the data communications industry consisted of telephone company equipment and services supported by the existing Bell telephone and American Telephone and Telegraph Company (AT&T) voice networks. Within the past several years, the need for data communications has grown to a point where the industry now has virtually thousands of independent suppliers and distributors. These vendors provide communications equip-

Extended Binary Coded Decimal Interchange Code

B4 B3 B2 B1

High / Low		0000	0001	0010	0011	0100	0101	0110	0111	1000	1001	1010	1011	1100	1101	1110	1111
B8 B7 B6 B5		0	1	2	3	4	5	6	7	8	9	A	B	C	D	E	F
0000	0	NUL	SOH	STX	ETX	PF	HT	LC	DEL		RLF	SMM	VT	FF	CR	SC	SI
0001	1	DLE	DC1	DC2	DC3	RES	NL	BS	IL	CAN	EM	CC		ITS	IGS	IRS	IUS
0010	2	DS	SOS	FS		BYP	LF	EOB	ESC			3M			ENR	ACK	BEL
0011	3			SYN		PN	RS	UC	EOT					DC4	NAK		SUB
0100	4	SP										¢	.	<	(	+	|
0101	5	&										!	$	*	)	;	¬
0110	6	-	/								\		,	%	_	>	?
0111	7											:	#	@	'	=	"
1000	8		a	b	c	d	e	f	g	h	i						
1001	9		j	k	l	m	n	o	p	o	r						
1010	A		~	s	t	u	v	w	x	y	z						
1011	B																
1100	C	{	A	B	C	D	E	F	G	H	I						
1101	D	}	J	K	L	M	N	O	P	Q	R						
1110	E	'		S	T	U	V	W	X	Y	Z						
1111	F	0	1	2	3	4	5	6	7	8	9						

Figure 2.6 EBCDIC Code Set

ment and support services to any organization requiring data communications capabilities. The equipment and services that were once exclusively provided by the telephone network companies are now widely available and can be acquired from a multitude of private concerns.

Today, data communications is a multibillion dollar industry with an annual growth rate in excess of 35 percent. Most computer manufacturers incorporate data communications capabilities into their hardware and software product lines.

The primary catalyst for the tremendous growth of the data communications industry has been the data processing industry and the need to access information in an online, interactive environment. The basic need has been to transfer data economically from one location to another over long distances.

Today's complex and diversified businesses must have immediate access to information regarding their operations in order to serve customers and control business activities. With operations, sales offices, and manufacturing spread all over the world, organizations must have methods to transfer vast amounts of data quickly and inexpensively.

The objective of data communications is to move data as quickly and as reliably as possible from one location to other locations that make up the network. From a hardware perspective, many different components are available for that purpose. Often, it becomes a time-consuming effort to evaluate and select equipment appropriate for a given situation.

Let's assume that we have nothing in place at this time and must start from ground zero in building our network. When we have determined what our data transmission requirements are and projected what the rate of growth will be, the first decision should be what type of computer processing capability is necessary for acceptable performance.

Needless to say, a large number of computer manufacturers exist to satisfy this need. We have a large task ahead of us in simply selecting the computer system. When making such a choice, you'll need to consider the following major areas.

CPU Processing Capacity

Planners should make a detailed evaluation regarding the breadth of applications the computer will have to support. Figure 2.7 shows a CPU capacity layout. As a planner, ask the following questions:

- Will this machine be dedicated to online processing activities, or will it need to support batch applications as well?
- Will the CPU directly support the terminal network, or will a communications processor be required to assist the CPU due to the number of terminals it must manage on the network?
- How many online applications must be supported simultaneously on the computer?
- Will we need to have a hot backup (another computer waiting in standby mode if the primary processor fails)?
- Will our online applications be transaction based or conversation and interactive?
- Will program development occur on this machine?
- What is an acceptable response time for users?

Network Handling Capabilities

To determine network handling capabilities, you must determine the network's physical and logical organization. Figure 2.8 illustrates a network layout. Carefully weigh the answers to the following questions:

- Will the network need to support local (hard-wired) terminals as well as remotely attached terminals and workstations (terminals connected via telephone or other commercial links)?

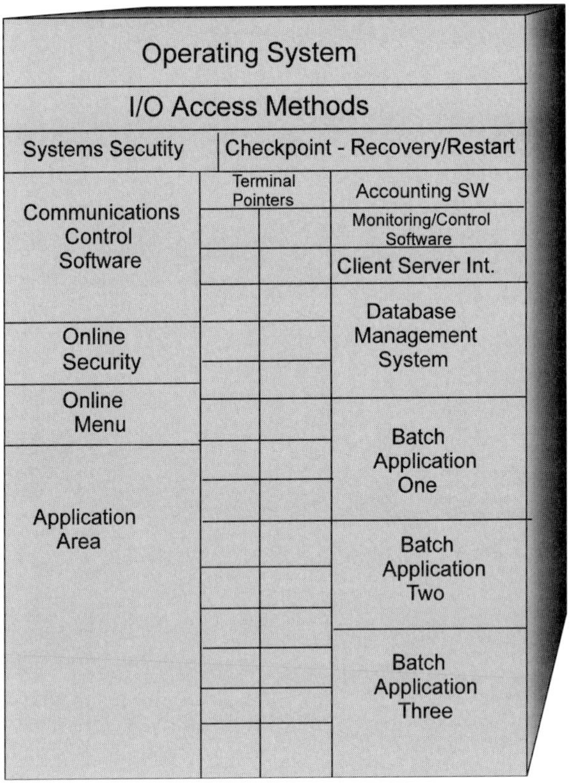

Real Memory Requirements

Operating System	8 + MB
Communications Software	4 + MB
Database Software	4 + MB
Accounting Software	1 + MB
Client Server Interface	1 + MB
Monitoring Control	1+ MB
Application Programs	20% +
Accounting Programs	1 + MB
Checkpoint - Rec/Res.	2 + MB

Support Requirements

Systems Software
Applications Software
Communications Software
Network Software
Local / Remote Processing
Client Server Software

Throughput Requirements

Terminal Response Time
Network Throughput
LAN Access
Batch Throughput

Figure 2.7 Illustrates CPU Capacity Requirements

- Are remote job entry (RJE) or network job entry (NJE) communications to be supported?
- RJE and NJE allow remote sites to use the remotely located computer as if it were a local computer. Should we plan for future growth in the equipment we purchase now?
- Will the network communications hardware support an interface to local area networks? (A local area network consists of computers restricted to a limited distance.)
- Will the system have to support a voice and video network?

Communications Hardware Considerations

Communications processors and multiplexers assist the CPU in performing data communications activities, including terminal management and internetwork communications. If you are responsible for planning, you should review these requirements to determine whether a front-end processor or communications processor (CP) will be needed. Does

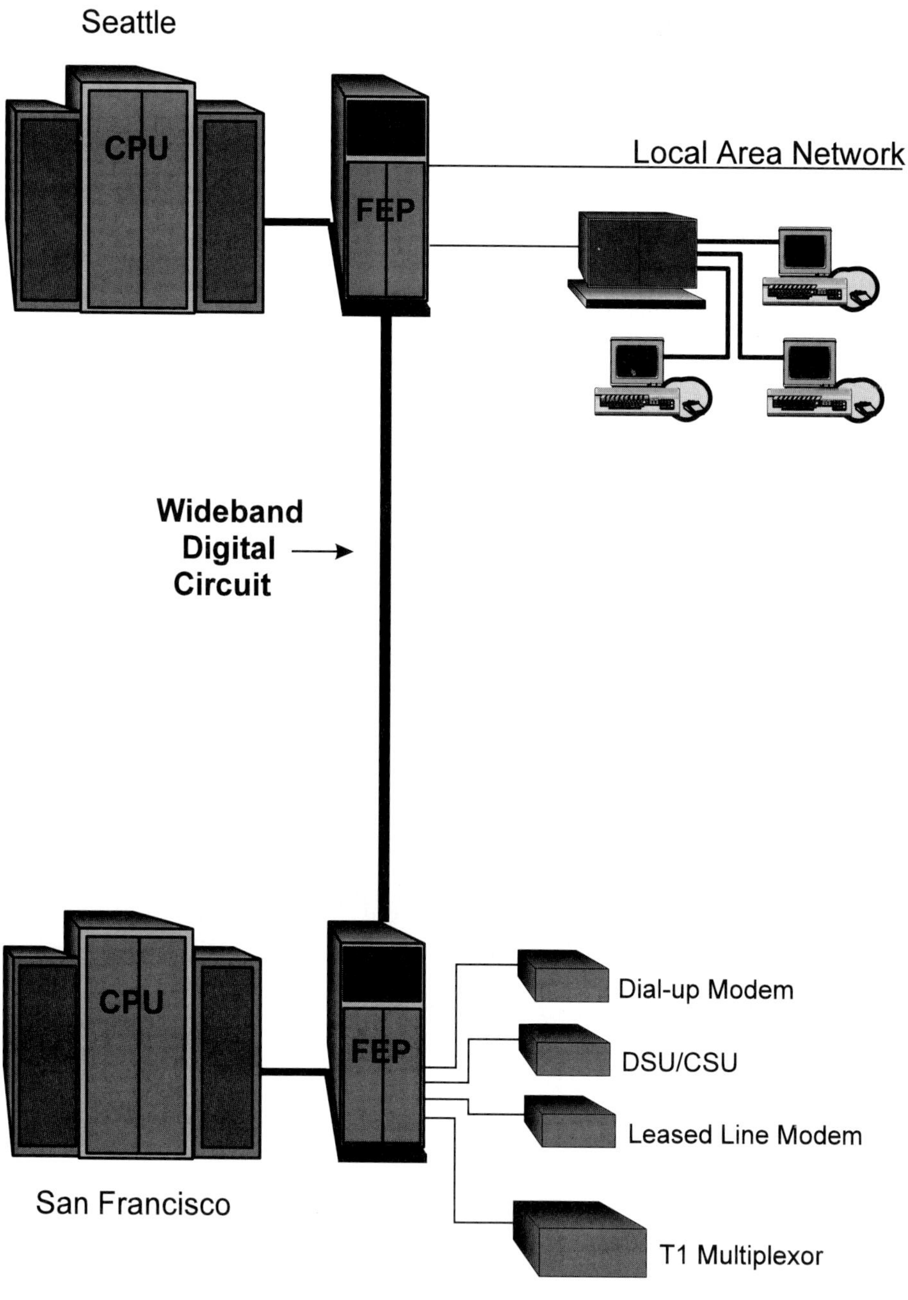

Figure 2.8 Network Connectivity

the network require a CP due to the large number of terminals and printers on the network? Can future growth justify the acquisition of a communications processor even though channel and port capacities currently exist for the attachment of all terminal devices to the CPU?

Will the network require a remote communications processor in order to attach clusters of remote terminals or LANs resident in the same location? Figure 2.9 depicts a communications processor. Multiplexers can be used in place of communications processors, but can the remote terminal network grow using multiplexers? Should a network control center that manages the communications network be installed and what type of capabilities are required to support it?

With respect to the transmission equipment, will a DSU/CSU (used for the movement of bits across the line) be required to support remote digital communications? Are modems (used for analog bit transmission over the telephone network) to be incorporated and, if so, what type and what transmission speed will be acceptable? Are multiplexing modems (which support multiple computers and terminals) suitable for this network so that modem pooling (sharing a single modem among several computers) can occur? When wiring the facility, what media or wiring types will the hardware support?

Transmission Links

Transmission link service options are also important to evaluate when selecting a data communications system. Costs associated with the selection of transmission link services should be carefully weighed. If a private microwave is considered, will the local laws support microwave transmission? Can FCC approval be obtained? Will enough capacity exist

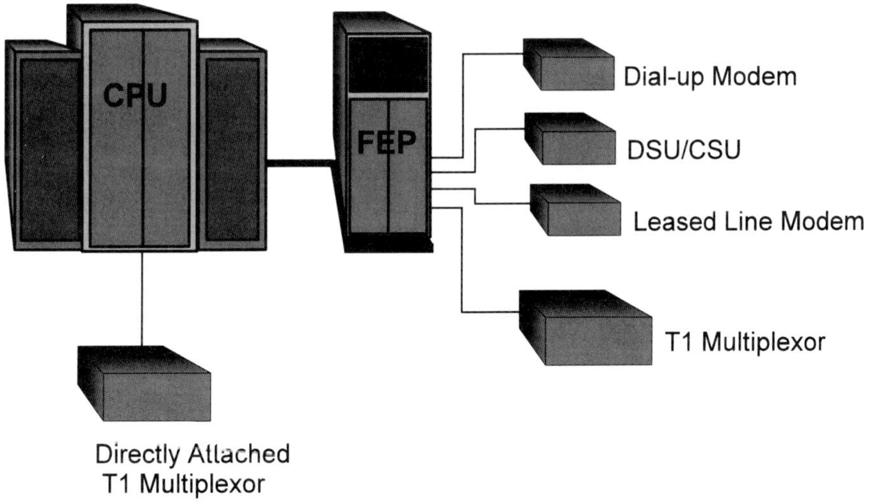

Figure 2.9 Device Connectivity

for future growth of the system? When using dial-up links, does the current public switched telephone network (PSTN) provide a high enough bit rate for reliable communications? Finally, what type of facility or service, if any, should be leased: satellite, wideband digital service provided by the telephone company, ISDN services also provided by the telephone company, or commercial microwave?

SOFTWARE REQUIREMENTS

At one time, the process of moving voice and data across a communications network was a mechanical as well as an electrical function. Eventually, telephone companies introduced electronic program control to the system through the use of computers. This became necessary due to the increased use of communications switching components required to manage data transfer. Because switches were provided with intelligence, software was needed for line control, multiplexing (line sharing) and management.

As the telephone networks were incorporating intelligence, computers were evolving to the next level of online processing. What was once a single application environment communicating with a few dedicated terminals, became a multitasking, online application enterprise linking thousands of networked intelligent devices. New hardware provided the development platform for the complex software we need today for network communications.

Aside from business application software, several other types of software are needed. These other software classifications include management software for the control of the overall computer environment, software for the control and management of the terminal network, and software for the management and control of internetwork communications. Monitoring and performance evaluation software for visibility into enterprisewide communications activities also is available, as are application program interfaces that manage terminal/program connectivity.

Decisions regarding acquisition of the software greatly depend on the hardware selected. Before selecting the software, you'll need to answer basic questions about how the network will function.

Application Control Software

This type of software controls the application-to-terminal processes to ensure error-free and timely access to data. Before choosing this software, consider these questions:

- Will the online environment require application control software to manage concurrent requests from multiple terminals? Or, will the communications activities be dedicated to a specific corporatewide function and have all terminals owned full time by a single application process?
- Does the control software need to provide password security, or will it be part of the application process?
- Should the control software support interactive or transaction-based activities?

Terminal Network Software

This software manages the physical terminal network and provides a method for controlling terminal access and availability. Depending on the size of the network, the expected performance and the type of hardware and software used, it may not be required.

Network Control and Interface Software

If the planned network is to be part of a larger network, network control and interface software will be required to support gateway and bridging functions. The corporation must decide what capabilities will be acquired to support these activities. Will the network control and interface software be required to communicate with other manufacturers' computer environments running different control software? What are the interface requirements? Will the host computer that uses this software be used for routing and relaying data to other computers in the network and will they have compatible software?

Network and Performance-Monitoring Software

Depending on the size of the network, network control may require a remote control center. When considering this type of software, ask yourself these questions:

- Will the network be large enough to support a network control and monitoring function at any level?
- Will funding and management support be sufficient for the purchase and use of the required software?

COMMUNICATIONS SERVICES

The need to move data between distant locations calls for a variety of equipment, software, services, and expertise. Many methods exist for the transfer of data from one location to another, including telephone services (both leased-line and dial-up) provided by the various public telephone companies and private bypass services provided by alternate carriers and network services providers. In addition, an organization may install and maintain an on-site private network or share facilities with other organizations. The following types of services for the movement of data need to be considered.

Carrier Services

A large variety of carrier services exists worldwide for the transport of data, but in many countries, the telecommunications facilities are owned by the government. Where a choice is available, consider the reliability of the carrier, the variety and cost of the services available, the available bit rates, and the mean time to install and upgrade to another class of service.

Specialized Carriers

If the installed network is to support worldwide communications, consider whether specialized carrier services will be required or available at each of the locations. These may include satellite services, trans-Atlantic or trans-Pacific fiber-optic service, and others such as switched 56 K/bps service or packet-handling services.

Communications Service Providers

If the planned communications network is to consist of a heterogeneous mix of equipment (components from different vendors and companies), the network may require the services of a communications service provider. Such a provider reformats the data into an understandable form for reception by incompatible computer systems and other networks.

CASE STUDY (CONTINUED)

At the present time, a manual processing environment exists at each of the branches of the Bank of Scotts Valley. The bank now has more than 30 offices in California, and this method of processing cannot support efficient handling of the bank's business activities. Manual processing means that each branch periodically must batch all financial transactions throughout the day, reconcile them to the teller ledgers and forward them to the service centers. This activity is required for capture by a computer.

The bank has two regions, Northern and Southern, and two service centers support the bank's operations; however, limited data communications capabilities exist between the service centers. This is because at the time of installation, it was determined to be an unnecessary expense. Each service center is responsible for processing batched data input from branches in the region it supports.

Interbranch banking is supported, but it is a slow, manual process involving the batching of transactions at the close of the business day. The transactions are taken to the service centers and then transported via computer tape delivery services between the centers. Transactions are processed overnight and returned, along with the appropriate listings, to each branch before the start of the next business day.

Because of the way the bank must handle its customers' activities, customer service is not competitive and suffers from a poor image. Customers wishing to make deposits must have a hold placed on the deposit until it clears. Withdrawals require a manual lookup and a telephone verification that the customer has sufficient funds to cover the amount requested. Account transfers, as well as automatic deposits, require manual entry by each of the tellers.

Each of the service centers has a Unisys mainframe and magnetic ink character recognition (MICR) equipment. This hardware is used to capture a variety of transactions that include checking account deposits and withdrawals, mortgage and loan payments and savings account deposits. Associated reports and tapes are prepared each night for the following day's activities and distributed to each of the branches before the beginning of the next business day. At the time the bank developed the initial data processing applications, data communications capability was not a requirement.

One objective for the new computer and communications environment is the use of teller terminals at each of the branches to provide employees with online access and update capabilities to the service centers. Additionally, automatic interbranch transfer activity for all accounts and transaction types will be needed. Another important objective for the bank is the elimination of paper and tape transfers between branches and service centers. Hardware required by the bank will include a new high-speed computer at each of the service centers capable of supporting a dynamic application environment for at least five years. Communications processors also must be considered to manage the overall

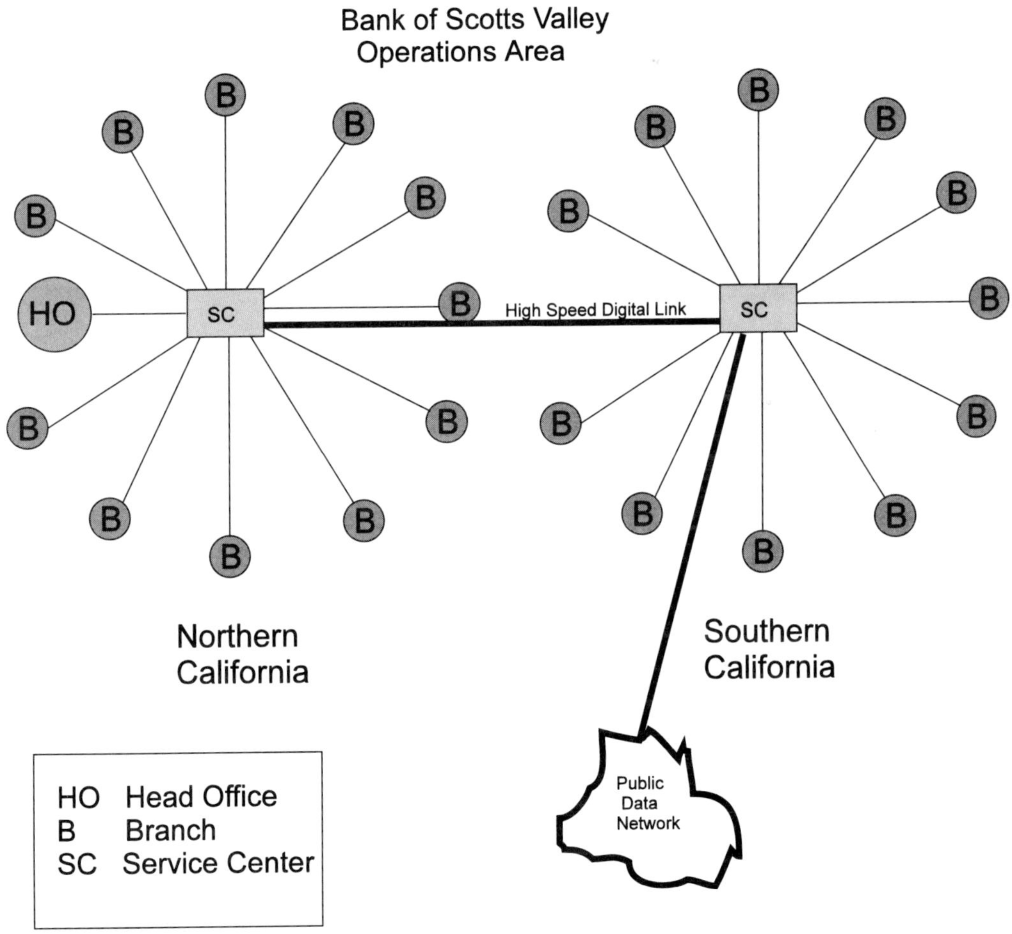

Figure 2.10 Bank of Scotts Valley Network Environment

network. The bank also must plan for communications transmission equipment to perform the data communications activities.

The bank requires a new software environment that will include a new operating system, specific banking application programs, communications control software and system programs for hardware and communications management.

A review of link service requirements will ascertain whether the data transmission capacity between the branches and the service centers will support existing and future business activities. Figure 2.10 depicts the Bank of Scotts Valley's planned network.

REVIEW QUESTIONS

1. What did Samuel Morse invent?
2. With what is Baudot credited?
3. What two code sets are currently used in data communications?
4. What was the prime catalyst for the data communications industry?
5. What factors must be evaluated when selecting data communications hardware?
6. What software needs to be considered for data communications?
7. What function does application control software perform?
8. What do communications service providers offer?
9. What role does the telephone company play in providing communications services?

PART II

SOFTWARE REQUIRED FOR DATA COMMUNICATIONS

CHAPTER 3

Communications Software

INTRODUCTION

Many types of software exist, and many applications are designed to perform end-user processing such as developing payroll checks and maintaining inventory control. Some have system support responsibilities such as control, placement and organization of data on a disk storage device.

Over the years, manufacturers have incorporated many software activities and functions into computer and communications hardware. This incorporation is intended to take advantage of the advances in software that can be hard-coded and contained in read only memory (ROM) semiconductor chips. There are still some activities better suited for inclusion into the software because these activities require modification of the code that direct hardware functions.

Data communications activities such as terminal management, network control, contention control, and error detection require extensive software support. This chapter will concentrate on the two major categories of software, system and application software, as they relate to data communications. Subcategories for system software include operating systems, communications systems, and network control software. Several subcategories of application programs exist; however, discussion is centered on online processing activities only.

ROLE OF SOFTWARE IN COMMUNICATIONS

Communications technology has rapidly evolved from early mechanical switching methods to the sophisticated fourth- and fifth-generation programmable digital computers we use today. However, due to the tremendous demand for speed and flexibility required by users, it is impossible to accomplish these communications functions solely within the hardware.

Communications and operating system software manage hardware resources so that all programs active in the computer can share them. This software has many functions, including activities such as the provision and management of all communications between the host computer system and its remote sites, monitoring performance of the computer communications network and the provision of utility functions such as control and error resolution.

Many different types of system software are available from either the computer hardware manufacturer or from independent software manufacturers. An example of the use of system software would be the simple task of adding or deleting terminals or LANs to or from an enterprise network. Without system software to accomplish this task, physical wiring of the hardware switch would be necessary to add or delete terminals, making it a much more time-consuming and difficult activity. System and communications software responsibilities include:

- Network configuration and maintenance
- Network control
- Application management

Network Configuration and Maintenance

Network configuration and maintenance software ensures that all communications lines are correctly specified and uniquely identified as to their logical addresses, bit rates, and types of service. These functions assist application programs in accessing specific terminals, networked workstations and lines when communications configurations are updated or changed. As an example, terminals must be uniquely specified in several software locations as to screen size, color and special features. Such specificity ensures that information transported to the terminal will be properly displayed. Another area requiring configuration and maintenance is network hardware. Many network components require information about the overall hardware environment to function properly. The role of network maintenance software is to manage communications and operating system software changes as well as to control update activities. Figure 3.1 illustrates line addresses and terminal address tables in memory.

Network Control

Control of the network is critical to acceptable network throughput and response time. Active network control software must be present to manage the overall computer and network resources. Network control software consists of many interlocking programs such as operating systems software elements, access methods and application program interfaces (APIs). These software components accomplish control by directing network routing and by providing error control and line congestion control for all processes requiring communications. Additional control activities include keeping track of active lines and activity in the overall network and resolving contention and congestion.

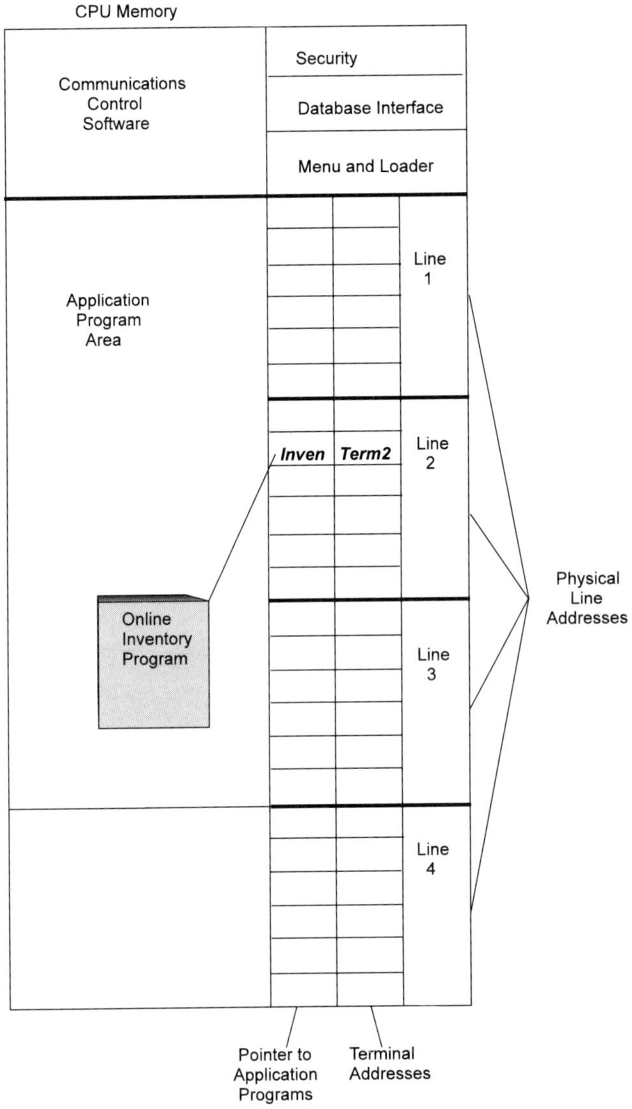

Figure 3.1 Illustration of Software-On-Line Connectivity

Application Management

In today's complex multitasking environment, the many active online programs need to be supervised to ensure that they do not interfere with each other or cause contention problems. Application management software's primary responsibility is to manage the hundreds of application programs that may be resident in memory. This management activity ensures that conflicts do not occur between applications for terminal or other hardware resources. In the absence of application management, multiple communications networks managed on the same host computer would compete for the same scarce resources. Such an environment would create contention between two or more application programs by attempting to address the same endpoint concurrently. Figure 3.2 illustrates a terminal-addressing conflict.

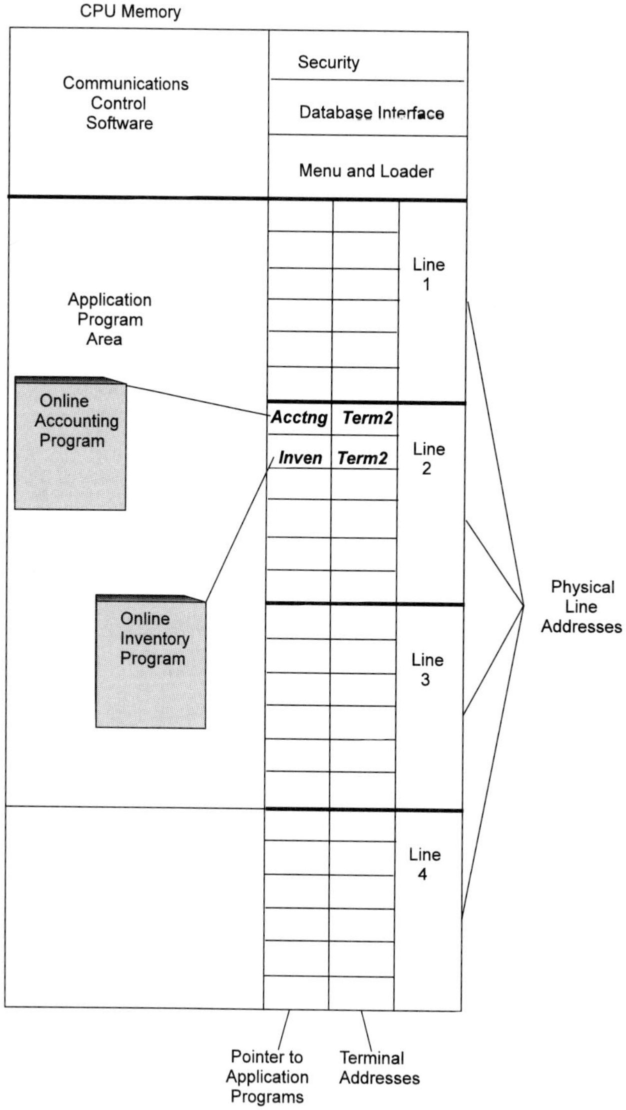

Figure 3.2 Example of On-Line Multitasking Environment

SYSTEM SOFTWARE

The term "system software" is frequently used to describe a vast number of general programs that perform a variety of functions. Basically, system software can be defined as software designed to perform interactively with application programs by managing them and providing a variety of services. The goal of system software is to provide acceptable throughput as efficiently as possible. As a rule, system software must be efficient in order to provide acceptable application program performance.

To achieve such efficiency, system software is typically written using a specific manufacturer's low-level computer language, generally the assembler language of the computer. The use of higher-level languages is less efficient because they often generate additional instruction code that adds processing overhead to the CPU. By using a low-level assembly language, the software programmer is given the most computer control because the assembly language enables the programmer to direct a specific action by the computer, as opposed to a more general one that may include nonessential instructions.

Most system software currently in use is provided by the manufacturer, although some system programs may be written by the programming staff within an organization. This is so, because it takes years of training and special skills to write system software. Almost all computer manufacturers provide communications network and application management software to support their hardware products.

Vendor-provided system software is most often used because of the computer's need to handle the complex tasks associated with management of the required communications functions. These functions include:

- Allocation of access time for all tasks and programs desiring to use the hardware resources
- Management of tasks resident and active in memory
- Control and management of communications links
- Resolution of error conditions originating either from the hardware or from the software
- Checkpoint, restart and recovery activity when required

Figure 3.3 shows major software system components in memory.

Operating System Software

The foundation for all computers is the operating system software. This collection of programs is responsible for the use of memory, the management of all communications hardware functions and the control of all input/output devices attached to the computer. Overall, operating systems manage the use and availability of hardware resources so that all tasks (batch, online, communications control, and system software) have access to the hardware in a timely and orderly manner. The operating system software performs several

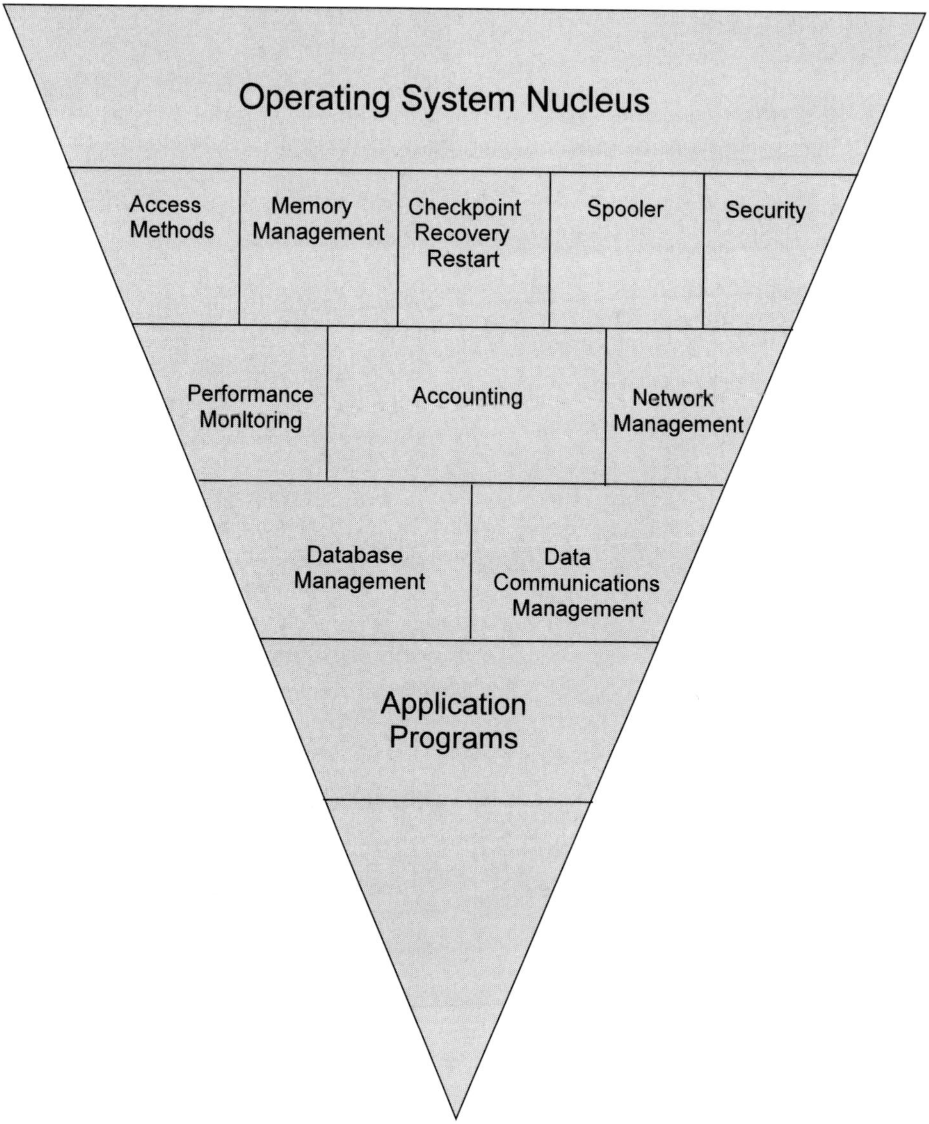

Figure 3.3

functions that are described below. Figure 3.4 illustrates an example of operating system components in an IBM processor.

The operating system must allow for interactive communications between operators and users. Operators require intervention to cancel executing programs, turn lines on and off, schedule output waiting for printer availability and a variety of other functions.

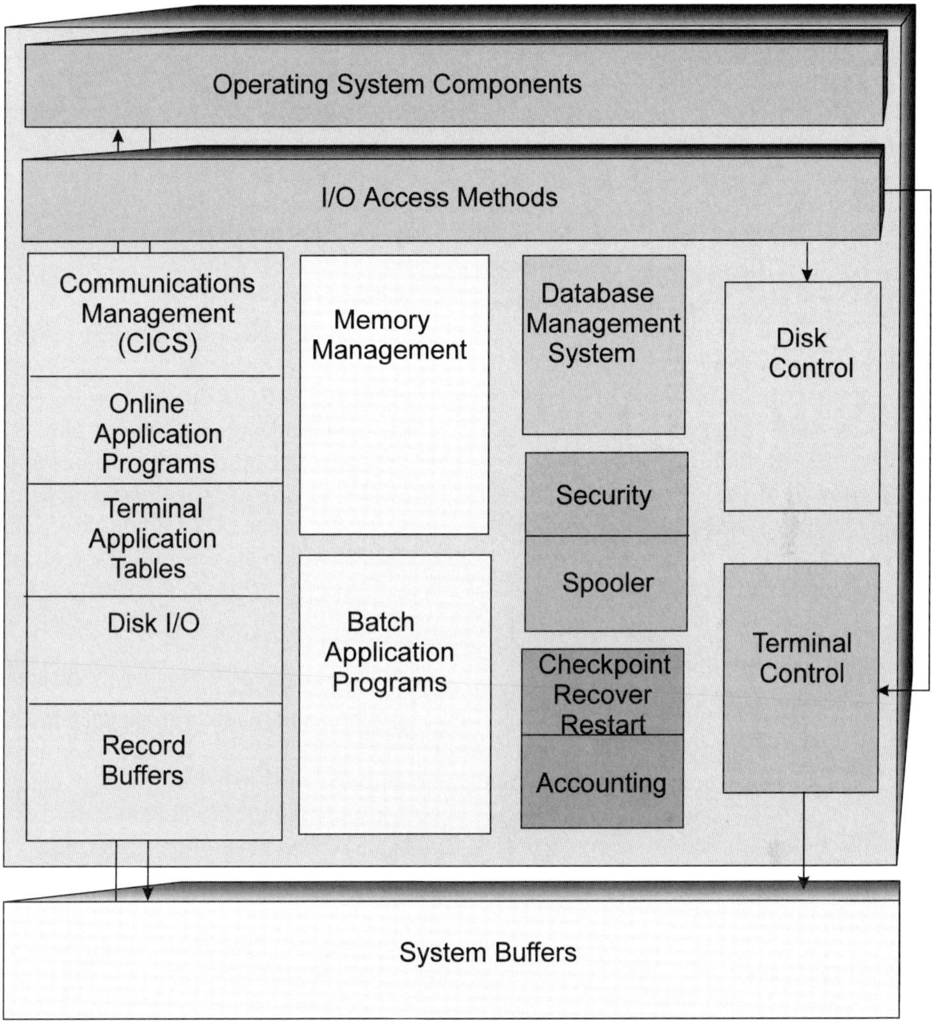

Figure 3.4 Operating System Components

Operating systems also must provide scheduling management for application programs waiting for execution. They accomplish this by continuously monitoring memory for availability. When memory is available, the operating system accesses a list of programs waiting for execution, obtains the name of the next program to execute, fetches the program and loads it into the available memory.

Another function of the operating system is to manage the hardware resources and make them available to programs that require them. The operating system is responsible for managing channel access to I/O devices and will temporarily assign access use on a queue basis to any application process requiring it. The operating system also resolves input/output conflicts created by multiple programs requesting the same resources at the same time, and it attempts to detect and correct errors.

Operating system software manages programs in memory and temporary storage. Because the size of memory is usually large, many programs can reside together. However, the programs must be directed to avoid addressing conflicts. Remember, unless using multiple processors, the computer is capable of executing only one instruction at a time.

The operating system must ensure that programs remain within their own boundaries. It also controls each program's active operations, limiting it to a specific number of instructions to be executed. Finally, the operating system ensures that all programs in memory have active CPU time.

If a failure or fatal error is detected, the operating system manages recovery and restart processes. The operating system attempts to identify and correct any error it can. The operating system may even be able to restart the entire system and each processing application. This is accomplished by frequently storing the contents of memory on disks or tapes and assigning identifiers to them. In the event of a failure, these snapshots of memory can be reloaded using a restart program within the operating system and processing can resume.

Communications System Software

Communications system software manages application and communications tasks, such as security and checkpoint activities that need system resources. This software provides control and access to the communications network and its terminals in an equitable manner. Many programs must interact to perform communications with terminals and other computer systems such as workstations on a LAN because of the complexity of the various activities that must be performed for successful communications.

In the IBM environment, programs that must communicate with terminals can communicate directly using logical access instructions written into the application program. (In other words, the code needed for physical access to the terminals is part of the application program.) This communications access code must interface to the operating system's access methods, which contain the specific code required to move data across the channels to the input/output devices or terminals. For other computer manufacturers, similar procedures must occur for accessing and controlling terminals and I/O devices. Two specific IBM access methods are the Basic Telecommunications Access Method (BTAM) and the Virtual Telecommunications Access Method (VTAM).

If application program code is to contain its own communications instructions, the program code must contain the hardware address of each of the terminals, as well as the instructions necessary to format messages to and from the terminals. The terminal display information along with the terminal address is passed by the application program's communications code to the correct operating system access method. Often, the access method code provided by the manufacturer is embedded in the application program code during its creation.

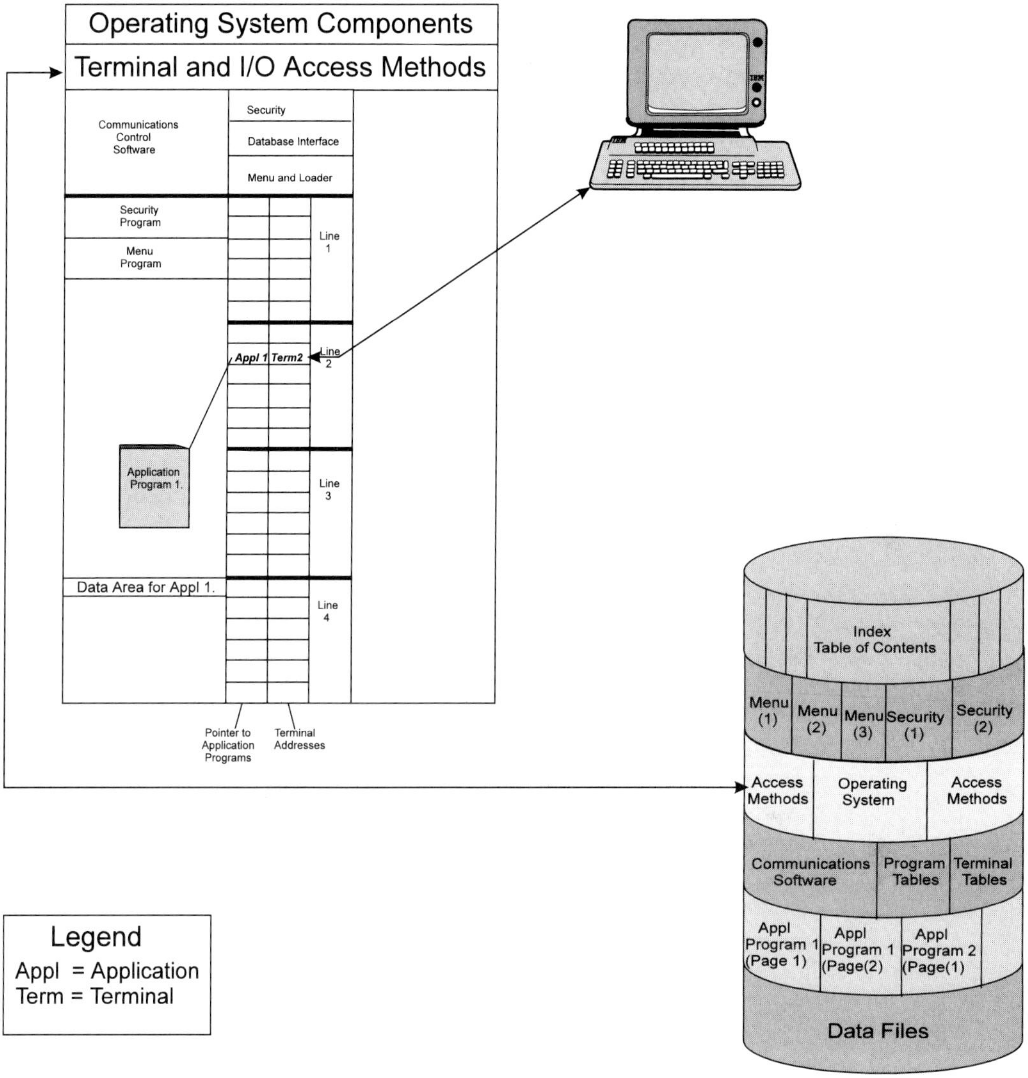

Figure 3.5 Interaction Between Hardware and Software in an On-Line Environment

This method of direct hardware control creates an efficient application program. It is important to bear in mind that a specific terminal can be active only to a single application program at any point in time. This may create a major restriction unless some form of virtual terminal service is utilized. A virtual terminal is a terminal that can display multiple windows, each containing an active process or having the ability to switch between active sessions. Using this method of direct control communications, terminals are normally assigned to a single program during a specific time period.

As an alternative method, organizations can use commercial communications management software such as CICS (Customer Information Control System), an IBM product. Communications management software is usually offered by the computer manufacturer and is designed to provide the same terminal access functions as user-developed application programs that contain embedded communications program statements. Several independent software companies provide communications management products that compete against software provided by the hardware manufacturers.

Communications products manufactured by Software AG, Cincom Systems Inc., Computer Associates International Inc., American Management Systems Inc., and other companies are designed to interface with the hardware access methods provided by the computer vendor. These products provide application program management, thereby making the program independent of the hardware. This means that the application program does not directly address a hardware device. Instead, it uses logical addresses that are translated into hardware addresses by the communications software product.

Most organizations choose to use vendor-supplied communications management software because of the complexities involved in writing their own. This software is responsible for the management of communications functions, such as polling the line and each terminal or workstation to determine if any activity is present, or to see if a terminal or workstation wants to initiate a communications session.

When a user depresses the Enter key, a poll bit is set within the terminal or workstation indicating activity for that particular device. One of the roles of communications management software is to detect the poll bit and establish a session with the device. The software also must be able to select (indicate to the terminal it has something for it), broadcast (send to all terminals), or send data or messages to specific terminals on the link.

Workstations connected to local area networks typically operate in an emulated terminal mode using resident software to establish sessions with the host or mainframe computer. Software running on a host sees the networked workstation as just another terminal so no special handling is required for communications to a LAN-oriented PC. Figure 3.6a through d illustrates the four stages involved in loading the CICS communications software and responding to an application terminal processing request.

Communications management software also schedules and loads online application programs into memory as requested by the terminal, workstation or another program. It is then responsible for transferring the message from the terminal or networked workstation across the link to the waiting program. It must be able to resolve any error conditions that may occur and to ensure that all users desiring host access have equal time.

Additional communications management software capabilities include the automatic broadcast of selection menus at start-up time or when sessions have been completed, and access-control validation to ensure that users have legitimate access to an online application.

Communications management software, when loaded, becomes resident in the computer's memory as a system task and provides services to all online applications requiring communications. Many vendors embed their communications support software into the

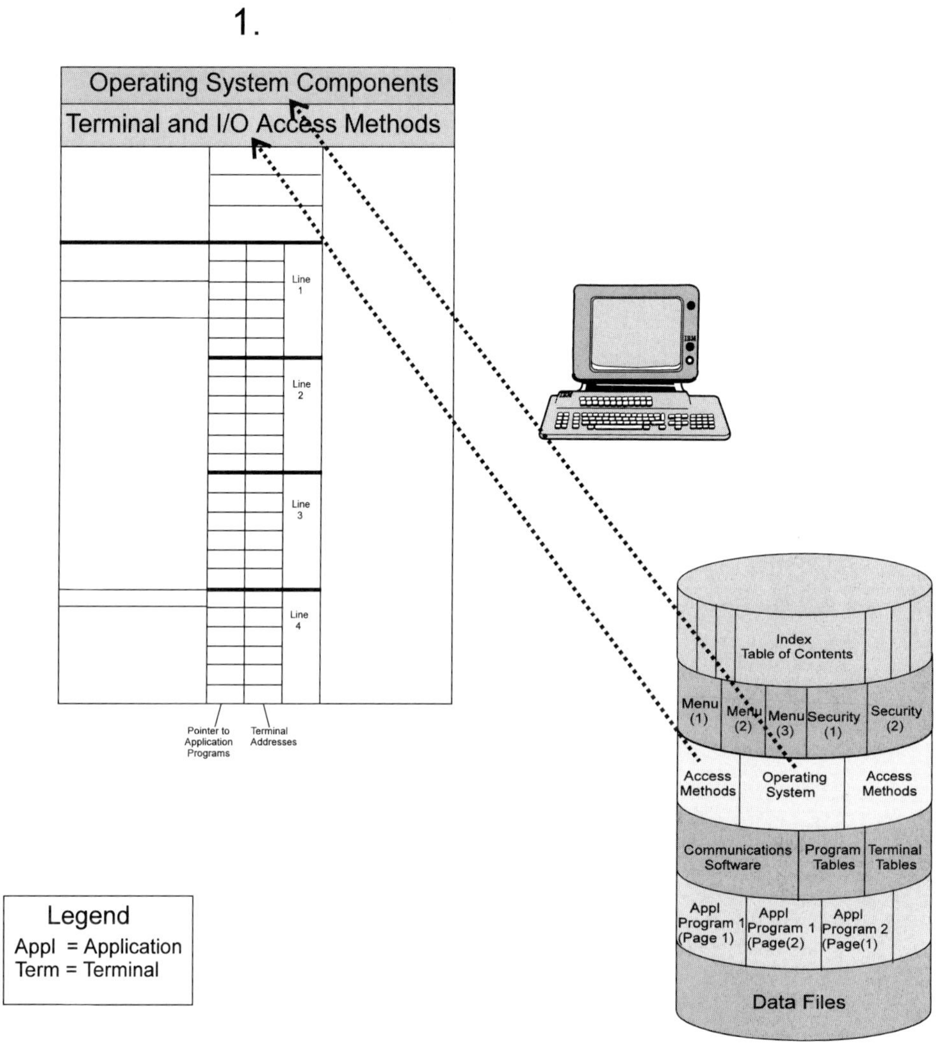

Figure 3.6A

operating system for this reason. UNIX is an example of the integration of communications capabilities into the operating system.

IBM's CICS and other communications management software products offer several distinct advantages. The software resolves contention between diverse applications without impact on programs or user devices. It handles polling, selection, and program interrupts to ensure minimum response time. It resolves error conditions on both the data and line levels. And it becomes a memory-resident part of the computer system, enabling much faster access time to a single copy of the control software.

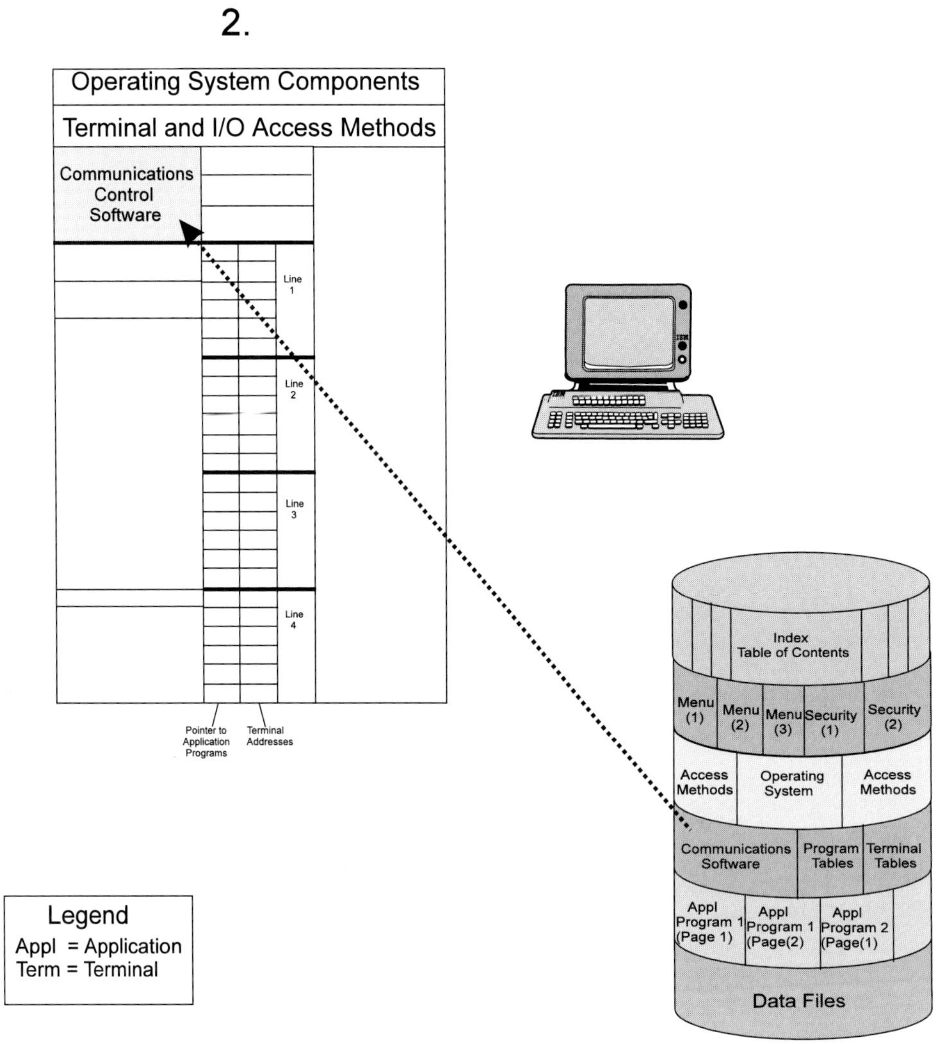

Figure 3.6B

NETWORK SUPPORT SOFTWARE

Depending on the hardware and the size of the network, network support software may be required to physically manage the network, either from the CPU itself or from a supporting communications processor. This software manages the physical lines, the physical terminals and all internetwork communications. The functions of network support software are many and varied and include communications control and network control. Many ven-

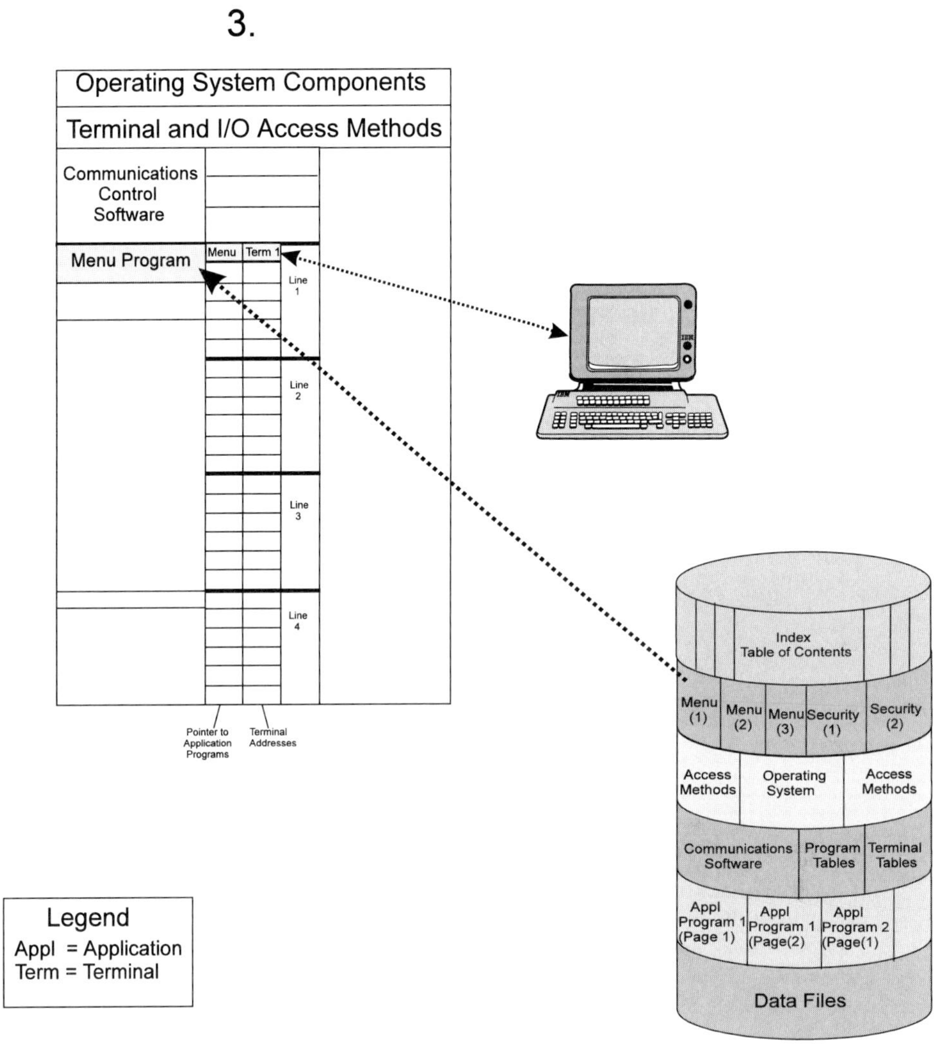

Figure 3.6C

dors, such as Digital Equipment Corporation, Hewlett-Packard Company, and Sun Microsystems Inc., have built these functions into their hardware and software as integrated components. IBM provides separate software packages to accomplish these activities.

Network communications controllers such as the IBM 3745 controller require software to send and receive data to and from local and remote terminals. Figure 3.7 illustrates some of the software components found in an IBM 3745 communications processor. Some of this software might be resident in the CPU, or it might be resident in the

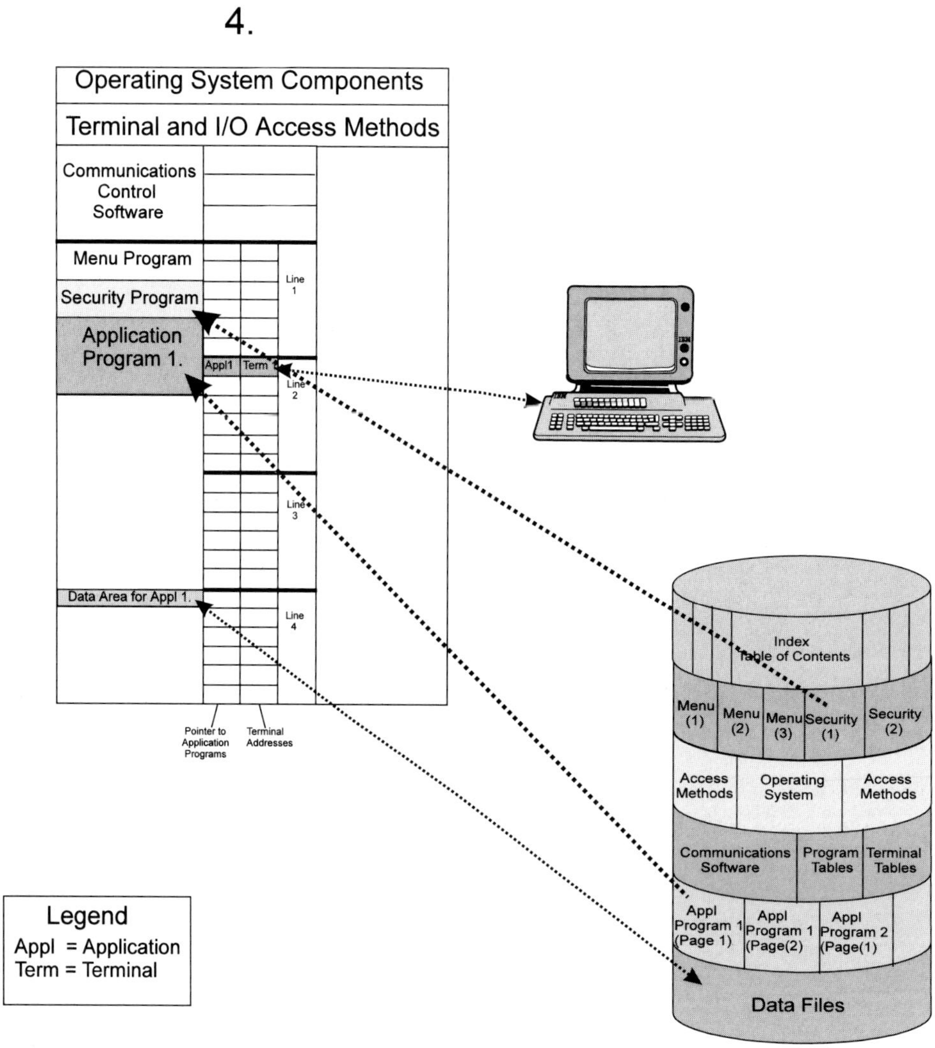

Figure 3.6D

communications controller depending on the type of equipment available and the network defined. It can be downloaded from the host (transmitted across the communications link), or resident as part of the hardware in memory ROM (read only memory); this is known as device software. It can also be referred to as network control software. Some of the following are examples of network control software.

Network Control Program (NCP). This program is resident in the front-end processor (FEP) and is part of the IBM Systems Network Architecture (SNA) environ-

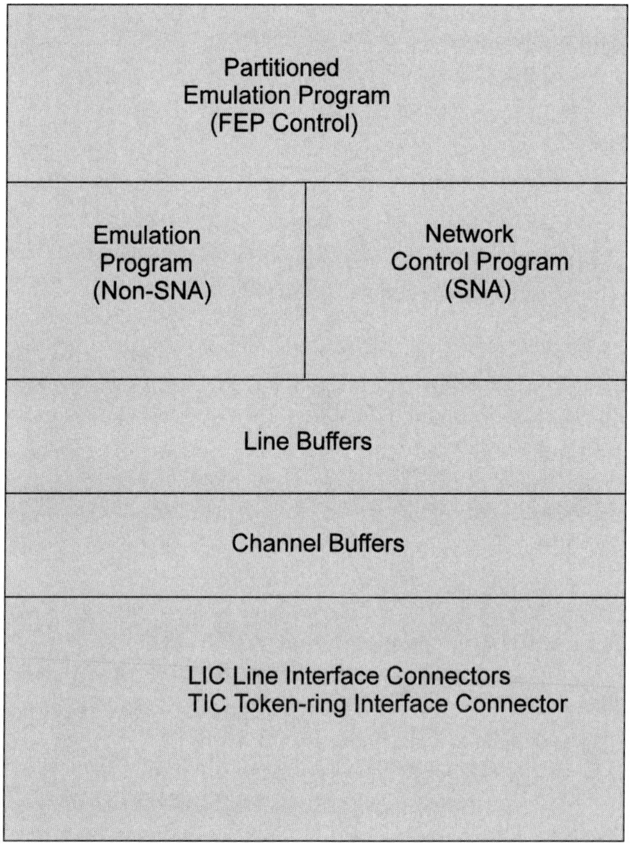

Figure 3.7 Network Control Software Example

ment. The NCP software offloads many of the communications functions that normally would be controlled by the host computer, such as line control, network control, and error resolution. This reduces the amount of processing time and memory that the host computer must allocate for communications activity, enabling it to better service the application processes.

Advanced Communications Facility (ACF). This IBM software assists VTAM in communications activity and enables the network operator or systems programmer to view and control the network. It can be classified as a monitoring and performance option.

Virtual Telecommunications Access Method. VTAM is an IBM access method resident in the CPU and is used to assist the communications monitor or other communications software to communicate across the network to the various nodes. It differs from BTAM in its ability to be terminal transparent. It can be modified (terminals added or dropped) without complex procedures, because it uses table entries that can be easily changed.

Basic Telecommunications Access Method. BTAM is similar to VTAM, but it is an older telecommunications access method. It is used primarily in pre-SNA networks and requires the hardcoding of terminal types and addresses into the program.

APPLICATION SOFTWARE

The function of application software is to address the business needs of an organization and to provide a data processing solution. This is accomplished by using the hardware, communications control software, and system software to process the data and produce a desired product.

Typical examples of applications software are payroll programs designed to calculate payments and print checks, accounts receivable programs used to keep track of monies owed to a company and produce billing statements, and automated teller machine programs designed to dispense cash and handle other ATM transactions.

Application software is often written by an organization's programmers using one of the languages provided by the hardware or software manufacturer. Alternatively, application programs may be purchased from the many software developers.

Application program responsibilities include editing and verifying numeric and character data, manipulation of the various fields to form units of information, mathematical calculations, and the preparation of the data into a format that the end user can understand. Another function of the application program is resolving error conditions that arise from the data or other programs with which it must interact.

Application programs, whether user-developed or vendor-supplied, employ one of two methods to provide online capabilities, depending on the needs of the user and how the applications are developed. Application programs used for data communications are written to use either the interactive-conversational or transaction-based method, as explained below.

INTERACTIVE-CONVERSATIONAL ONLINE APPLICATION PROGRAMS

This type of program allows the user to establish and maintain a running dialog with the program. The interactive-conversational method is employed when the systems designer wishes to allow the user to converse interactively with the application (i.e., maintain a conversation and keep the application active until it is intentionally released by the user). For example, financial application programs often require long periods of active time with the user. Conversational interaction is used when a number of options, such as questions answered on a screen display, determine what actions the application will perform.

In the conversational mode, the user controls the application until he or she wishes to release it. This type of access is used for continuous access to a specific application or task, such as a customer inquiry activity in an insurance or utility company.

Transaction-Based Application Processing

Transaction-based application methods are used for short, predetermined actions such as those found in automatic teller machines. This activity may include a variety of functions that are indirectly related to each other. The specific transaction is completed and the terminal is returned to an inactive status until the next user wishes to employ it.

Neither of these two methods is superior to the other. They simply apply to different operational environments and each one has advantages for specific applications.

CASE STUDY (CONTINUED)

The move to a data communications networking environment by the Bank of Scotts Valley must be very carefully orchestrated. The implementation of new services and the methods for providing them represent a major change in the way the bank operates. These planned changes also represent a major financial investment and therefore must be cost justified to the stockholders. A key issue is that the hardware and software required to support the environment must be implemented in such a way as to allow the bank to continue operations during the transition period. This necessitates that the bank maintain two independent service center environments for both the Northern and Southern regions so that the conversion process is smooth and trouble free.

Because the bank has a Unisys system already installed, the bank performed an extensive evaluation to determine whether Unisys can provide solutions that will not require a total conversion process. At the conclusion of this evaluation, it was determined that the existing systems located at the service centers do not have the capabilities to meet the new requirements. Both the hardware and software are inappropriate and cannot be upgraded to meet the objectives set forth in the evaluation study.

After careful review, the bank has decided to implement an IBM solution using an IBM S390 mainframe and 3178 terminals with a carefully planned migration to a local area network environment in the future. This decision was made based on the upgrade paths they think they will have to take, the availability of off-the-shelf components that will interface to the processor, the impact on the user community, and the availability of systems, communications, and applications software.

The bank has decided to employ IBM's CICS communications software because it does not have the resources to develop custom communications software. CICS is a mature product in operation at thousands of locations around the world, and a large variety of banking applications run under its control. This last point is particularly important to the bank because it does not have a large or experienced programming staff to create either application or system software. The operating system will be IBM's MVS because it is a mature operational environment for CICS and supports several database management systems. Database management is a requirement for online real-time data access.

The bank's management wishes to use commercial applications and system software as often as possible to expedite implementation of the network. Additionally, all applications must be written in the COBOL programming language because the bank will need to maintain the applications after a period of time.

It also has been determined that, because a communications processor initially will not be required, network control software also will not be necessary. This is not to say that at some point during the growth of the network, network control will not be needed.

REVIEW QUESTIONS

1. What functions does communications software perform?
2. What functions does system software perform?
3. What activity is application software responsible for?
4. Name four major functions performed by the communications control software.
5. What does the term "conversational mode processing" describe?
6. What functions do BTAM and VTAM perform in the IBM environment?
7. What functions do network control programs perform?
8. True or false: Some overlap of functions exist between communications software and system software.
9. What is the difference between a terminal and a LAN workstation to the host application software?
10. Communications and system software are usually purchased from the hardware manufacturer or a software vendor as opposed to the facility developing the software itself. Why?

CHAPTER 4

Application Development

INTRODUCTION

Chapter 3 covered data communications software requirements. This chapter expands on that information and provides a more detailed view of the various types of software required for online activities. It describes the different types of data communications management programs that can be employed for data communications activities and explores the differences between them.

One critical concept that must be understood is the detailed interaction needed between application programs and the controlling communications software. Interactions between multiple application and system programs are necessary because of the tremendously complex operational environment of online systems. It is essential to understand the activities that occur within this environment in order to plan, create, install, and operate a data communications network.

Often, a communications environment requires millions of lines of code contained in several hundred programs to perform communications processing. Multiple programs are necessary because it is extremely difficult if not impossible for a single programmer to develop and maintain all the required code in a reasonable period of time. Many man years are often necessary to create the required application and system software.

To develop a data communications network, software systems must be designed, developed, and put in place. The application processes (programs) also must be designed, tested, and implemented. Rules for program interaction and operation are also important. Detailed rules governing program design must be followed for communications between online application programs and system software components. Rules are needed because all online programs must follow specific sets of procedures to accomplish their processing functions.

Application programs used in an online environment follow specific interactive activities in sequence when performing terminal processing activities. These activities provide orderly processing of the transactions and meaningful responses to requesting users.

Online application programs operate under the control of system software (the operating system and communications control software) when executing in the communications environment. As described in Chapter 3, this software controls and manages the hardware so that many programs can run concurrently. Remember, concurrent processing does not imply simultaneity, but the speed of the computer makes it appear so. Figure 4.1 illustrates the interface between a typical application program and the communications system software.

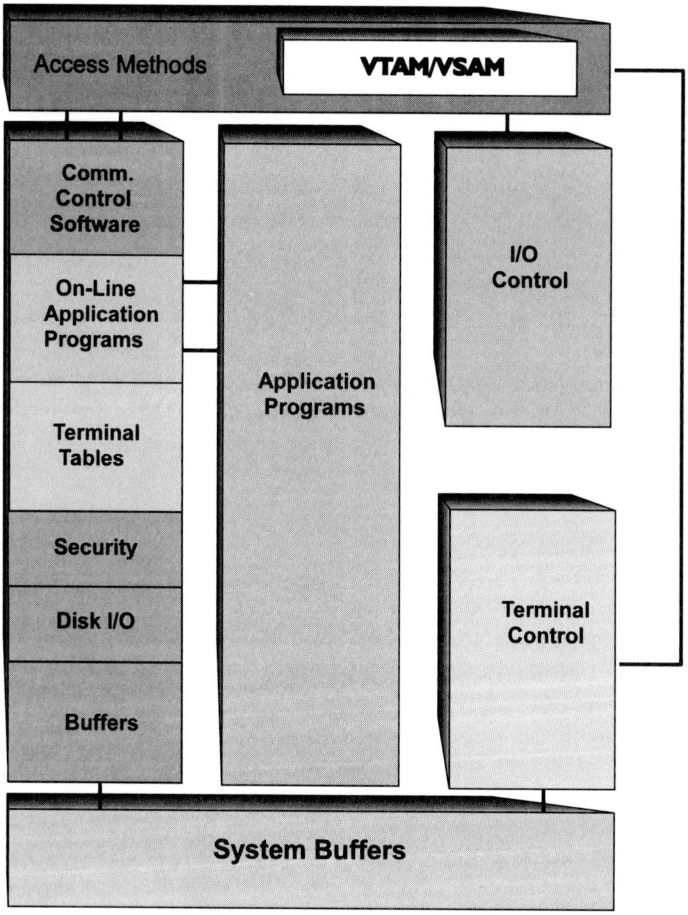

Figure 4.1 Software Map

Interaction between the application program and the communications control software is mandatory for data to be processed and transferred across the network to its destination in the correct order and format. This interaction is also needed because of the complexities of the hardware and operating environment. Online application program interaction with system software must be performed in a specific sequence due to the large number of activities that must take place. Interaction is also needed for error-free data communications in a multitasking environment where many programs must coexist.

MODES OF COMMUNICATIONS PROCESSING

Many methods can be used for creating an online environment. The simplest is to dedicate the computer to a specific set of application processes. It is important to note, however, that unless there is a budget to support this type of environment, the computer usually will have to support a mixed environment consisting of multiple batch processing activities, program development and testing, and one or more interactive or online environments.

Two types of online processing environments are *session-based processing* and *transaction-based processing*. Session-based processing is an interactive conversational type of activity requiring continuous online interaction between the user and the program. This typically involves interactive conversational time-sharing management. Session-based communications is needed when long periods of time are required for interactive activities (e.g., program development).

Transaction-based processing is usually represented by interactions of short duration from many communication points. Transaction based processing can be illustrated with products such as IBM's CICS communications programming system or Tandem Computers' Transaction Processing Facility (TPF). These products are designed for short, high-speed, burst-like types of multiple communication activities such as those associated with ATM processing.

TRANSACTION-BASED COMMUNICATIONS

Online programs that require data transmission services to terminals and end users typically will use the services provided by the transaction-based communications management software. In some cases, custom communications software is used for the online application systems processing because of throughput demands. An example is the Programmed Airlines Reservation System (PARS). The design of PARS provides high-speed airline reservations processing using a subsystem called Transaction Processing Facility (TPF). Several banks also use TPF for their ATM networks because of the number of terminals that require support.

Online application programs running under the control of transaction-based system software use special commands to specify the activities the communications software needs to perform. These commands are communicated in the form of parameters passed on to the communications management software. The management software examines the values in the information passed from the application program to determine which type of activity is necessary. This information includes commands for passing display screens back to a specific terminal, retrieving a screen and other parameters. Before exploring the online application program requirements for communications with the system software, let's examine the activities of typical communications management software.

Transaction-Based Communications Management Software

The communications management software is invoked (executed in the computer) by the computer operator at some point during the day. This activity usually occurs at the beginning of the work day. On systems that must provide 24-hour accessibility, the communications management software is resident and active 100 percent of the time.

One of the functions of the communications management software is to bring up or activate the communications lines that connect the terminals or network(s) to the processor. It also monitors and maintains the lines to ensure that links are operating correctly. Additionally, the control software monitors each line and terminal for requests; this is known as *polling*. When the software encounters a line that has an activity indicator set (when a user enters a message at his or her terminal), the following events occur.

First, the communications control software identifies which online application program the terminal and user require. The control software accomplishes this using some code or value located in the incoming message that arrives from the active terminal.

The next step is to transfer a message block received from the terminal into the communications management software's record buffers. A record buffer is an area of the computer's memory reserved for storing information from terminals and disk storage devices. A transaction code that identifies the terminal and application program will be contained in the received terminal's data block (part of the received message). The communications software uses this transaction code to look at an internal program name list that was made resident in memory when the system was activated. The communications software uses the input transaction code in order to determine which online program should be fetched from the disk-resident program library and loaded into memory.

The communications management software loads (fetches) the online program identified in the code table from its location on a disk device and places it into memory. If this is a highly active program, the management software might make the online program resident in memory for better response. Other programs with lower activity will be loaded when required and erased from memory when no longer needed. The communications software then informs the memory-resident online program where the terminal message is located and turns control over to that program so processing can begin.

In some cases, communications networks such as those used for ATM activities provide limited functions and usually require fast response times because of the number of terminals that require support. To provide acceptable response times, additional functions are often contained in the terminal or in the computer if it is dedicated to that one activity. In a dedicated environment, all application programs that service the ATM network would be made memory resident. A dedicated environment is one in which the terminal is designed for a specific activity, such as those used for airline reservations or bank teller activities.

Online application programs written for communications activities can transmit information using either transaction-oriented processing or conversational-oriented processing. Let's explore the differences between the two types by examining typical activities required for each method.

Transaction-Based Application Program Activities

The online application program accesses the transaction resident in the communications management software buffers and performs whatever action is necessary to respond to the message. That could include calculations, access to databases, and data manipulation. Often, several interactive display screens are transferred between the computer and the terminal to complete the transaction. The application program either completes the activity or requests additional input from the terminal. It does this by passing another command back to the management software and requesting that the response be routed back to the sending terminal.

When the application program has completed processing the transaction, it informs the communications software so the management software may use that memory space for another transaction-processing application if the space is needed. The application program must follow certain logical steps in processing transactions and messages.

Upon receipt of the initial request from the terminal, the application program begins to process the transaction. It acquires records to produce the information required as output to the terminal. Before the application requests a transfer of data across the link, the program must identify the templates, called mask and maps, it wishes the communications management software to display on the terminal screen for that data layout. Figure 4.2 is an example of a typical terminal map used to identify display data.

Unless screen formats are to remain the same, this activity must take place for each formatted data layout to be returned to the terminal. The communications software transmits the screen map and then lays the data on top of it. The mask and maps create display formats for the data that make it comprehensible to the end user.

Prior to virtual systems concepts, when memory was limited, it was common for application programs to modify themselves to save memory. This was done by changing some of the branching or other instructions in the code that was already resident in memory. Another method for running large programs in limited memory was to overlay portions of the original code with new code. While this worked well, if the original code was required again, a fresh copy had to be brought in from disk. With the advent of multithreading methods, several terminals were able to access the same program. Of course, access must be at different times and in different program code locations. This allows the program to service several transactions and terminals concurrently. Multithreading allows several users to access the same program without conflict. The communications management software keeps track of the specific location of each user's program thread Figure 4.3 illustrates an example of multithreading.

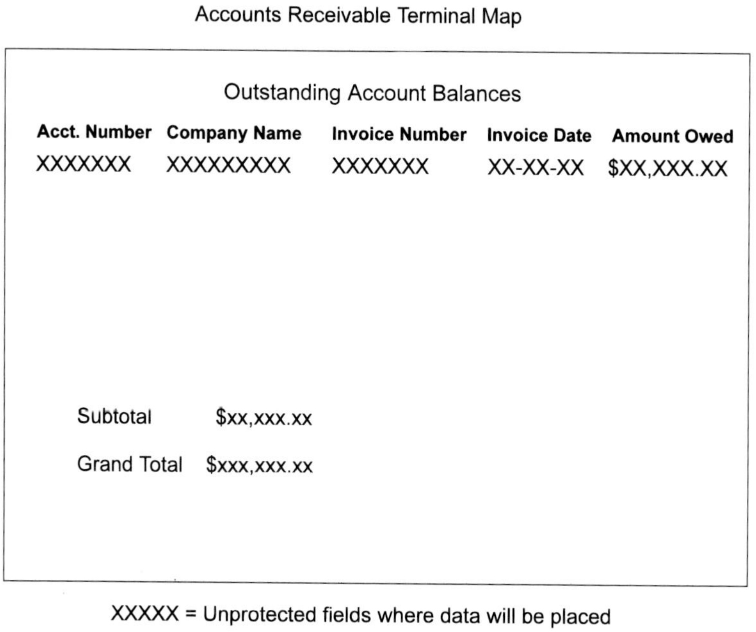

Figure 4.2 Screen Layout Showing Protected Fields

In this environment, each of the active transaction context data blocks (this is the specific terminal data being worked on) must be saved by the communications software and be made available for restoration on demand by the online application program. This activity must take place whenever the online program switches from one active transaction or terminal to another. Switching of program control normally takes place when an online program needs system services such as requesting the retrieval of a record from disk storage. The program cannot continue processing the specific terminal transaction until it receives the requested records from the communications control software. At this point, the application program can begin processing another transaction if another terminal has an outstanding request. (This is analogous to placing someone on hold on the telephone while another call is serviced.) As a result, the application program must not modify itself.

A second reason that an online program must not modify itself is that it could be paged out (removed from memory to disk storage) while it is inactive and memory is needed by other programs. When the online program needs to become active in memory again because a new transaction requires processing, the communications management software will bring in a fresh copy of the program. Any counters, switches or total accumulators within the online program used for placement or mathematical activity will have a zero value. This program must be written in what is referred to as re-entrant code.

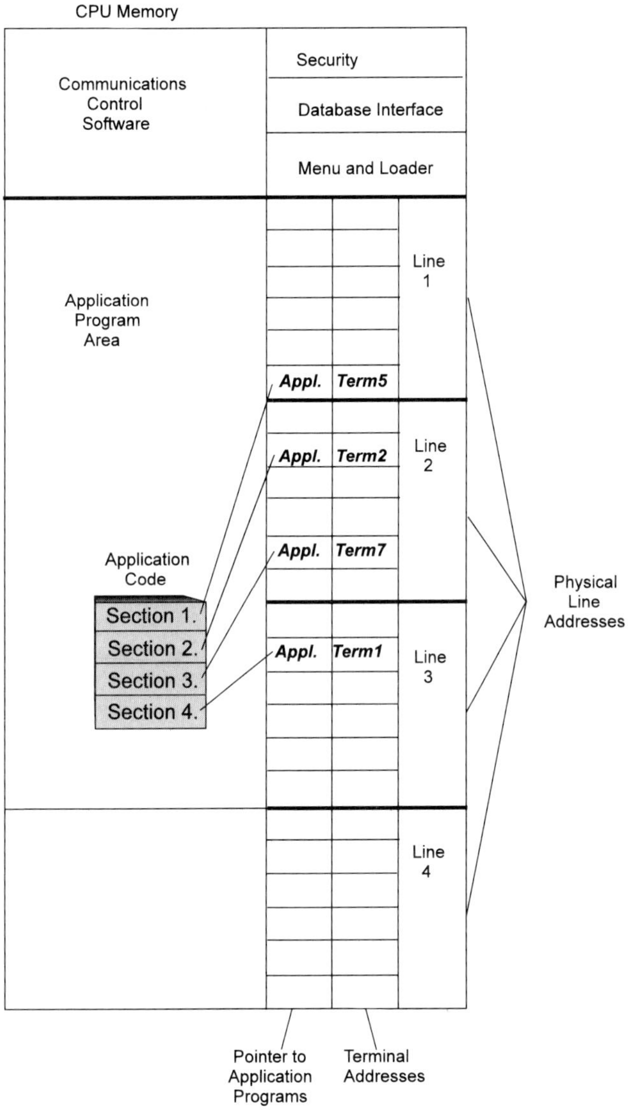

Figure 4.3

The online program also should not engage in long record searches because system programs view the communications management software and its online application programs as a single activity. Long searches by one online application make all online programs, including the management software, wait. It also makes the end user wait an excessive amount of time for a response. This delay is nonproductive for the end user and it tarnishes the image of the computer department.

If the application program is a transaction-based program such as those used in ATMs, it normally will be designed to handle single transactions and may remain resident for the next transaction. This is opposed to low-volume transactions that do not need rapid response time. Typically, the latter will be dynamically loaded when required and then disappear until the communications software schedules it again.

Most communications control software allow the application to be written in any programming language supported on the machine (COBOL, PL/1, BASIC, C, etc.). It is important to keep in mind, however, that processing efficiency is extremely important and throughput efficiency may require programs to be written in a low-level language such as Assembler.

CONVERSATIONAL COMMUNICATIONS

Conversational communications programs are written like transaction-based applications and interact with the communications management software in an identical manner. The major difference between transaction-based and interactive-based processing lies in the area of end-user interaction. Transaction-based programs are designed to service any terminal on the network in a surprise or random environment (i.e., the communications management software doesn't know when a transaction is going to arrive or what application program it is destined for until it is received). This creates a dynamic environment whose constant state of flux governs the way the online program is written. Transaction-based application programs break the connection to the terminal when finished processing one request and await requests from other terminals.

In the conversational environment, the application programs remain active and connected to the terminal until the end users at each of those terminals indicate they are finished. Typical conversational environments include customer service departments where customer information needs to be accessed on a continuous basis, as well as inventory look-ups, sales order look-ups, and payroll. With these types of activities, a department's terminals remain online to the specific application until the user signs off.

Time-Sharing Communications

Most computer manufacturers base their primary communications software on time-sharing concepts. One example is the VMS operating system and conversational communications management system developed by Digital Equipment. Another time-sharing monitor is the VM operating system and conversational management system (CMS) provided by IBM.

Time-sharing communications management software is considered to be an integrated component of many operating systems. It is designed to offer a variety of user services on a time-sharing basis. You could look at the use of a time-sharing system as an alternative to the personal computer because the user interacts with the computer on a one-to-one basis using the software and services provided. This is not to imply that you cannot perform batch processing or provide interactive transaction-based processing in a time-sharing environment.

Many computer manufacturers offer specific time-sharing software in support of their hardware. Time-sharing software is usually user friendly and requires little or no knowledge of the hardware or software environments for many of its basic functions. Users can create private disk storage for their exclusive use or create public storage files that can be accessed by all users. Additionally, the user can employ community printers and access electronic mail software for communications.

Time-sharing management software is similar to transaction-based management software in that it provides hardware services and data communications to the end user on request. Time-sharing software is normally used by programmers and engineers for application development or utility activities, including file transfers, word processing, text editing and electronic mail. Transaction-based communications software is used to process data and communicate information to and from the end user.

The major difference between the two is in the method used to provide the hardware and software access by a user program. Most transaction-based communications programs operate on what is known as an *interrupt-driven* basis. In an interrupt-driven environment, the method of transferring control from one program to another (application to operating system or application to application) is governed by the specific processing program having control of the system at that time. The program in control voluntarily gives that control up when it needs operating system services or processing assistance from another program. This is normally done when a program needs to perform some type of I/O request.

An example of this is a transaction-processing program receiving a request from a terminal for a customer look-up. The program takes the customer number from the incoming transaction and passes on an I/O request to another program designed to perform the record retrieval function. When the request for data is issued, control of the system is transferred to the data retrieval program and the transaction processing program is placed into a wait-state until the data is available.

The program performing the requested service (usually the access method) passes the request to a channel control program. On many machines, this is an independent activity operating within its own microprocessor-based environment. The channel control program then takes the request and tells the access method to come back in a little while to get the record.

Input/output activity is normally an independent program process in most computers. The requesting transaction-based application program is placed in a wait-state pending receipt of the retrieved data from the input/output access program. At that time, the communications management software can transfer control to another program to allow it to begin or continue processing. The communications management program also may have passed a record back to the waiting program, which will then be given control. The same process is repeated for each program operating under control of the communications management software. Although this may seem like a time-consuming activity, bear in mind that the hardware is capable of executing millions of instructions per second. A well-designed system can actually process thousands of transactions per second.

Time-sharing communications management software, on the other hand, operates using a technique known as time slicing, whereby the communications software and operating system allocate to each application a preset amount of time. This permits an application to process a portion of its data. When the time expires for that application, the software passes control to the next application for a slice of time. All active users are serviced rapidly because the amount of time allocated to each user is minimal in the interactive environment. Remember, between communications activities, the user must review and act on the data transferred. Time-sharing techniques are also referred to as round robin techniques using quantums of time.

Selected Communications Software

Most communications software products are geared toward the IBM market because IBM and other companies making IBM-compatible computers require additional software for online processing. Following are some of the time-sharing communications products available for use on IBM systems.

IBM's Time Sharing Option (TSO). TSO operates as a system task under the Multiple Virtual System (MVS) operating system. TSO also can operate as a subsystem under the Virtual Machine (VM) operating system. TSO actually has its own subsystem, referred to as the Interactive System Productivity Facility (ISPF). These systems are designed to assist the programmer in program development, the computer operator in monitoring and managing the computer system, and the system programmer in maintaining the various software systems.

TSO with its ISPF subsystem provides many features and has an excellent text editor facility for program development. This software also allows users to access the computer system for status and program inquiries. It also enables programmers and users to access many utility functions. A unique feature of TSO and ISPF is that they provide an interface capability that allows other vendors' software to operate under its control. An example of this would be third-party software designed to monitor and tune the system's performance.

TSO and ISPF allow programmers to write small pieces of code to perform special functions, and they enable users to access and create data files. Both are extensively used by system support staff for software maintenance and provide a user broadcast and message-switching activity between various terminals on the network. Finally, TSO and ISPF provide a batch capability that allows programmers and users to execute batch programs and receive the program output on their terminals. This type of monitor is not used for transaction processing.

VM and Conversational Monitoring System (CMS). VM/CMS incorporates several time-sharing options and treats every user as though that user were on the machine alone. VM provides user services by allowing users to access a number of subsystems that it controls. VM provides an editor (XEDIT) similar to the ISPF editor sup-

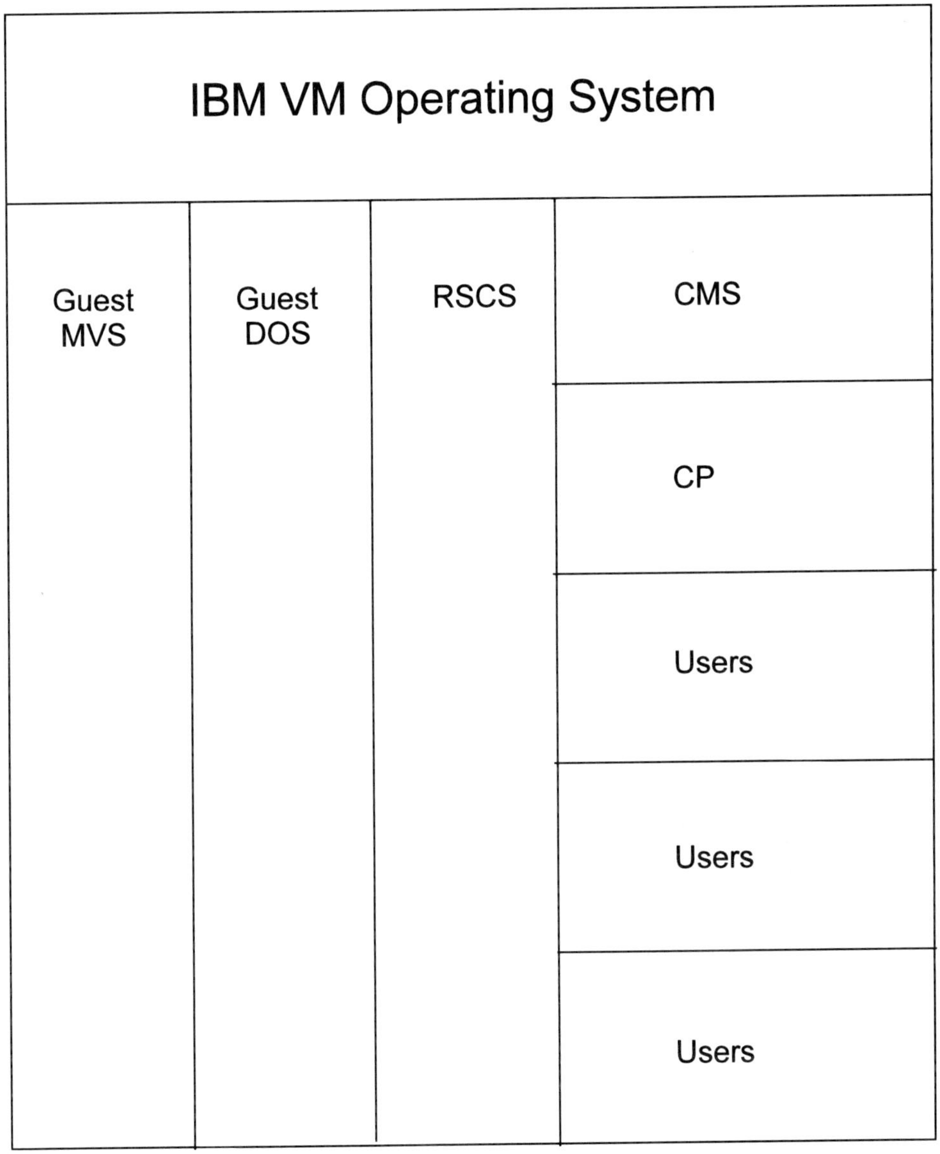

Figure 4.4 Map of VM Operating System Modules

ported in TSO; this is part of CMS. Figure 4.4 illustrates an example of the VM operating system environment.

VM also provides for communications between end users on the same computer and end users on other computers connected to that computer. It does this by using a subsystem known as the Remote Spooling Communications Subsystem (RSCS). VM allows terminal users to execute programs from their terminals and permits the sharing of data and program files between multiple users.

In addition, VM and its subsystems create an environment where each user is provided with his or her own virtual machine. This allows users with little or no knowledge of computer hardware or software to use the facilities of the computer. One last point is that VM and its subsystems are user friendly and are not normally used for system programming activities.

CASE STUDY (CONTINUED)

For the Bank of Scotts Valley to meet the initial goal of providing online interactive systems to support its basic activities, it either must develop or acquire applications to provide the services required. These activities include demand deposit checking and savings, automatic teller machines, loans, mortgages, credit card transactions, and other financial services. The necessary data will have to be accessible from both a teller branch online environment and a service center batch environment. Figure 4.5 shows the Bank of Scotts Valley's CPU configuration.

One of the bank's major objectives is to develop applications that will exploit data as a corporate resource. The bank wishes to use a database management system (DBMS) to manage and support the total operational environment. A DBMS will allow applications to share resources while reducing the time it takes to enhance or create new programs and systems.

The system planners believe that end users (tellers) will not need a free-form query capability because of the nature of the processing activity. Based on this information, the IBM Information Management System (IMS) DBMS is selected for managing the data.

The majority of the bank's data processing/data communications activities will take place in a transaction-based interactive manner; as previously stated, IBM's CICS communications software will be used for online processing. CICS was selected because it is a transaction-based communications system that provides acceptable transaction processing throughput and is used extensively within the banking industry.

The bank's data processing organization also has determined that a conversational interactive communications environment is a critical requirement for access to systems and program maintenance. To satisfy this requirement, TSO and ISPF from IBM also have been chosen.

The bank also hopes eventually to incorporate a production and development environment. However, the bank cannot, at this point in time, afford a separate development environment for programming and testing applications. To solve this problem, the bank has decided to install the IBM VM operating system on the computers in each of the ser-

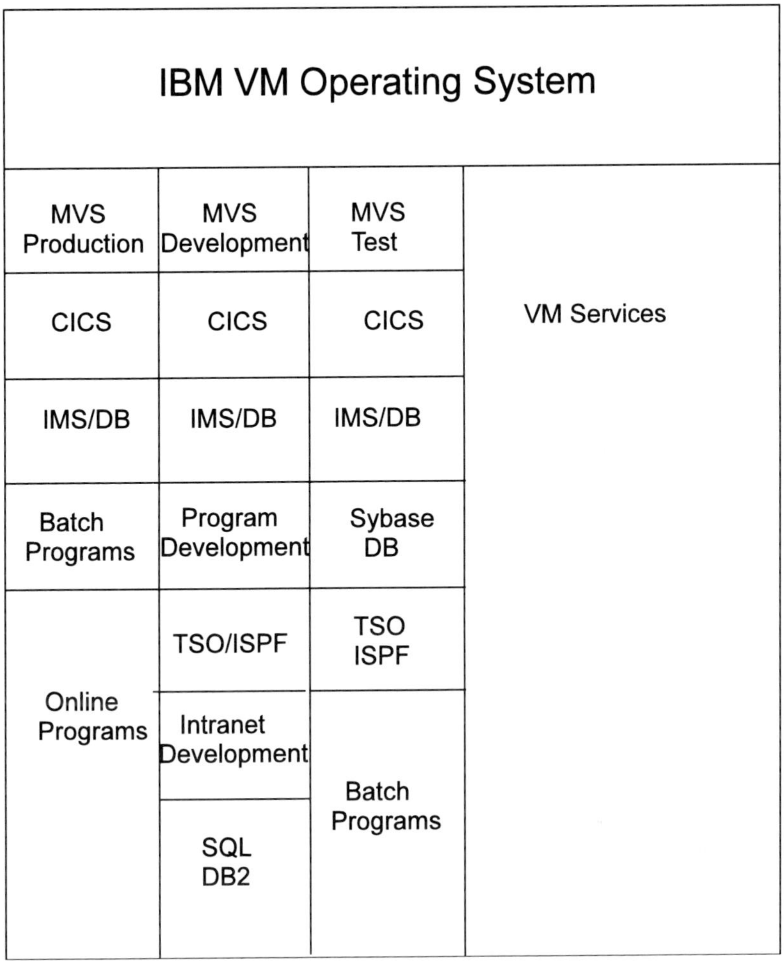

Figure 4.5 VM OS Acting as a Host to MVS

vice centers. This system will be used to create multiple MVS environments on each machine, one for production and one for testing and development. While this creates a somewhat complicated environment, it will provide a safety net should a development activity get out of control and accidentally terminate the production environment.

REVIEW QUESTIONS

1. What are the differences between transaction-based communications and time-sharing based communications?

2. Can time-sharing software be used in place of transaction-based communications software, and why would one want to do so?
3. What special considerations do application programmers need to take into account when writing programs that perform online functions?
4. What is the major difference between a conversational online system and a transaction-based online system?
5. What is a time-sharing system primarily designed to do?

PART III

DATA TRANSMISSION CONCEPTS AND COMMUNICATIONS SERVICES

CHAPTER 5

Transmission Modes and Methods

INTRODUCTION

Establishing a network that is cost-effective and provides acceptable performance requires extensive planning. Important factors to consider include the bandwidth of the data transmission media (wire or cable) and the type of line required to support the desired bit rate. You also should consider the degree of reliability needed and the level of security required. Finally, the type of directional transmission necessary to accomplish the communications activity should be planned appropriately.

Although the cost for high-speed data communications has been falling rapidly, it is still expensive for the small user to incorporate and maintain high-speed links to remote sites. Small users either pay for capabilities they do not need or do not have the capabilities required due to high costs.

This chapter introduces transmission modes and grades of services to help you to determine what type of service, mode, and facility are required for a given communications activity.

BANDWIDTH CONCEPTS

Data communications concepts are built on a number of specialized terms used to describe certain functions and activities. An understanding of the following terms is a prerequisite to a solid data communications foundation.

Circuit. This term generally applies to an established medium such as wire, fiber-optic cable, or radio frequency waves. A circuit can be impressed with a light or electrical signal to carry analog or digital transmission information.

Channel. This term describes the path for one-way transmission across a link. A circuit may utilize one or more channels.

Carrier. A carrier can be a constant signal on a channel or circuit at some fixed amplitude and frequency. This can be combined with an information-bearing signal such as a data signal of ones and zeros from a terminal. A modulation process within a modem creates a combined signal (carrier plus data signal) to produce output signals suitable for transmission.

Modulation. The term "modulation" refers to the variation of the amplitude (power), frequency, or phase (angle) of the carrier in accordance with an input source to represent information, sound, or video on a transmission medium.

Baud. Baud is an old term that denotes the number of sine waves or signal changes per second. It is frequently and incorrectly used interchangeably with bits per second to represent the number of bits that can be transmitted across a data communications link per second.

Bits Per Second (BPS). Bits per second is a measurement of how many bits can pass across a data communications link in one second. The terms "baud" and "bits per second," although representing different activities, may in some cases be used interchangeably Figure 5.1 shows the proper placement and use of baud and bps. Keep in mind

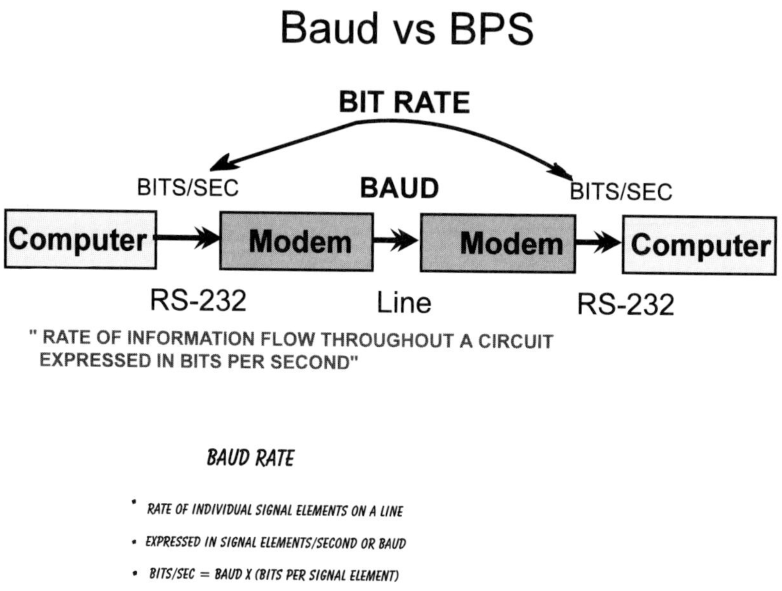

Figure 5.1 Baud vs BPS

that they do not mean the same thing. Baud represents the number of signal changes per second or cycles per second (frequency of the link), whereas bps represents the bit rate of the link.

Decibel Measured (Dbm) Levels. Decibel measured is the strength or intensity of a transmitted signal placed on the communications circuit. Various standards govern the Dbm level for equipment compatibility.

Bandwidth. Bandwidth as it relates to modulation is the total frequency on the carrier available for modulation of the signal. Figure 5.2 shows the total bandwidth available for the information signal and the signal overhead. The signal does not completely overlay the entire available bandwidth but does use most of it.

The term bandwidth implies some measured amount of frequency available on a channel. Two general bandwidth categories have been defined for communications: broadband and baseband. Figure 5.3 compares baseband and broadband transmissions.

Broadband Transmission

Broadband transmission is also referred to as wideband transmission because the bandwidth is usually wide on the transmission medium and can reach well into the gigahertz range (billions of cycles per second) depending on the media type. The available bandwidth is split (multiplexed) into a number of frequencies or channels by intelligent devices. Broadband transmission is employed in Community Access Television (CATV) cable systems and can carry analog as well as data (digital) signals. CATV cable is usually capable of handling a 500 Mhz bandwidth broken into 6 Mhz channels. A single television channel requires that amount of bandwidth for video, audio, and control information.

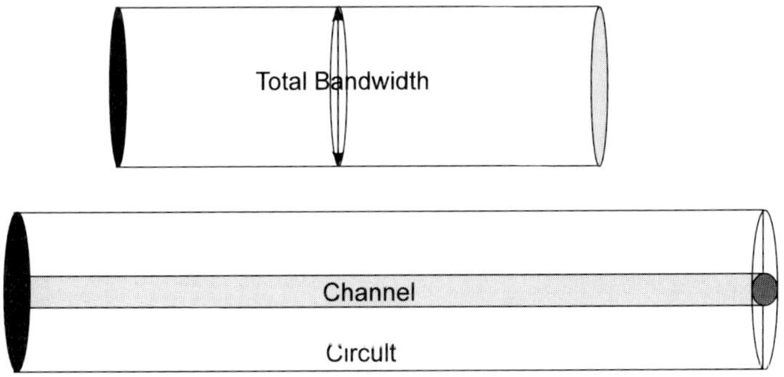

Figure 5.2 Illustration of Channel, Circuit, and Bandwidth

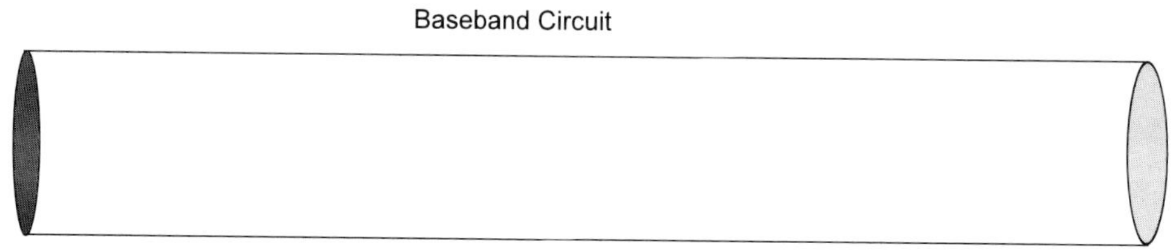

* Baseband is single channel digital
* May be Time Division Multiplexed

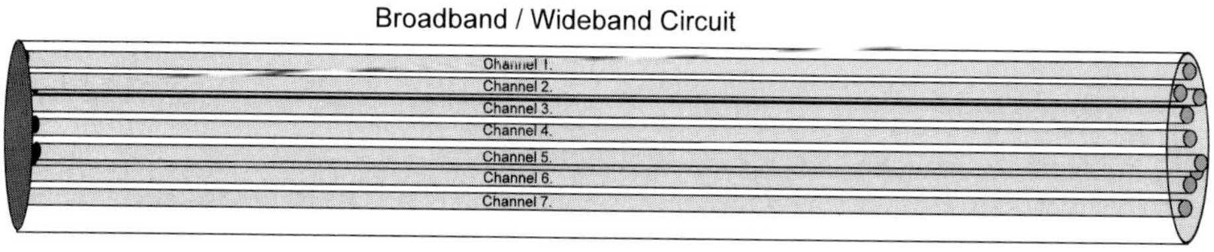

* Is channelized for better utilization
* May be analog or digital
* Space is allocated between channels for noise control

Figure 5.3 Examples of Baseband and Broadband Circuits

Baseband Transmission

Baseband transmission refers to one unit of information (a unit of information is data and control bytes) transmitted over a single channel one message at a time. Baseband transmission is extensively used in local area networks and consists of a single-channel, high-speed data path for the interconnection of workstations and computers. Because baseband transmission is used primarily for digital transmission under controlled distances, data transfer takes place at very high speeds, often well into the range of megabits per second.

MODES OF TRANSMISSION

For wired, cabled, or radio frequency communications, three transmission modes exist for the transfer of information across a network link: simplex, half-duplex, and full-duplex. Use of a specific mode depends on hardware requirements and budget constraints. Most installed lines are full duplex because carriers primarily support only full-duplex configurations. (Communications hardware and software also operate using these concepts.)

Simplex Transmission

In the simplex transmission mode, the operation of a channel or link is limited to one direction only with no capability of reversing the directional flow. The terminal will either be transmitting or receiving. Simplex transmission is analogous to one-way streets, escalators, doorbells, and home security systems.

An example of the simplex transmission mode used in data processing and data communications is the remote station printer. This device is used to create printed output such as that used in airline ticketing systems. Most other types of printers require other than a simplex link to communicate status information back to the sending device. Bear in mind that some remote printers have large enough buffers to accept the entire transmission and do not need to communicate status information back.

A badge reader is another example of a device that can use simplex transmission. In some badge readers, the identification card is read and the data is transmitted to the intelligent device on a simplex link. The intelligent device validates the card and sends a signal to the door lock to open using another simplex circuit.

Card readers also use simplex links. CRTs without keyboards, like those found in airport terminals, are excellent examples of simplex transmission, because they are output devices only.

In modern data processing/data communications, simplex transmission is not frequently used because the receiver normally needs the ability to send error or control signals to the transmitter. Returning signals back to the transmitter is usually required for acknowledging receipt of the data or indicating that a problem has been encountered.

Half-Duplex Transmission

Half-duplex is a transmission mode in which the channel operates in either direction, but not both directions at the same time. In most communications environments, half-duplex is the predominant transmission mode in use today. It is important to note, however, that with the speed of lines rapidly increasing and the costs correspondingly decreasing, most lines are in fact full-duplex. Half-duplex operation is largely due to the capabilities and costs of the communications equipment and the type of data communications taking place. In a conversational environment, in which one side will be listening while the other is sending, this probably would employ the use of a half-duplex link.

A major disadvantage of half-duplex transmission is the time required to change the direction of the data flow so that the other end can respond. In a typical half-duplex operation, line reversal can add up to 500 milliseconds (ms) of delay time on some types of links. This is referred to as line turnaround time and can range from 50 to 500 ms, depending on the distance the signal must travel.

An example of the use of half-duplex transmission in data processing and data communications is the line printer. Most printers operate using half-duplex transmission because they must send control and status information back to the sending device. For example, a printer with a paper jam needs to inform the CPU it cannot accept any more data at the present time. Terminals also typically operate in half-duplex mode because they normally either are sending or receiving but do not perform both functions concurrently.

Some modems actually transmit in full-duplex mode using half-duplex facilities by splitting the channel in half. Each side of the link uses one of the subchannels. The only problem with this technique is the reduced speed of the link in each direction. It is usually half of the total available speed because only half of the bandwidth is available for each direction.

Full-Duplex Transmission

Full-duplex transmission is considered bi-directional in that data communications occurs in both directions at the same time. Two simplex channels are required for full-duplex transmission; each channel carries signals in one direction. One might look at full-duplex channels as two paired simplex channels.

One example of full-duplex transmission in data processing and data communications is the channel-to-channel interface between two processors. This is required when the processors are in continuous communications with one another. The Tandem fault-tolerant computer system and its communications network operate in a full-duplex environment. Terminal controllers that interface to multiple terminals also may operate in full-duplex mode, which would allow them to send and receive data to and from their terminals and the processor at the same time.

Full-duplex lines are often used to support half-duplex terminals so that no turnaround time delay is encountered when reversing the direction of the data flow. While this is more expensive because of full-duplex line charges, it reduces the amount of time required for a device to acknowledge receipt of the transmission, which can be considerable over long distances.

GRADES OF SERVICE

An incredible number of link service providers support the world of data communications. Standards have been established by both the industry and international standards organizations to ensure worldwide compatibility between communications providers. Various types of services are offered by American Telephone & Telegraph (AT&T), Microwave Communications Incorporated (MCI), the regional Bell operating companies, and the independent operating companies. In addition, other private companies also offer competitive services for data transmission.

The communications line, whether it be wire, fiber-optic cable, microwave circuit, or other types of circuits, is the physical portion of the communications link. The physical link normally is divided into two or more channels depending on its rated bandwidth. A channel is a logical path through the physical link over which information and data flow.

Channels are described in terms of transmission type or transmission mode. Grade of service, or the link's information-carrying capacity, refers to the volume or speed in bits per second that a channel can handle. Grades of service as discussed here refer to the various types of service offered by most telephone operating companies.

Bit transmission speed is determined by the available frequency bandwidth of the channel. Bandwidth is one of several important characteristics that you must consider when configuring a communications link. Bandwidth is the difference in hertz (or cycles per second) between the lowest and highest available frequencies of that channel. This available frequency bandwidth determines the overall carrying capacity of the transmission link.

Be careful to analyze bit rate requirements when ordering and installing lines. It would be foolish to install a very fast megabit capacity wideband line for personal computer communications; that would be comparable to a truck driver who uses an 18-wheel truck to transport two trash cans to the dump. Capacity planning is vital to the design and implementation of a data communications network. It often can take extended periods of time and a great deal of money to install or upgrade line capabilities if future usage projections were not accurate.

The major telephone companies generally offer many grades of telephone service—they are broken down into two categories: voiceband and wideband services. Specialized wideband services such as Digital Data Service (DDS) and T1 are usually offered to organizations that need bulk or high-speed data transmission capabilities. Grades of service are generally defined in terms of available bandwidth or digital bit rate:

- The voiceband grade is considered a low-medium-speed service with a frequency of 300 to 3300 hertz, for a total bandwidth of 3,000 hertz. This is also the bandwidth of a standard voice-grade circuit used between homes and central telephone offices. Bit transmission speeds range from 60 bits per second up to 33,300 bits per second. This grade of service is primarily used for dial-up switched facilities but can also be leased as a dedicated line.
- The wideband or broadband grade is considered a high-speed service and can be either analog or digital. Analog wideband service extends from 3,300 hertz into the megahertz range. Analog transmission speeds range from 19,200 bits per second to greater than 1.544 million bits per second. It may be implemented for multiple voiceband channels that are multiplexed into groups, supergroups, or portions of line groups. It is used for the transmission of large amounts of data over high-speed links. If the circuit is digital, it will support bit rates well into the billions of bits per second and standard multiplexing techniques are used which are described in the chapter on hardware.

Not all hardware links support all grades of services, so be sure the lines selected support the entire geographic network. This is particularly true for the faster links because they are designed more for high-speed multiplexed service that may not be available in certain locations. Slower links, associated with most grades of service, are usually more widely available.

TRANSMISSION METHODS

As was previously stated, transmission modes apply to the direction of the transmission and the channel. Additionally, there are two basic transmission methods—asynchronous and synchronous—for moving the information across the link. Each of these methods is used with specific network architectures, hardware, transmission speeds, and transmission protocols.

Asynchronous Transmission

Asynchronous transmission is really an offshoot of the old Baudot telegraph system and the equipment that was used. Because of the variable number of pulses associated with each character to be transmitted, some method was needed to indicate to the receiving device, the end of one character and the beginning of another. This was accomplished by inserting start and stop bits in front of and behind the bits representing the transmitted character. This was necessary because of the way Baudot's telegraph machine was designed. It was basically a rotating drum that needed to be returned to the 0 degree position for the beginning of each character. The insertion of one start bit and one or two stop bits was used to synchronize the starting position of the drum to the start bit inserted before the character.

This asynchronous transmission method was adapted to the data communications environment because it was simple to implement and designed for low transmission speeds. It is also referred to as stop/start transmission. This is because start and stop pulses (bits) are used to separate each character thus allowing random transmission to take place. Figure 5.4 shows the placement of start and stop bits before and after the data bits in a byte. Today asynchronous transmission is primarily used for low and medium speed dial-up connections such as with PC-to-PC communications or PC-to-mainframe communications. This is because the modems are inexpensive and this type of communication typically does not involve the movement of large amounts of data. Note that the addition of start and stop bits adds a 20 percent to 30 percent overhead to each byte transmitted, which reduces the number of data bits transmitted. Also as we move into an era of all digital communications and Internet access, asynchronous communications will be replaced by synchronous communications.

Synchronous Transmission

Because of the inefficiencies associated with asynchronous transmission, another method allows for higher transmission speeds with better error detection and correction capabilities. The synchronous transmission method is the principal transmission method used for high-speed communications. It uses a clocking method to establish timing synchronization. Figure 5.5 illustrates an example of clocking and synchronous transmission.

The synchronous transmission method does not use start and stop bits to denote the beginning and ending of each character being transmitted. The data is framed (blocked),

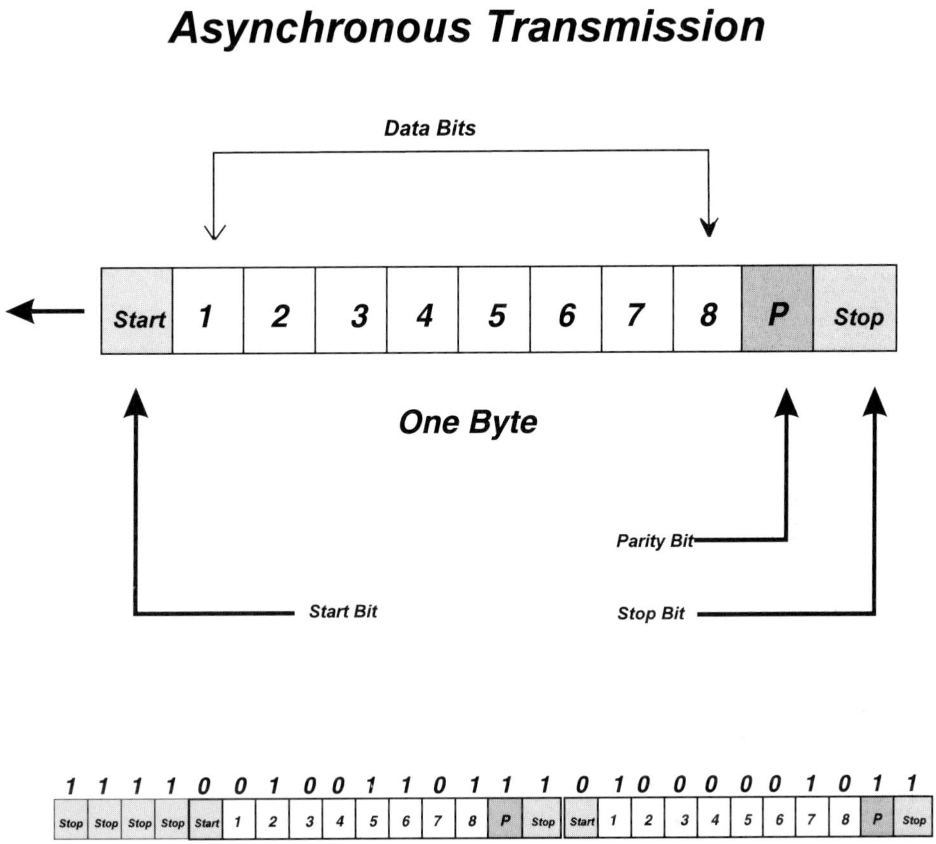

Figure 5.4 Asynchronous Transmission Example

and a synchronization character or bit pattern is placed in front of the frame. This allows for very efficient communications because much of the overhead associated with asynchronous communications is eliminated.

Synchronous transmission requires the synchronization character or bit pattern be transmitted as the first byte of the frame. This byte or bit pattern is used by the receiving side to establish timing. Synchronous communications require that the frame be transmitted at a fixed clock rate. The first byte is used by the sending and receiving modems to establish synchronous timing. This ensures that the transmitting circuits are sending bits out at the same time that the receiving circuits are taking them in. This is known as *bit synchronization* or *clocking*.

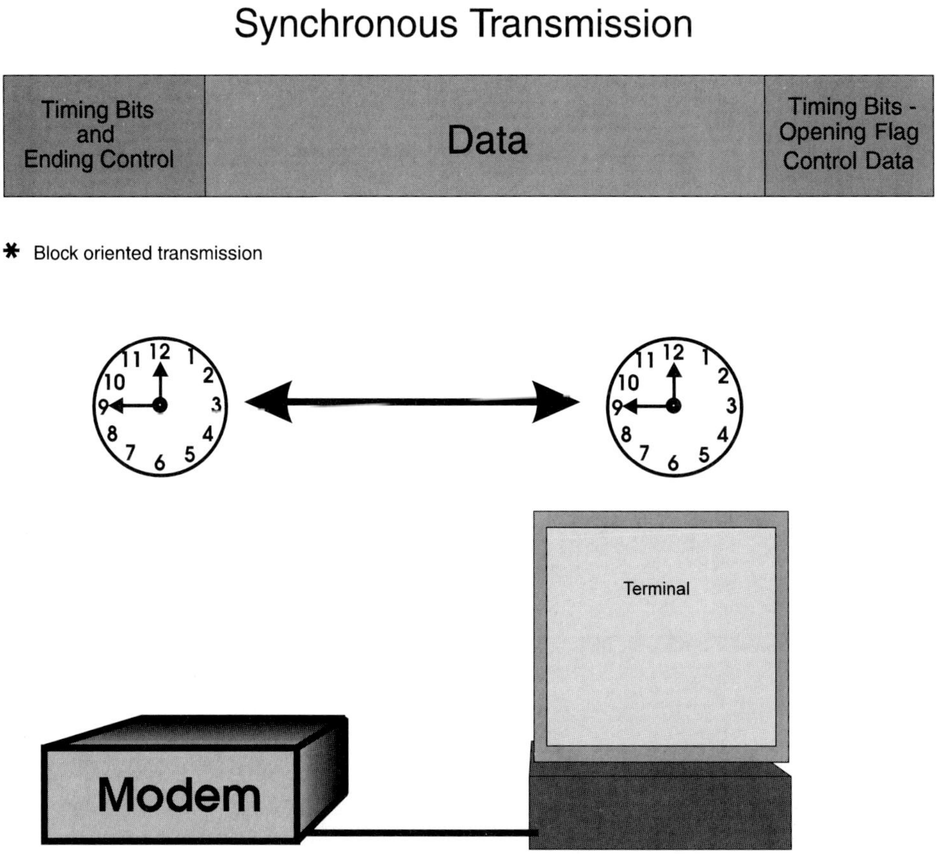

Figure 5.5 Example of Synchronous Transmission

CASE STUDY (CONTINUED)

Up to this point, the bank has not had to support a communications network other than to perform batch file transfers between service centers. It has accomplished this by using standard, low-speed telephone facilities. The link that connects the two service centers is a half-duplex, low-speed line that operates asynchronously because there was no need for full-duplex, simultaneous transmission. The bank uses a voice-grade link and asynchronous half-duplex transmission method because of the cost associated with the transmission components and telephone facilities. Figure 5.6 shows the evolution of the bank as it implements a data communications network.

The new implementation plan for the bank mandates extensive communications capabilities to support all of its operations. This calls for a variety of high-speed wideband

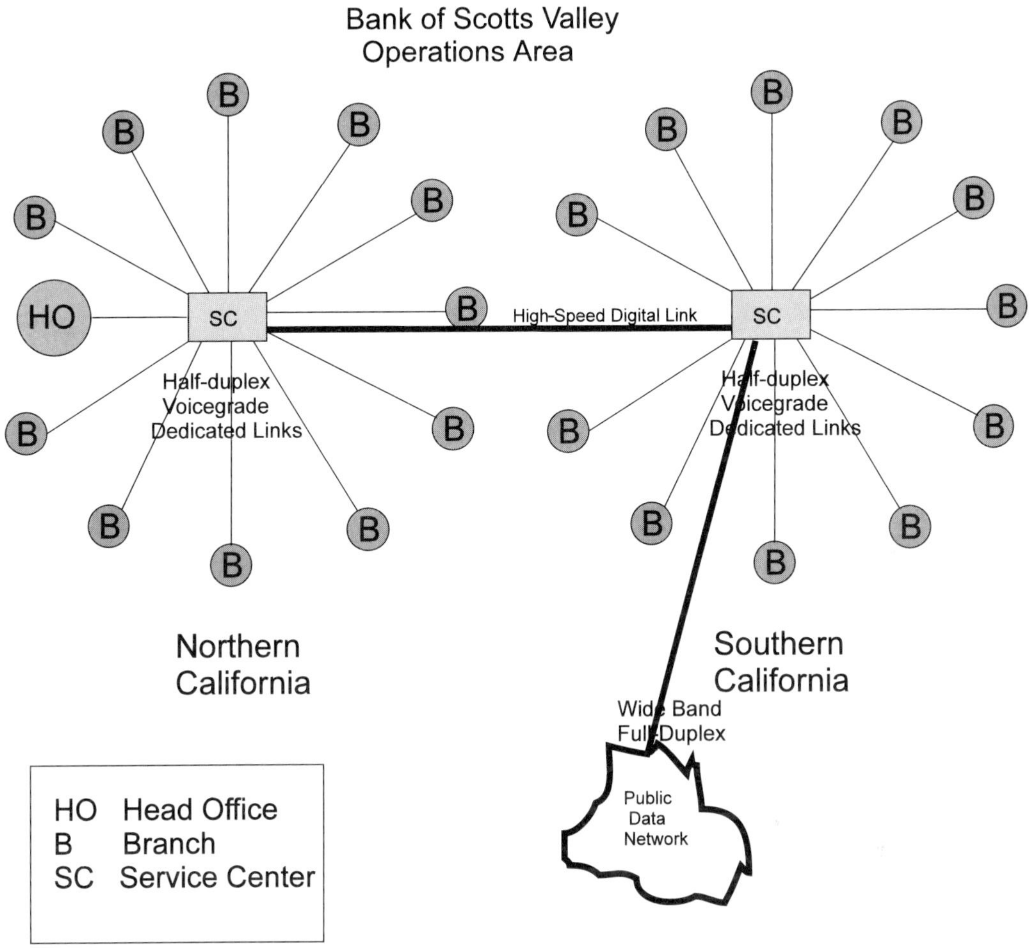

Figure 5.6 Bank of Scotts Valley Operations Area

and low-speed voiceband links to interconnect the branches and service centers. Each branch will have a full-duplex, leased voiceband link to its regional service center, and a second link that will be used for backup in the event of a failure. All communications will use the synchronous transmission method.

The service centers will be synchronously interconnected using high-speed, full-duplex wideband links. Multiple lines will be installed to ensure that back-up capabilities are available in the event of a line failure. Within the branches, each PC workstation will be hard-wired to a central transmission point within the building using a Local Area Network (LAN) and high speed connections.

REVIEW QUESTIONS

1. Define the term "channel."
2. What is baseband transmission?
3. True or false: Simplex transmission is transmission in both directions but not simultaneously.
4. True or false: Full-duplex transmission is the most common transmission mode.
5. True or false: Half-duplex transmission allows two-way communication at the same time.
6. What are the two grades of transmission service available for data transmission?
7. What type of transmission, analog or digital, are current telephone transmission networks designed to support?
8. What type of transmission mode are most types of data communications equipment designed to support?
9. What are the two transmission methods available for data communications?
10. Which transmission method requires clocking information?

CHAPTER 6

Transmission Concepts

INTRODUCTION

Several methods are available for moving data across a communications link, including radio frequency broadcast transmission, infrared radio transmission, and modulated land-based transmission over a wired media such as coaxial cable or twisted-pair wire. The process of converting the digital ones and zeros into a form suitable for transmission and placing them onto a link is known as modulation.

This chapter introduces basic modulation as it relates to wired media and radio frequency transmission. Several standard radio frequency modulation techniques are currently in use by international modem manufacturers. These modem standards, which are governed by the CCITT, permit compatibility between different vendors' products.

For low-speed communications, modems use simple modulation techniques because they work well, do not require extensive error detection and correction logic, and are inexpensive to manufacture. High-speed modems use more complex modulation techniques because the circuitry provides better error detection and correction and is more efficient.

The radio frequency spectrum is the principal vehicle for transporting modulated information. Figure 6.1 shows the frequency spectrum associated with hearing and the human ear. This graphic represents the bandwidth, or range of frequencies, associated with human hearing. Although the audible frequencies are limited, there is no real limit to the overall frequency spectrum available for use in communications, except that of the equipment.

Initially, the limitation imposed by the audible frequency spectrum dictated the frequency range available for transmission, because it was governed by the bandwidth of the voice-based telephone network. While this narrow range of frequencies no longer governs the available bandwidth for electronic communications, the voice frequencies did estab-

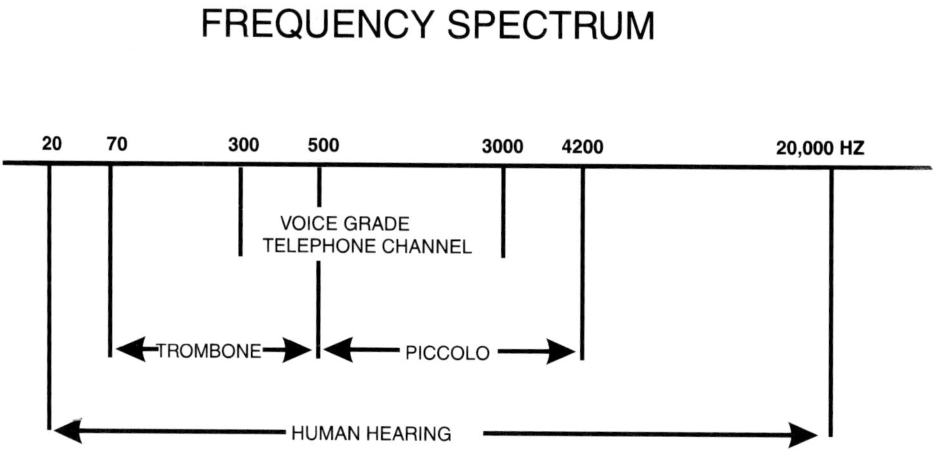

Figure 6.1 Voice Grade Frequency Example

lish the basic foundation for modulated transmission. Data communications bit rates originally were limited to the voice bandwidth (300 Hz to 4000 Hz) of the telephone network, but with the wideband circuits currently available, no practical limitations exist.

Modulated transmission bit speeds are directly related to several additional factors, including the physical properties of electrical transmission, network limitations, and hardware. A correlation between data rate and available bandwidth governs overall transmission capabilities. For data communications, a frequency spectrum (channel bandwidth) broad enough to handle the required data rates must be available to support the bit rate requirements.

For many years, two theories governed radio frequency transmission activities: the Nyquist theorem and the Shannon theorem.

Nyquist Theorem

In 1933, a theory was developed by Harry Nyquist that stated that a maximum bit rate cannot exceed twice the frequency bandwidth of an allocated channel. This suggested that a frequency bandwidth of 2,400 Hz could handle a maximum bit rate of 4,800 bits per second. The theory applies to a simple binary signal applied to two different amplitudes or two different frequencies.

Shannon Theorem

C.E. Shannon developed a formula that relates transmission speed to signal power and error rates. Very simply put, Shannon stated that as the power level increases, so does the noise level. This combination created a limitation on the speed at which the link could operate.

Both theories are still valid today, with some modification. However, the derived modulation techniques that incorporate error correction assume a much different technological approach. This approach mitigates the impact and restrictions set forth by Shannon and Nyquist in their respective theories.

Table 6.1 defines the general frequencies allocated worldwide for radio and communications activity.

Table 6.1 Frequency Allocations

Frequency	Band Designation	Usage
3000G HZ	Unassigned	Unassigned
300G HZ	EHF Extremely High Freq.	Satellite Communications Microwave Transmission Amateur Radio
30G HZ	SHF Super High Frequency	Satellite Communications Radar Microwave Transmission
3G HZ	UHF Ultra High Frequency	Microwave Transmission Amateur Radio Pub. Svcs. UHF Television
300M HZ	VHF Very High Frequency	Amateur Radio (HAMS) VHF Television Broadcasts FM Radio 88M HZ–108M HZ
30M HZ	HF High Frequency	Citizens Band Radio Diathermy Machines
3M HZ	MF Medium Frequency	Radio Navigation AM Radio 500 K HZ–1.6 M HZ
300K HZ	LF Low Frequency	Radio Navigation Maritime Communications
30K HZ	Very Low Frequency Upper end of hearing range	Navigation, Weather Comm. Piccolo
3K HZ	Voice Frequencies	Human Voice Bass violin
300 HZ	Extremely Low Frequency Low end of hearing	Power lines Submarine Communications
30 HZ		

MODULATION

Modulation is the mechanism or technique that "pushes" the data-bearing signals across the link. Modems modulate the digital data into a form suitable for transmission over analog lines. Technically, modulation is the process of varying the characteristics of a wave in accordance with another wave or signal. The modulation of a carrier (a carrier is an unmodulated electrical signal on the link that can be modified to carry data bits) on a channel produces an information signal that can be sent from one location to another.

A carrier is used to establish a logical connection from point A to point B. The carrier is created by applying voltage or current to the communications link (transmission circuit) at a specified frequency. Figure 6.2 shows an unmodulated carrier and a modulated carrier. Modulation of that carrier can be accomplished by one of the following techniques:

- By keeping the voltage or current constant on the link from point A to point B and then varying the signal timing to create different frequencies representing one or zero bits
- By keeping the frequency (signal timing) constant and varying the amplitude (voltage) up or down in order to represent one or zero bits
- By varying both the frequency and the amplitude of the signal at the same time

Figure 6.3 illustrates a signal as it attenuates outward from its source point.

BASIC MODULATION TECHNIQUES

Three basic analog modulation techniques (amplitude, frequency, and phase) are available for data transmission. Figure 6.4 shows the sine wave for the three basic modulation techniques. Several derived modulation techniques are also available, but because of their

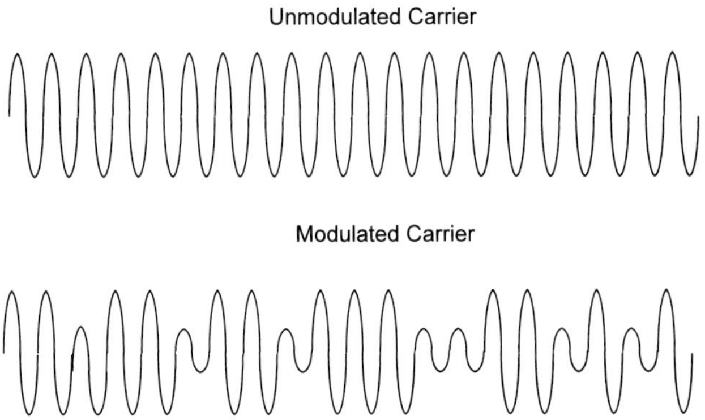

Figure 6.2 Unmodulated and Modulated Carriers

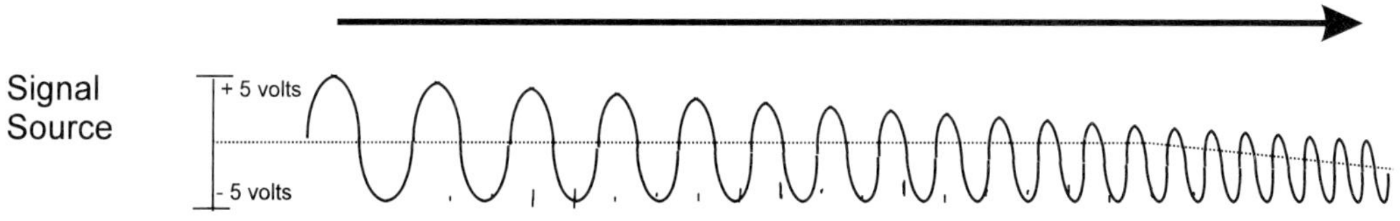

Figure 6.3 Signal Attenuation

complexity, they will not be discussed in detail in this book (though most international modems in use today employ derived modulation techniques). A derived modulation technique is one that combines the three basic techniques in order to form a derivative. The derived technique has the ability to carry more data than the basic technique. Additionally, some modulation techniques use encoding schemes to pack more bits into the transmission and then decode them at the receiving side.

Amplitude Modulation

Amplitude modulation (AM) is a descendent of the telegraph system in which a circuit attached to a battery was keyed on or off. This keying activity created a current that represented dots and dashes defined as characters of the Morse Code character set.

A basic method of transmitting analog signals, amplitude modulation operates by varying the amplitude or magnitude (voltage) of the carrier signal according to the input data source. Figure 6.5 shows an analog waveform. To put it simply, the voltage is modified up or down depending on whether the input digital signal is a one or a zero bit. In amplitude modulation, the carrier signal timing remains constant; that is, 1,200 cycles per second, 2,400 cycles per second, and so forth. Only the voltage (amplitude) is changed.

On a 2,400-baud line, the amplitude or current will change 2,400 times to indicate zero and one bits coming from the digital source. This results in a communications rate of 2,400 bps sent over a 2,400-baud line.

Because amplitude modulation increases the noise level as the signal level increases, it is no longer used extensively by itself. However, amplitude modulation is still used in combination with other modulation schemes.

Frequency Modulation

Frequency modulation (FM) is a method of data transmission in which the frequency changes to correspond to the changes in the information signal. Figure 6.6 illustrates a frequency modulated waveform. In frequency modulation, the number of cycles per second (hertz) varies according to the source (more cycles per second for one bits and fewer cycles per second for zero bits). Unlike amplitude modulation, the bits per second and the number of cycles per second are not the same with frequency modulation.

If you were able to listen to the analog signal of ones and zeros in frequency modulation, one bits would sound a higher tone than zeros bits. Generally speaking, two sets of

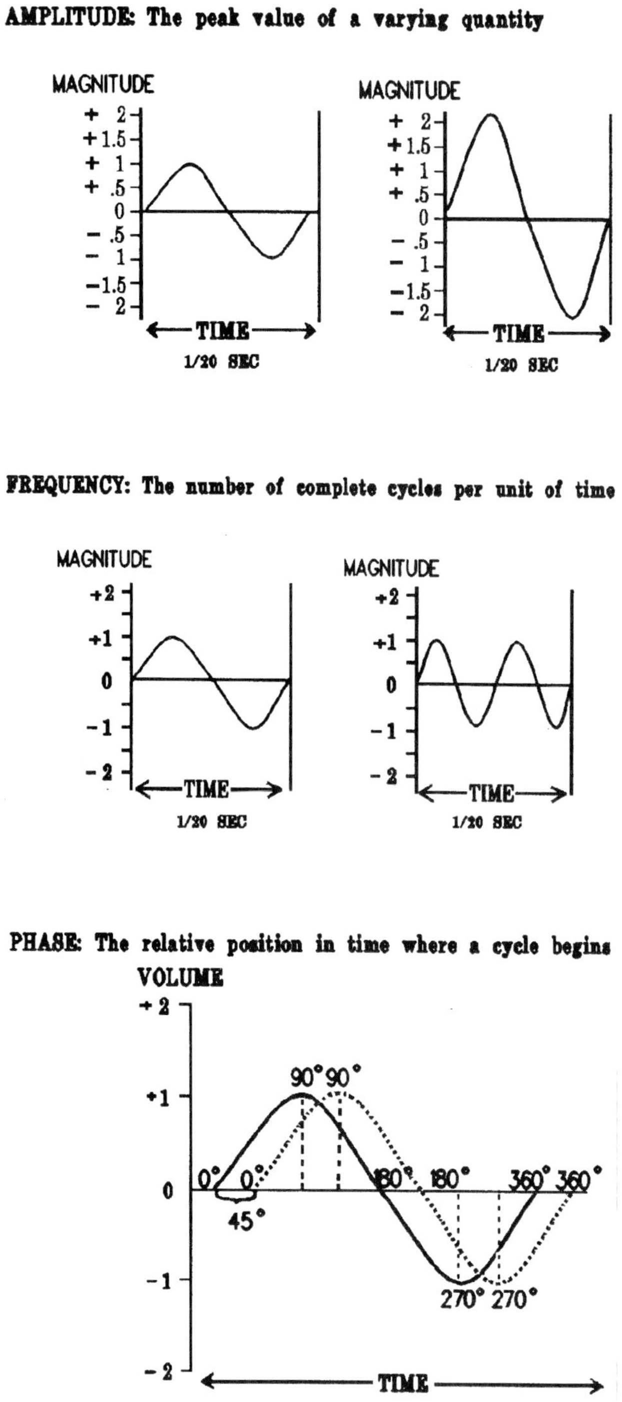

Figure 6.4 The Three Basic Modulation techniques

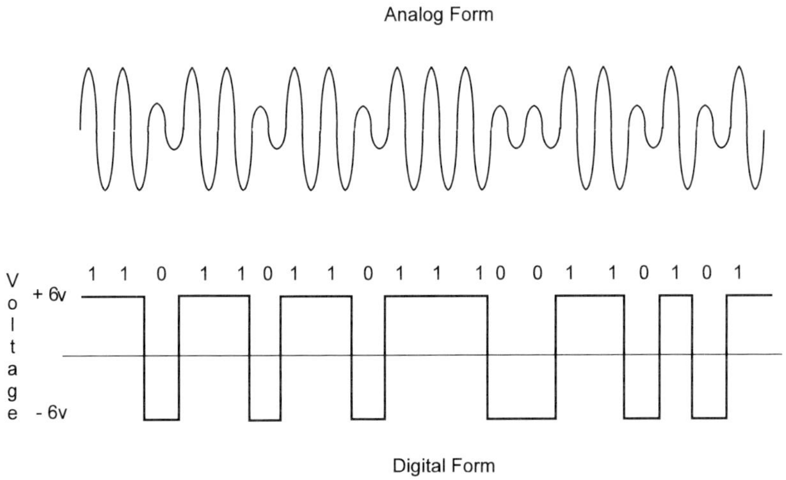

Figure 6.5 Amplitude Modulation

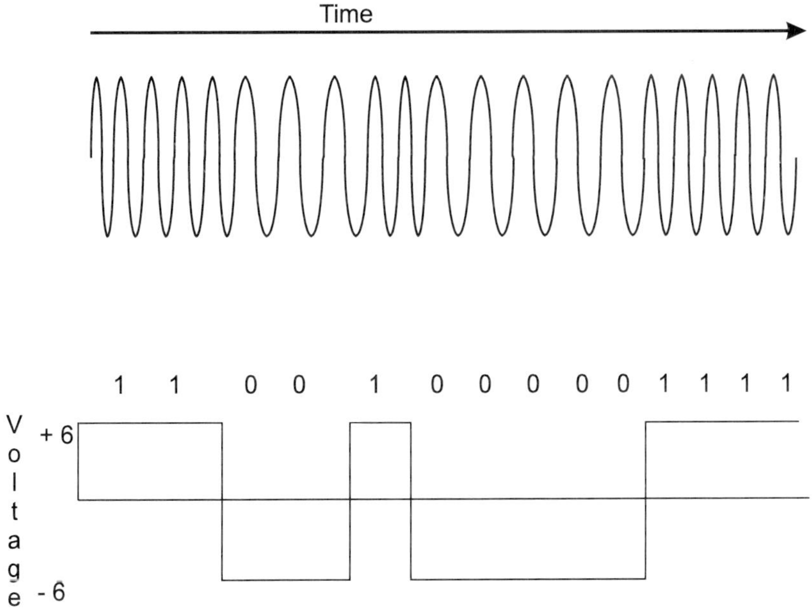

Figure 6.6 Frequency Modulation

frequencies are used for FM transmission; 2,025 Hz and 2,225 Hz are used for the first set of carrier frequencies, while 1,025 Hz and 1,225 Hz are used for the second set. Note that although these are the commonly used frequencies, some manufacturers elect to use other frequencies.

Until recently, frequency modulation was the most popular modulation technique because it is less prone to interference than amplitude modulation techniques; however, modem manufacturers have replaced FM with newer derivative modulation techniques. This is because of the derivatives ability to handle higher bit rates on the same type of lines. Nevertheless, frequency modulation is still used in lower speed modems because of the low noise aspects of the signal.

Phase Modulation

Phase modulation (PM) is often referred to as multibit modulation because it is used in modems that have the ability to produce more than one bit per cycle of signal (see Figure 6.7). In PM, the angle or phase of the carrier signal can vary (change its shape). This change occurs based on the ones and zeros of the digital signal. There are many types of phase modulation, but phase shift keying (PSK) is normally used for data transmission. Phase modulation is considered to be a derivative modulation technique and has many variations. This discussion addresses PM in its simplest form; explanations of more complex variations are beyond the scope of this book.

As already discussed, the carrier signal transporting data on a communications circuit operates at a specified set of frequencies or cycles per second. If one cycle of a carrier is viewed as a 360 degree circle, that cycle can be divided into degreed sections. Phase modulation is accomplished by shifting the phase at certain degrees within the circle to represent one and zero bits.

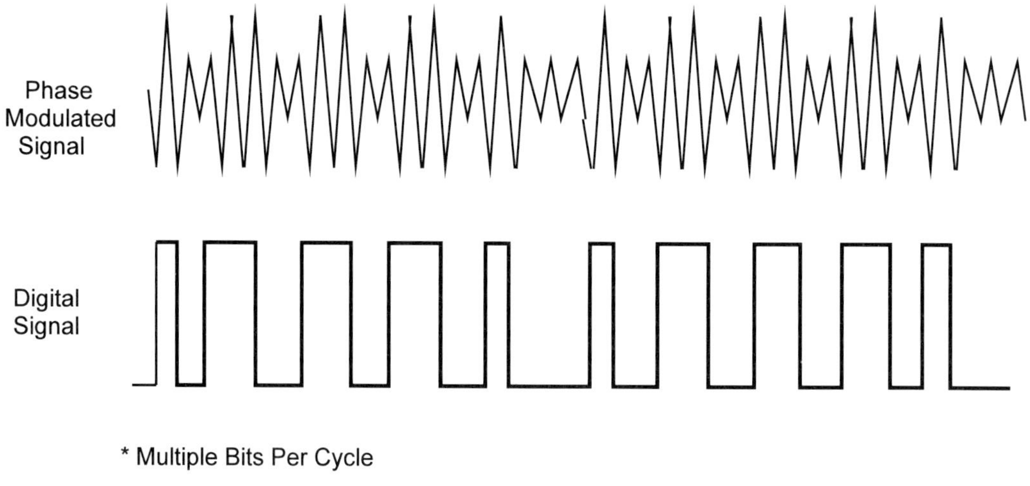

Figure 6.7 Example of Phase Modulation

Phase modulation is accomplished by interrupting the amplitude (voltage) for a duration of time that corresponds to the degreed position defined for combinations of zero bits and/or one bits assigned to and predefined in the 360 degree circle. When the phase is shifted, the carrier begins again at zero degrees. By controlling the timing, one bits and zero bits can be represented at the specific degreed locations in the circle.

Thus, it is possible, using phase modulation techniques, to create two bits per cycle (dibits), three bits per cycle (tribits), or four bits per cycle (quadbits). With this modulation technique, the rate in bits per second can exceed the baud rate.

Variations

There are many derived modulation techniques that permit more efficient transmission of data across analog communications lines. Examples include double vestigial sideband (DSB/VSB), frequency shift keying (FSK), phase shift keying (PSK), quadrature/differential phase shift keying (QPSK/DPSK), quadrature amplitude modulation (QAM), Trellis coded modulation (TCM), and pulse code modulation (PCM). All of these derived modulation techniques use combinations and variations of frequency, amplitude, and phase modulations.

Trellis Coded Modulation. TCM is used by higher-speed modems because it employs forward error correction techniques and provides a more reliable high-speed data transmission rate. Because of the complexities of TCM, specific technical information is not presented here; however, this modulation technique uses an extra bit per cycle per second (cps) and encodes the bits into a transmission pattern for decoding at the other end.

Quadrature Amplitude Modulation. QAM uses two carriers of the same frequency that are 90 degrees apart and added together. It is a derivative of phase modulation and has the ability to create multiple bits per cycle.

The use of derived techniques increases the bit rate of the line over that provided with basic modulation techniques. At higher bit rates, trade-offs must be considered due to variables such as line quality and signal-to-noise ratios.

Several modulation schemes using phase encoding techniques are designed for the conversion of voice signals into a digitally encoded analog form. The two most common are *pulse-code modulation* (PCM) and *adaptive differential pulse-coded modulation* (ADPCM).

PCM functions by sampling the voice signal 8,000 times a second, quantizing each signal sample into a numeric digital value and creating an eight-bit word for each sample. Figure 6.8 illustrates an analog to digital conversion process. This creates a 64,000-bps signal that encodes the entire voice spectrum (300 Hz to 3,330 Hz) .

ADPCM operates similarly to PCM but encodes only a four-bit sample that represents the difference between the serial samples (current and next waveform sample). If there is no difference between samples created by pauses or silences, a two- or three-bit

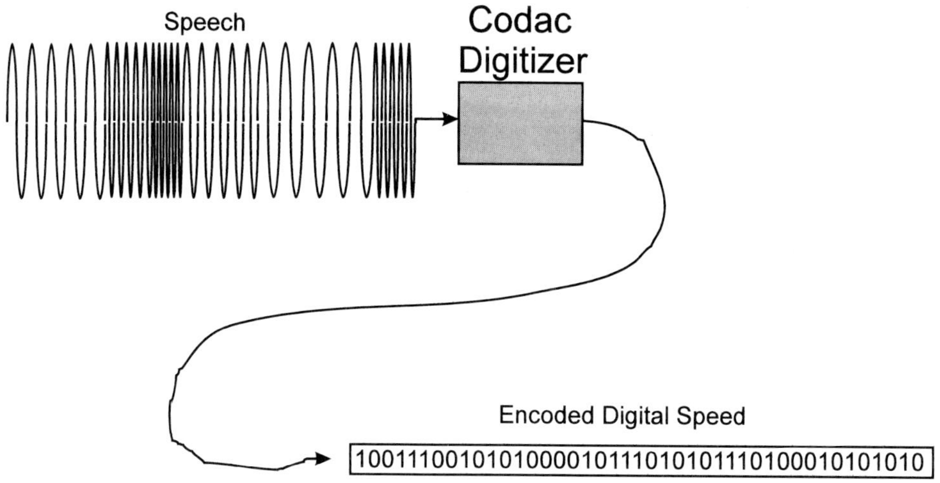

Figure 6.8 PCM Digital Process

sample is formed. This technique allows ADPCM to create a full spectrum signal in 32,000 bits per second.

DIGITAL TRANSMISSION

Pure digital transmission is represented by changes in voltage on the communications media. Figure 6.9 depicts a digital transmission signal. Some digital signals are nothing more than the voltage turning on and off. Other digital methods use a constant voltage having positive and negative values that represent one and zero bits. At the transmitting side, the voltage polarity changes from positive to negative depending on whether one or zero bits are coming from the computer. At the receiving side, the voltage is monitored

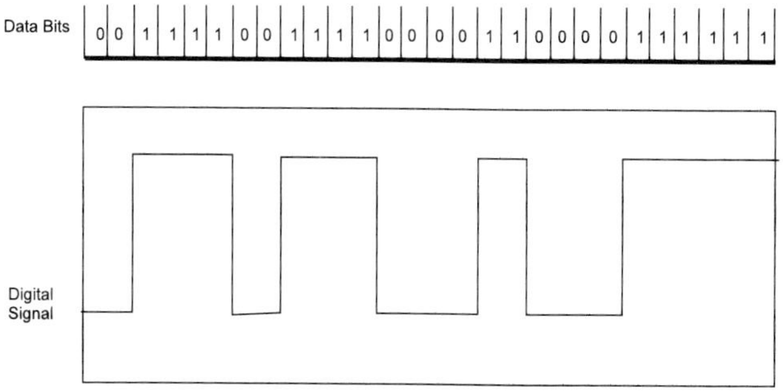

Figure 6.9 Digital Transmission Signal

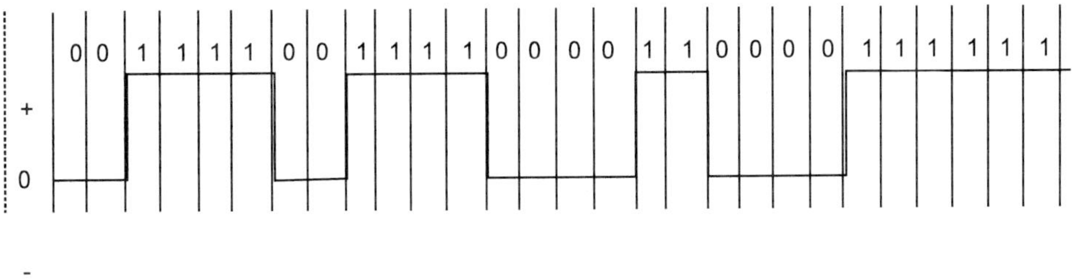

Figure 6.10 Bipolar Transmission Signal

and, when it makes the transition from a positive to a negative value, one and zero bits are derived.

Typically, digital signals from a computer or terminal are not in a form suitable for direct placement onto the media partly because of the low power levels used in the computers. Even though the signal from the computer is a pure digital signal, it often must be changed or shaped to be transmitted. This requires a DSU/CSU.

Some of the digital transmission techniques currently used are bipolar, return-to-zero (RZ), and non-return-to-zero (NRZ).

Bipolar Current Loop

Bipolar digital transmission uses positive and negative voltages to represent digital information. Figure 6.10 illustrates a bipolar signal in which one bits are always a negative voltage and zero bits are represented by a positive voltage.

Return-to-Zero (RZ)

RZ digital transmission uses a self-clocking transmission method. Data bits operating at a given bit rate begin on the link at a zero reference level regardless of value and go through a transition during bit time or signal period. Figure 6.11 presents an RZ digital signal. Zero bits begin at a zero reference level and go negative for half a bit time or signal period. During the second half of the signal period, the signal turns positive and then returns

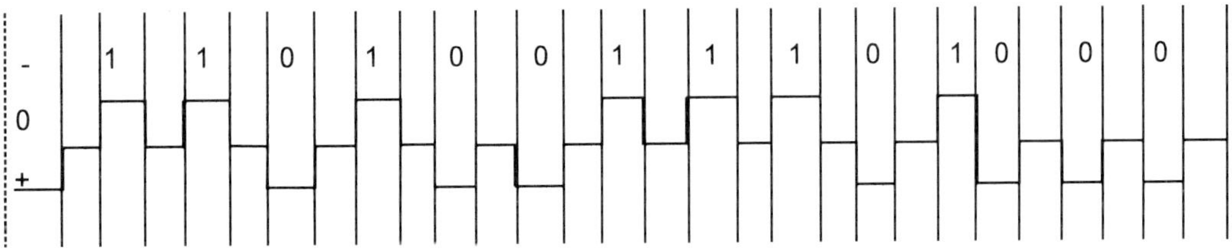

Figure 6.11 Return-to-Zero Transmission Signal

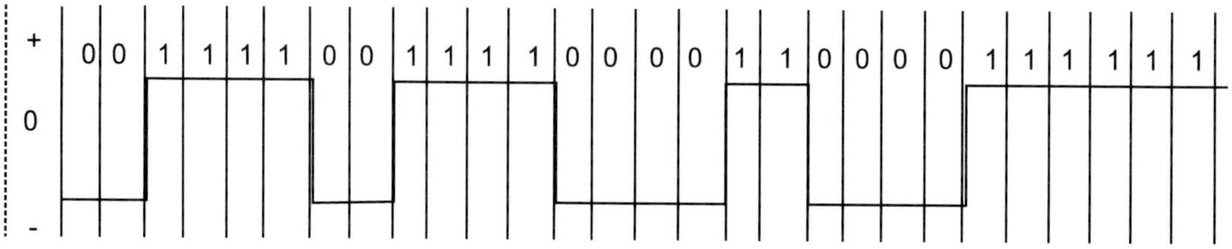

Figure 6.12 Non-Retun-to-Zero Digital Transmission Signal

to a zero reference level. For one bits, the exact opposite signals occur. Two different RZ signals can be used and care must be taken to ensure that both devices are using the same form of return-to-zero transmission.

Non-Return-to-Zero (NRZ)

NRZ also can be referred to as differential encoding and is unique in that transitions only occur when a one bit is present. Figure 6.12 illustrates an NRZ digital signal. The first one bit encountered causes the signal to go positive; the signal remains positive until the next one bit causes the signal to go negative and remain there. When the next one bit is encountered the signal will go positive. One bits may be represented by either a positive or negative signal; therefore, synchronization must occur between the communicating devices because it is the transitions in the signal that represent bit values, not the signal value itself (positive or negative).

CASE STUDY (CONTINUED)

The Bank of Scotts Valley uses many types of data transmission equipment on its data communications links. As stated earlier, the bank will use both dial-up and dedicated links for communications with the service centers. Modems using TCM and operating at 19.2 Kbps or 9,600 bps will be used over the dial-up or voice-grade dedicated links. The bank's digital links used between the service centers will employ the telephone company's signaling and digital transmission formats.

REVIEW QUESTIONS

1. Name the three basic modulation techniques used in data communications.
2. Describe frequency modulation.
3. Describe amplitude modulation.
4. What is the difference between baud and bps?
5. What are dibits, tribits and quadbits?

6. What is phase modulation?
7. Does TCM include a coding technique?
8. A pure digital signal can be defined as?
9. What is PCM?
10. Name two digital transmission techniques.

PART IV

HARDWARE, MEDIA, AND INTERFACES

CHAPTER 7

Hardware Links

INTRODUCTION

Various types of wiring are available for video, audio, and digital information transport. A point of emphasis is that most communications systems today are designed to support voice transmission or analog data communications. This is rapidly changing, however, due to new technologies such as ISDN, low-cost DDS links, and new equipment that supports pure digital transmission. Nevertheless, traditional cabling that maintains analog network facilities is still the most predominant media in use.

The type of cable link best suited for a user's application is usually dictated by specific requirements, which include but are not limited to environment, security, bit rates, and availability of transmission and support equipment. These should be carefully evaluated by both data communications planners and personnel responsible for the facilities prior to cable link selection. The availability of network connections and the need for high-speed or low-speed capabilities also play an important role in the decision-making process.

Another consideration is the integration of voice and data onto the same transmission link. The types of services to be combined (e.g., digital voice) may dictate what is required. In reality, it is common practice among most carriers to incorporate a variety of different links in situations where distances are great. This is not necessarily by the user's choice, but rather because of the various telephone circuits that are available from the carrier.

WIRE-BASED TRANSMISSION

Two- and Four-Wire Twisted Pairs

Two-wire and four-wire twisted pairs used for telephone circuits are referred to as single-ended or unbalanced circuits. This is because the signal strength is not balanced from end to end. The signal is carried over one wire with the ground return on the other wire of the pair. Two-wire twisted pair circuits are usually located between the home and the central office or between an organization's facilities and the central switching office. This type of link is a local loop between the location of the handset and the local telephone office.

Four-wire twisted pairs are the wiring scheme of choice for many organizations that need wiring for data communications. This is primarily because newer communications equipment can better resolve error conditions associated with the transmission media. Four-wire twisted pairs also allow higher data rates and a wider bandwidth than two-wire twisted pairs and can provide full-duplex transmission.

Twisted-pair wires are manufactured in many configurations, including unshielded wire, shielded wire, and multiple conductors. Twisted-pair wire found in the telephone network varies in size from 19 gauge to 28 gauge, but the most common size used is 24 gauge. While the telephone companies generally limit bandwidth (mainly due to the equipment used), twisted pairs can accommodate much higher bandwidths than those required for telephone communications. Many companies reduce wiring costs by installing twisted pairs and using them to transmit data at rates in excess of 10 megabytes per second.

Recent technology breakthroughs are now allowing very high bit rates to be transmitted over shielded and unshielded twisted-pair wires ideal for use with local area networks (LANs). LAN bit rates can go as high as 100 Mbps using twisted pairs.

Four-wire circuits are commonly found between many of the older central offices (telephone offices) for long distance transmission. This is rapidly changing, however, as a result of the rapid conversion taking place among many telephone carriers to fiber-optic cable. There are usually 24 twisted pairs of wire grouped into each twisted-pair cable, but this may vary from four to 28 conductors. The use of twisted pairs for long distance circuits is rapidly diminishing with the introduction of the newer high-speed transmission methods mentioned earlier.

Four-wire transmission is also used for most business telephone systems because four-wire circuits have superior full-duplex capabilities compared to two-wire circuits. It is also possible to create several channels over a four-wire circuit by using a form of multiplexing.

The conversion of two-wire to four-wire transmission in order to transmit data over long distances must take place at the telephone company's central office. This is done using hybrid coils that transform the signal on the two-wire circuit into one that can be placed on the four-wire circuit.

Standard telephone lines (twisted pairs), when used in the public switched network, are considered to be switched lines that are unconditioned. Data transmission using the

PSTN is generally limited to a maximum of 33,300 bps for long distances. Speeds greater than 33,300 bps over standard telephone links are prone to high error rates. This is not to say that future communications components will not be able to detect and correct errors at higher speeds. Newer modems use variant modulation techniques such as Trellis Coded Modulation so that higher speeds are now much more common on the switched network.

The term "switched" simply means that the connection is established manually or dynamically. Manually implies the act of a mechanical switch actually moving and making contact. Dynamic switching is accomplished via an electronic device and no mechanical activity takes place. Figure 7.1 is an example of twisted-pair wiring.

The type of circuit (the medium) and line availability are not guaranteed by the telephone companies. Line busy/circuit busy signals may be encountered at any point in the network. This is due in large part to contention between users who want to use the network at the same time. In addition, variable noise levels may be frequently encountered.

Switched lines operate on a point-to-point basis. This means that routing is determined by the availability of lines from one telephone central office to another. Line contention can occur because a much smaller number of circuits exist between central offices than between central offices and homes or businesses. The central-office-to-central-office circuits are available on a first in first out basis which means that circuits are assigned sequentially based on incoming requests. Line contention usually results in a busy signal or reduced signal quality due to the excessive distance the signal must travel as a result of alternate routing from CO-to-CO.

Contention is frequently encountered on holidays such as Mother's Day or Christmas due to the sheer volume of traffic on the network. Because limited capacity is dictated by two-wire and four-wire connections, busy signals are often encountered at the area code, city code, or country code entry points. The lack of available circuits is responsible for this overload situation.

Several additional types of twisted-pair wiring are available for dedicated point-to-point connections with an eight-wire configuration being the most popular. This type of wire known as category three or category five wire is commonly used for wiring local area networks. It uses 24 or 28 gauge strands in an unshielded configuration and depending upon use, can have runs up to 28,000 feet.

Coaxial Cable

Coaxial cable consists of an unbalanced pair of wires, an inner conductor, usually copper, surrounded by a grounded outer conductor, which is either solid or braided. The inner and outer conductors are held in a concentric configuration by a dielectric material, usually Teflon or PVC. Figure 7.2 illustrates two types of coaxial cable.

Coaxial cable can handle more than 10,000 voice-grade channels (300 Hz to 4,000 Hz) using a variety of multiplexing techniques. It is used extensively for terminal-to-terminal controller connections, LAN connections, and cable television connections because of its superior noise control.

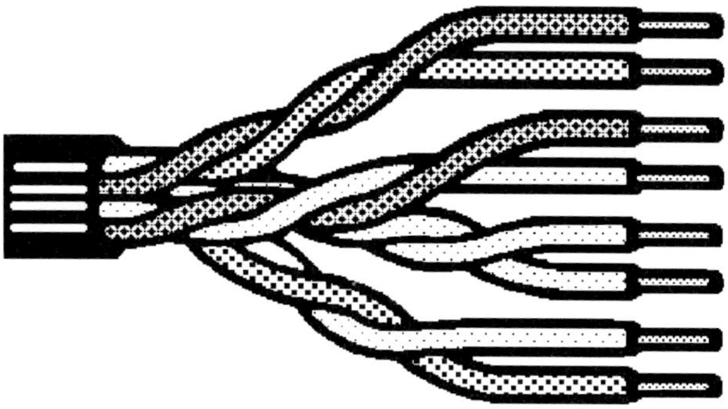

* Two- or Four-Wire Twisted-Pairs
* Two-Wire used for Subscriber Loops to Central Office
* Four-Wire used for Central Office Trunk Connections
* Subject to Noise and Line Restrictions

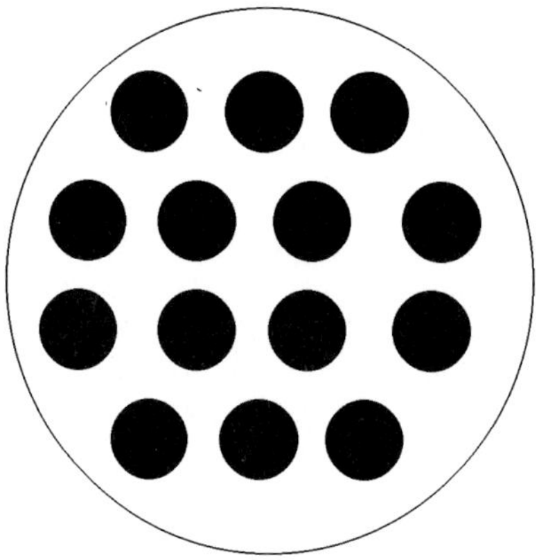

Twisted-Pairs Cable or Trunk

Figure 7.1 Twisted-Pairs

Coaxial cable requires signal repeaters (a type of line amplifier) for transmission over long distances. It can support bit rates well in excess of 1.544 Mbps and is superior to two- and four-wire circuits with respect to noise control due to the shielding placed around the signal conductor. Coaxial cable has a wide frequency bandwidth in excess of 500 MHz.

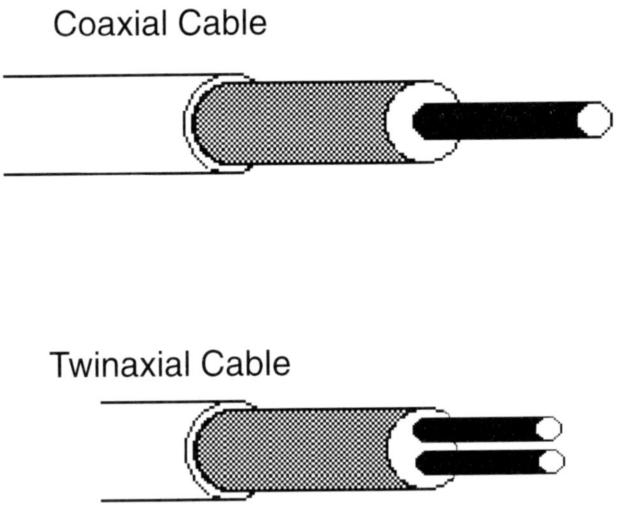

Figure 7.2 Coaxial Cable

Coaxial cable can be configured with multiple center conductors for increased transmission capability. Care must be exercised when using PVC coaxial cable because it may violate some fire regulations.

Fiber-Optic Cable

Fiber-optic cable is rapidly becoming common for digital transmission. Virtually all of the telephone companies—local, long distance, or government operated—are implementing or expanding fiber-optic cable networks each year. The availability of fiber-optic local area networks and the emergence of the Fiber Data Distributed Interface (FDDI) standard have made it desirable for users to incorporate fiber-optic cable into their wiring plans.

The concept of using light pulses transmitted from a laser or light-emitting diode (LED) source is really nothing more than an extension of the concept employed by Samuel Morse and the telegraph system. The major difference is in the technology, using pulses of light instead of electricity. Moreover, fiber optics represent a major advancement beyond electronic transmission because a single fiber has the potential to carry incredible amounts of data in the future.

When fiber-optic cable was introduced, it was very expensive because of rigid manufacturing requirements and the special handling needed. As new and improved splicing techniques and mass production of the cable have evolved, prices have rapidly decreased. The reduction in costs has enabled fiber-optic cable to be competitive with other types of transmission media.

Fiber-Optic Technology

Fiber cables are composed of hair-thin fiberglass or other light-conducting fibers. The fibers are covered by a light-reflecting opaque material referred to as cladding. The multiple fibers are then wrapped in an outer protective sheathing, usually Kevlar, which is an extremely tough material (used in the construction of bullet proof vests). This rigid protective covering protects the cable from stretching.

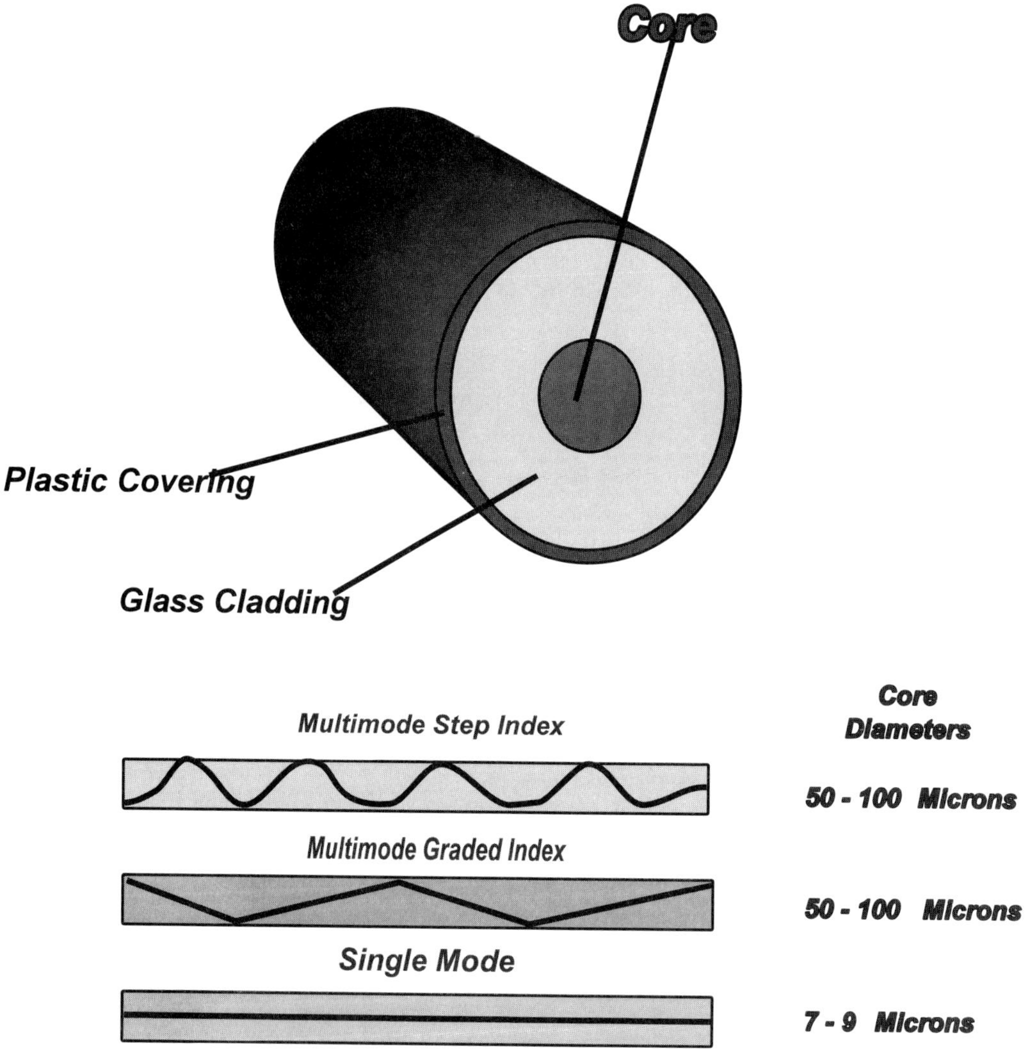

Figure 7.3 Cross Section of a Typical Ortical Fiber Cable

Three fiber-optic technologies are now in use (Figure 7.3 shows the three major types of fiber cable). The oldest is multimode step-index transmission, which works by bouncing the light along the fiber cable path to its destination. This technology is primarily used within a building because of the limited distances normally encountered. A second fiber transmission technology is multimode graded index, which is typically used in a campus or multibuilding environment because it performs most efficiently within a limited distance. The fiber-optic technology with the greatest potential is known as single-mode transmission. This technique uses a very narrow core and causes the light pulses to travel in a straight line down the center; it is suitable for greater distances. Various core sizes are used in all three methods. Typical core and cladding sizes are 8.3/125 microns for single-mode fiber and 62.5/125, 50/125 or 100/140 microns for multimode fiber. Each type of cable is designed for specific applications based on the distances and bit rates supported. Fiber transmission uses one cable for each direction of transmission because fiber cable operates in a simplex mode.

Nominal transmission speeds are in the range of 300 to 500 Mbps using a single-frequency laser. Much higher data rates can be achieved using the newer coherent communications transmitters because the coherent technology provides multifrequency laser capabilities (parallel lasers transmitting laser light pulses of different colors at the same time). Recent tests by Bell Laboratories using multifrequency laser capabilities have achieved speeds in excess of 50 billion bytes per second over a single fiber. At the present time, distances are limited to less than 50 miles without repeaters. Repeaters are placed enroute and reamplify the light pulses received on the input side. The repeater then retransmits the amplified light signal across the next segment.

For long-distance fiber applications, the fiber cable contains an electrical wire that carries power for the repeaters. Ongoing testing at Bell Laboratories and other organizations has achieved distances of more than 3,000 miles without the use of repeaters, but these experiments have been conducted in highly controlled environments using special test equipment.

Many regional telephone companies, long-distance companies in the U.S., and various private and government telecommunications companies in Europe are switching to fiber-optic networks because of their speed, capacity and reduced error rates. Ongoing research and development of new LED and laser technologies will continue to enhance fiber-optic capabilities.

One of the many advantages of fiber-optic cable is that it eliminates crosstalk commonly found in twisted-pair transmission. It also is not subject to electromagnetic interference (EMI), and it provides a high level of security because it is difficult to tap without detection.

Extensive fiber-optic networks have been installed in North America as well as overseas with trans-Pacific (Haw-4/TPC-3), trans-Atlantic(TAT-8), and trans-Caribbean (TCS-1) undersea cables. (These cable designations are defined by the carriers for the Pacific, Atlantic and Caribbean circuits.) Fiber-optic potential is nearly unlimited; the only real restriction is the transmitting and receiving speed of the equipment.

WIRELESS TRANSMISSION

Microwave Transmission

Microwave is a form of radio transmission using ultra high frequencies in the gigahertz (GHz) range. Microwave is considered to be a line-of-sight transmission method because the transmission and reception facilities must be within a line of sight. Distance is usually limited to less than 50 miles because of the curvature of the earth's surface. Figure 7.4 illustrates an example of a microwave transmission environment.

Microwave towers and transmission facilities act as amplifier stations to refresh the signal and retransmit it. Microwave transmission can provide multichannel transmission in excess of 2 Mbps. Thousands of channels can be integrated into a single microwave circuit because microwave transmission can make use of multiplexing techniques.

This type of transmission is used extensively for short-haul, point-to-point communications. It is, however, expensive due to the equipment required to handle the transmission. Microwave may be subject to interference by passing airplanes, rain, and other weather-related phenomenon. Microwave transmission and reception sites must employ extensive error correction technology to ensure acceptable performance.

Satellite Transmission

A satellite link is a channel connection using radio frequency waves between an earth station transmitter and a device on a satellite called a transponder. Satellites act as relay sta-

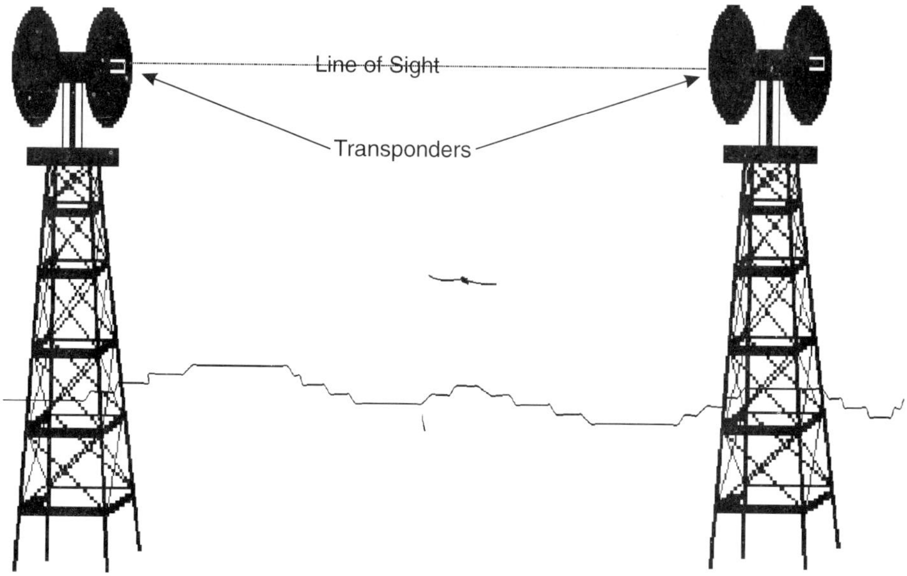

Figure 7.4 Microwave Transmission Towers

tions for very high-speed communications from one point to another. Typically, a satellite consists of multiple channels, each having the capacity to handle in excess of 1.544 megabits per second. The transponder is a receiver and amplifier coupled to a transmitter which receives the incoming signal, amplifies it, and then retransmits it to another receiving earth station or satellite. Figure 7.5 illustrates a geosynchronous satellite in orbit around the earth.

Data communications connections using a satellite link require ground stations for transmission and reception of the high-speed signals. Like microwave, satellite transmis-

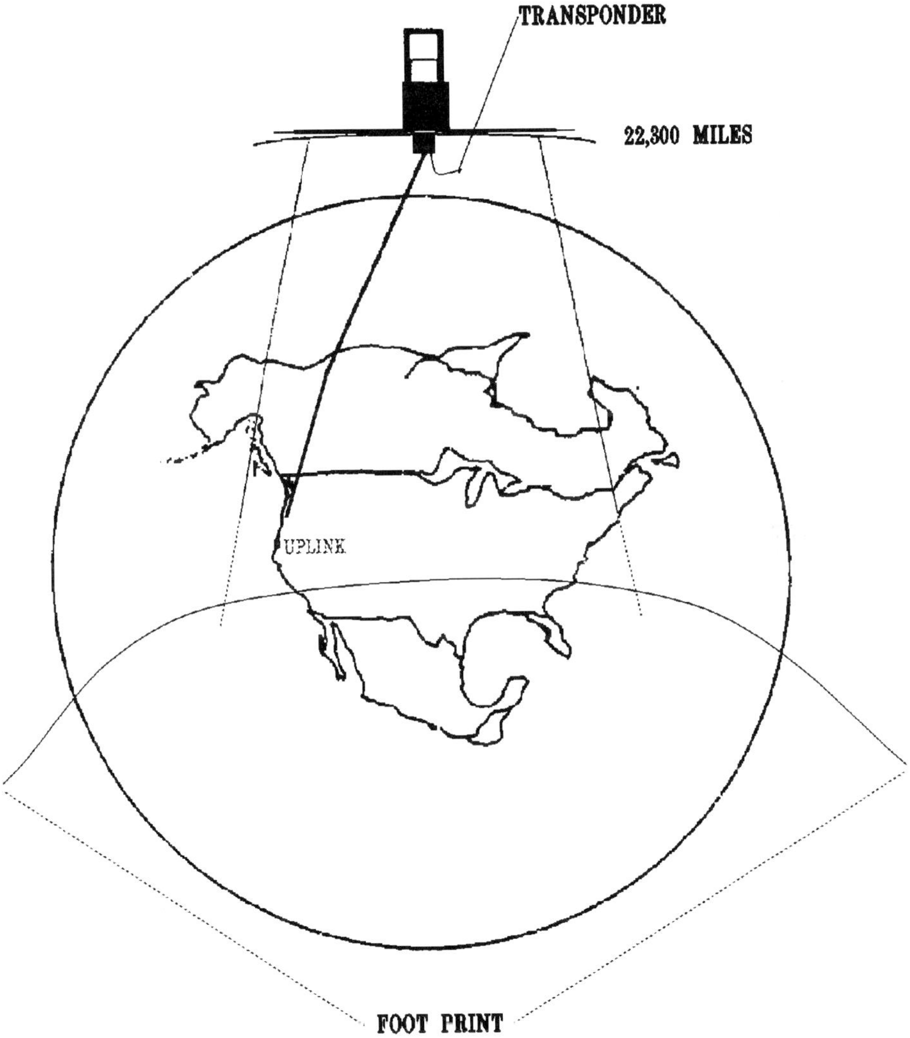

Figure 7.5 Satellite Transmission Example

sion is considered a line-of-sight transmission method. However, because of the distance involved and satellite location and placement, it's signal can cover a wide area. This wide area is referred to as its footprint and is determined by the altitude and position of the satellite.

Two sets of frequencies are used, one set for uplink activity and the other for downlink activity. The CCITT has designated two frequency bands for satellite communications. These bandwidths are allocated by the various governing agencies; the C band uses the 4 GHz to 6 GHz range and the KU band uses 14 GHz to 16 GHz. A typical user on the KU band would use 16 GHz for uplink activity and 14 GHz on the downlink.

The cost of a satellite link was expensive in the past, but with the placement of additional satellites in orbit and new technology to increase the speed of transmission, the cost appears to be dropping rapidly. One disadvantage of satellite use is the propagation delay of the signal (it must travel 22,300 miles up and then return the same distance), which causes potential timing problems with certain transmission protocols and frequencies.

Equipment may be expensive for companies that want on-site links because both uplink and downlink transmission facilities must be acquired and installed. Typical costs for leasing equipment for a private ground station range from about $3,500 per month to over $8,000 per month. Most companies that want to use satellite transmission use a satellite ground station service for link accessibility. Data to be transmitted using this service is carried over dedicated high-speed links from the customer site to a ground station called a satellite port. For improved performance, some vendors provide a digital technique called Demand Assigned Multiple Access (DAMA), which results in better service and more efficient circuit use than the standard analog links. Time-share satellite service is available from a number of value-added suppliers of data communications services.

Radio-Wave (RW) Transmission

Radio-wave LANs typically connect nodes located in different buildings. They basically work in the same way that wired networks do. Signals are carried on electromagnetic waves for both wired and radio-wave LANs. Radio networks use frequencies between 902 MHz and 928 MHz. Because signals in the air cannot be shielded from interference, radio networks require more elaborate error-prevention techniques than wired networks. To connect a wired network via radio waves, a special network driver and interface card in a gateway use a digital-to-analog converter to translate the networks binary signals into analog waves. The analog waves are sent to a radio transceiver (transmitter/receiver), where they are modulated into radio waves that are sent to the other nodes on the network. Table 7.1 lists several wireless technologies.

Spread Spectrum

Using a transmission technique known as spread spectrum, the signal is sent across a broad bandwidth of frequencies to minimize the danger of data loss. In one type of spread-spectrum transmission, signal hopping, transceivers send data for a few milliseconds on one frequency, then change to another frequency. In a further attempt to reduce data loss, this type of spread-spectrum transmission uses redundancy whereby the same

Table 7.1 Wireless Services

Service	Usage	Network	Benefits	Drawbacks	Representative
Cellular	Circuit-switched roaming modem and telephone connections	WAN	Mobility; mature technology; existing infrastructure; voice and data support	Expensive if connection is poor; not yet available; penetration may be poor	CDPD
Infrared	Line-of-sight data transmission	LAN, peripheral-sharing devices	Ease of installation and use; mature technology; permits faster transmission than many other wirefess LANs	Line-of-sight transmission only; units must be aligned and data path kept clear	BICC, Photonics, Windata
Microwave	Short-range data transmission	LAN	5–7 Mbps data rate; ease of installation; good compatibility	Signal cannot penetrate many walls; inappropriate for highly mobile sites	Motorola
Mobile satellite	Highly mobile messaging and location services	WAN	Saturation coverage; reliable connections	Expensive; mediocre data rates; limited capability	American Mobile Satellite Corp., Orbital Communications, Qualcomm
Packet radio	Packet-switched roaming data transmission	WAN, messaging services	Mobility; user pays only for packets transmitted	No current voice support; coverage limited to large metropolitan areas	Aridis, EMBARC. Mobile RAM Data
Spread-spectrum radio	Short-range data transmission	LAN	Signal capable of penerating a number of supporting walls; flexibility	Unreliable near heavy electromagnetic transmissions; requires one transmitter per linked PC	NCR, O'Neill Communications, Proxim

data is sent on several different frequencies in case interference on one end corrupts the signal. The receiving transceiver strips the signal of its carrier wave. An analog-to-digital converter in the network card converts the signal to binary code. In the LAN gateway, the data is reconstructed and undergoes statistical checks to ensure that it has arrived uncorrupted. If the checksum used for error detection is incorrect, the receiving gateway requests that the data be sent again.

Direct Sequencing

Transceivers that cannot switch between frequencies in milliseconds must use a different kind of spread-spectrum transmission called direct sequencing. This method uses a special chipping code that breaks data into chunks and transmits it across various frequencies. The receiving gateway pieces the data back together using the same chipping algorithm.

DATA TRANSMISSION ERRORS AND IMPAIRMENTS

Many types of data transmission errors and impairments apply equally to wire transmission and radio frequency transmission. However, certain types of error conditions apply specifically to wired forms of data transmission, while microwave and satellite transmission are especially subject to others, most notably propagation delay. Detailed explanations for certain error types are not provided here due to their extreme complexity. Figure 7.6 shows the effects of distortion on a digital signal.

Wired Switched Networks

Distortion. Signal distortion is a general term resulting from multiple conditions. As a current travels through a circuit, it encounters a resistance or opposition known as impedance which acts as a partial barrier to the signal's flow. Distortion, also referred to as attenuation, is caused by loss variations (power, speed) at different frequencies. Two of the primary causes of distortion are signal attenuation and delay.

The installation of smaller gauge wire will reduce some distortion, but this solution is good only for short distances and is expensive. Another method for controlling distortion is to add electronic components called inductors uniformly spaced along the line to ensure that the circuit remains stable. This process is referred to as loading. This technique however, will increase delay distortion. Insertion of attenuation equalizers in conjunction with hybrid coils will partially correct this problem.

Reflection (Echo). Reflection is produced by impedance (resistance at certain frequencies) due to irregularities along a transmission line or at its end points. This irregularity causes a portion of the signal's energy to be reflected back toward the originating end. Reflection can be corrected by the insertion of echo suppressors into the line. Be cautioned that although echo suppressors work well and are required with voice transmission, they have an adverse effect on half-duplex data transmission.

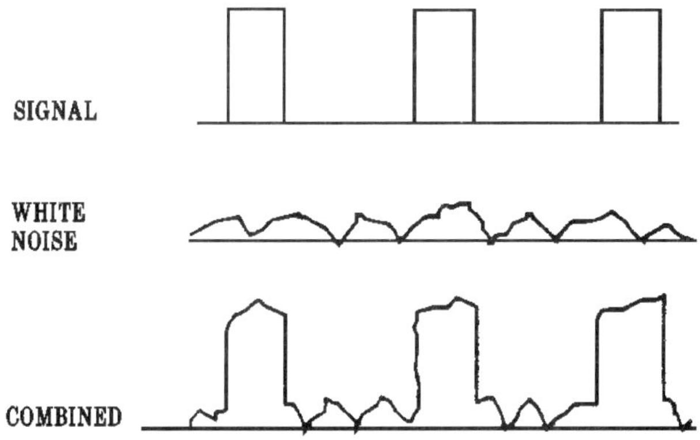

Figure 7.6 Digital Signal with Noise

In data communications, echo suppressors located in the satellite or central office can be disabled by the user via the answering modem, which can transmit a high-pitched tone at a frequency of 2,000 to 2,250 hertz. This must be maintained for a duration of 400 milliseconds. Echo suppressors will remain disabled provided that the carrier signal is not interrupted for more than 100 milliseconds. Echo suppression is required in data communications because an echo distorts the forward bound signal and creates data errors.

Noise. Background, white or steady-state noise is the average noise power on a voice channel and usually maintains a constant level. The presence of this type of noise often causes problems in voice transmission and, to a lesser extent, data transmission. Another type of noise, referred to as impulse noise or line hits, is characterized by very short bursts of high amplitude (i.e., power surge). Impulse noise produces major problems for data transmission because the pulses are of very short duration. This activity can alter bit values and even block out the entire data signal for short durations of time. At the present time, little can be done to eliminate noise on switched lines.

Propagation Delay. Propagation delay is a variation in the time it takes for a signal to travel from the transmitter to the receiver. Such a delay can be viewed as a slowing down or speeding up of the transmission signal. On switched networks, the propagation delay is not controlled and creates problems at higher speeds. In addition, propagation delay in switched networks can vary greatly depending on the routing within the network. This can cause problems because the delay may exceed line turnaround time (the time it takes to reverse the line in a half-duplex environment in order for the receiver to acknowledge a response). This type of delay can disable echo suppression circuits in the central office switch and can prevent the line turnaround from occurring. Often, the only solution to data transmission problems of this nature is to terminate and re-establish the connection.

Wired Dedicated-Line Networks

In addition to the transmission errors discussed for switched links, data transmission over leased or private facilities can also experience signal degradation. In addition to attenuation distortion and envelope delay, which already have been discussed, several others also create problems. These include nonlinear distortion, phase jitter, line transients, poor signal-to-noise ratios, and impulse noise.

Nonlinear Distortion. Nonlinearities in amplifiers and other electronic devices can cause the generation of signal components (distortion) along with the original data signal. This distortion is harmonically related to the fundamental frequency of the base signal. These added components can reshape the transmitted signal to such an extent that the data signal is difficult to detect.

Phase Jitter. This random distortion of the transmitted signal's phased sine wave results in the intermittent lengthening and shortening of time related to the transmitted signal.

Line Transients. Transient phenomena (unexpected and very rapid) are a major cause of transmission problems in data communications. They include impulse noise, phase hits, gain hits, and dropouts. Phase and gain hits are similar to voltage spikes that momentarily will increase the carrier voltage to a higher level. This can change the value of the data bits being transported at the point of the hit.

Poor Signal-to-Noise Ratio. The signal-to-noise ratio (SNR) represents the difference between the noise level on the link and the information signal, expressed in decibels measured. If the SNR exceeds the rated specifications for the link, the receiver will not be able to distinguish between the noise and the signal.

Impulse Noise. As stated previously, impulse noise is characterized by short bursts of high amplitude. This can be caused by a variety of things, including telephone switching equipment, lightning, and other electrical surge activities. Impulse noise creates problems in data transmission because it may cause bits to change their value during transit. A one bit could become a zero bit, thereby altering the composition of the data. If this problem occurs, the only solution is often to retransmit the entire block of data after the receiving station returns a negative acknowledgment.

Wireless Networks

Wireless transmission experiences many of the same problems as wired media, however, because wireless communication is a line-of-sight methodology, transmissions may be blocked by obstacles such as walls, mountains, or water. Wireless networks are also more prone to distortion because of environmental or weather related conditions.

CASE STUDY (CONTINUED)

The Bank of Scotts Valley uses several types of communications links between the two data centers and the various branches. This is because most of the dedicated links are provided by the telephone carriers contracted for link services.

Within each branch, the bank planners have decided to use twisted-pair wiring for terminal connections. This decision was based on the fact that distances are limited and unshielded twisted-pair wire will support standard terminal connections at rates in excess of 2 million bytes per second. Twisted-pair wiring also was selected because, at some point in the future, the bank probably will convert its terminals to personal computers

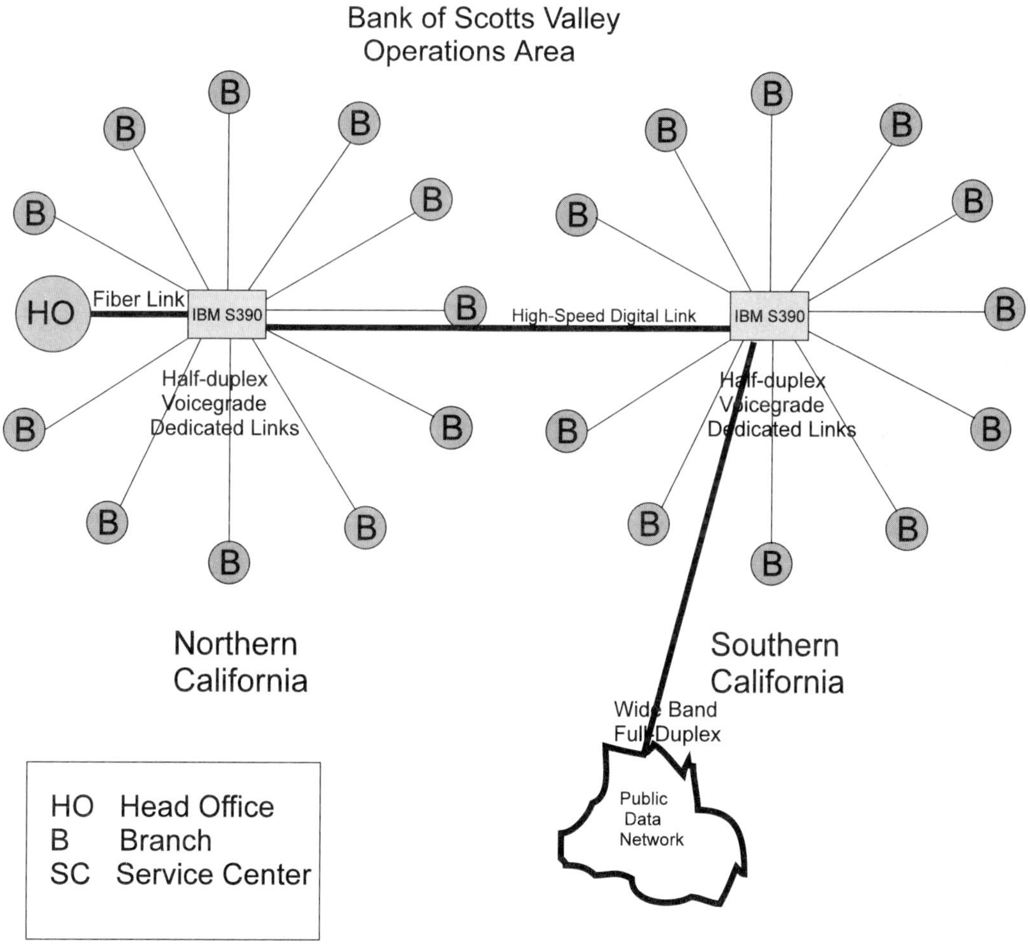

Figure 7.7 Types of Links used by Bank of SV

with LAN capabilities. Twisted-pair wiring is ideal for this type of connectivity. Fiber-optic cable also will be placed into each of the bank's branches, because it will be quite expensive to convert to fiber in the future.

For connection between the branches and the service centers, the local telephone carrier has provided a coaxial cable link from each of the branches to its central offices. The telephone carrier is in the process of installing fiber links throughout its access area but has not yet completed the conversion.

Communications between each of the service centers through a dedicated telephone link employs fiber for its primary link and coaxial for its secondary links. Figure 7.7 shows the links used by the Bank of Scotts Valley. The fiber links terminate at a telephone company central office and are then placed on a digital, long-haul microwave circuit.

REVIEW QUESTIONS

1. Name some of the basic transmission media used in data communications.
2. What is the main advantage of coaxial cable over twisted-pair wire?
3. Why is fiber optics considered the wave of the future?
4. Are microwave and satellite transmission considered line-of-sight transmission methods?
5. To what transmission medium does "cladding" refer?
6. What function does a transponder perform?
7. For what function are twisted pairs primarily used?
8. To what does line-of-sight transmission apply?
9. In what manner is distortion a problem in data communications?
10. True or false: Distortion is really a general term for many types of errors.

CHAPTER 8

Data Transport Formats and Connections

INTRODUCTION

In the real world, we are filled with a richness of sound, color, and the written word, which are all used to convey limitless amounts of information. Adapting this to the world of the binary computer requires that we take and convert it into a form and format that is compatible with a digital environment.

Regardless of the type of information we wish to transport within a network or between networks, a standard format is necessary. This is required so that communications software processes do not have to be concerned with data formats, but only with its movement from one point to another. To the computer, communications lines and telephone switches, audio, video, and data files are nothing more than sequences of on-off pulses representing ones and zeros. In order to handle this efficiently, the data to be transported is segmented into packets or frames.

A packet is a self-contained entity that includes a variable amount of data as well as any addresses required to deliver it to its destination. Figure 8.1 shows a typical Packet/Frame format. This allows packets to be mixed (multiplexed) from multiple sources onto a shared link. Each packet contains sending and receiving addresses that are used by the network hardware and software for routing and delivery.

RECORD AND PACKET CONCEPTS

As stated at the beginning of this chapter, data must be broken up into logical and manageable pieces so that communications software can provide equitable service to all users. Placing the data into a packet can be likened to placing a letter into an envelope and putting addressing and routing information on the face of the envelope. Another reason

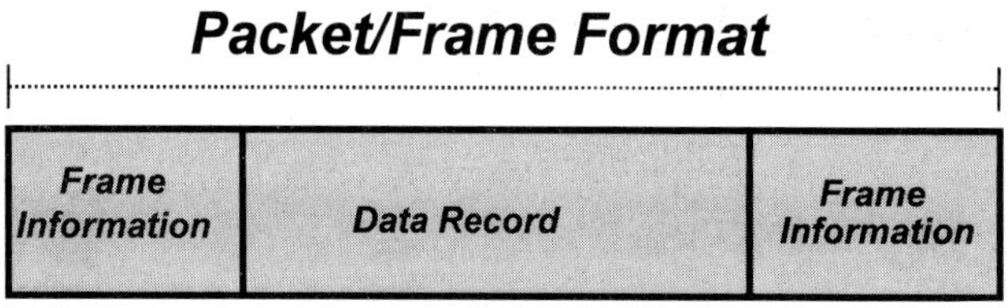

Figure 8.1 Example of a Packet or Frame of Data

for placing data into packets is one of control. A packet's control information includes routing instructions, the sending address, the receiving address, and error-detection information. Almost all networks today use the packet concept and supporting services.

Packet services originated in Europe to support transborder data communications. In the 1970s, a transborder packet switching network called X.25 was defined to support public data communications provided by many of the European telecommunications agencies. Because telecommunications was a government controlled function in many European countries, the telecom authorities required a common format for sending data between their networks and a concept known as datagrams became the standard. Figure 8.2 illustrates the X.25 Network installed in Europe.

Today X.25 is a International Telecommunications Union (ITU) standard designed to provide a packet delivery service across multiple networks. The ITU is an United Nations organization chartered with establishing and maintaining worldwide communications standards and was formerly known as the CCITT. The original standard covers basic services and does not guarantee that the packet will arrive intact if corrupted during transport. Arriving packets are not delivered in any predefined sequence and no capability exists for acknowledgment of the received packet to the sender.

A datagram is a single packet that has no relationship to previous or future packets. Each datagram packet is a separate entity routed independently and may arrive in any order from which it was entered into the network. Depending on priority and network traffic, each datagram packet is on its own and could take a different route toward its destination. Datagrams are delivered to their destination with a high probability of success, but can be lost or discarded.

Datagram services have improved as they have matured. Now it is possible to obtain error correction and routing services, which are provided by other supporting software in the network. Basic datagram service was primarily designed for videotext (video information service) activities, but today because of widespread use of X.25 types of networks around the world, two types of datagram packets now exist: unreliable datagrams and reliable datagrams.

Unreliable Datagrams

An unreliable datagram is one in which end-to-end confirmation of packet delivery does not occur. In other words, the packet may be discarded or lost. No connection setup or acknowledgment from the receiver is used with unreliable datagram service. This is a plus

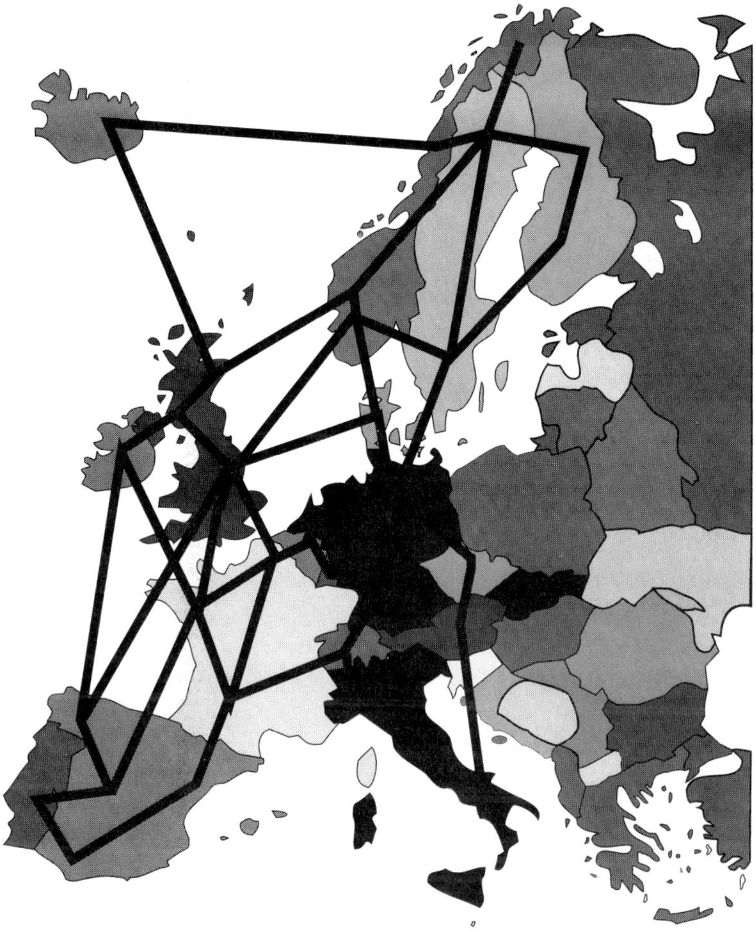

Figure 8.2 Example of the Links in the European X.25 Network

for datagram operational efficiency, because no overhead is required in establishing a connection before transmission can occur.

The unreliable datagram service does not guarantee delivery of the data packet to its destination. If end-to-end confirmation of delivery is desired, it must be provided by other communications software. This level of packet service does not ensure that delivery will be without error or that packets will be delivered in the same sequence in which they were transmitted. Unreliable datagram service was originally developed to provide an inexpensive delivery service for data, text, and videotext information. Videotext is a television or computer-based service provided to subscribers. Because videotext data is transmitted on a repetitive basis, a damaged or lost packet would be resent automatically on the next transmission cycle.

Reliable Datagrams

Reliable datagram service assures that a packet will always be delivered to the destination address. This can be accomplished by end-to-end controls within the network software.

Reliable datagram service provides improved capabilities over unreliable datagram services. It ensures that data packets will be delivered to their addressed destinations. Additional levels of service include improved routing and error detection, as well as notification back to the sender that the packet was received intact. Most networks, including the Internet, provide reliable datagram service.

SWITCHING TECHNIQUES

Data communications typically involves the movement of packets in short transmissions between many points of a network or between networks. Because of the nature of this type of activity, economies can be realized by sharing circuits (links) between many computers on the network rather than establishing dedicated circuits (creating a telephone type of connection) for each connection required. Two methods are available for moving data between computers or workstations on a network, they are called circuit switching and packet switching service.

Circuit Switching

Switched circuit service allocates the entire bandwidth on an existing circuit and is identical to having a dedicated link for the duration of the communications activity. One advantage with circuit switching is that it eliminates the need for source or destination addresses after the connection has been established. This type of switching is usually found in networks that are self-contained (no internetwork communications activity is supported) and is not really considered a network service. Figure 8.3 shows an example of circuit switching.

Circuit switching requires a dedicated path between each end, and it is usually quite expensive in terms of wasted bandwidth. In other words, when a connection is established and no activity is taking place, the idle bandwidth cannot be used for any other activity. It can be compared to a telephone call when no conversation is taking place. Telephones are linked using established circuits that are switched into place mechanically or electronically when a call is placed. The circuit is allocated to that call regardless of any activity. When the call has ended, the circuits are made available for other calls. Typically circuit switching designed for data communications is employed in areas where it is practical to have dedicated links to each terminal or computer in the network.

Packet Switching

Packet switching transmits data packets over a shared transmission channel or circuit. Packet transmission is a very efficient data transfer method and was primarily designed for low speed communications, however, it is also used today for high speed computer

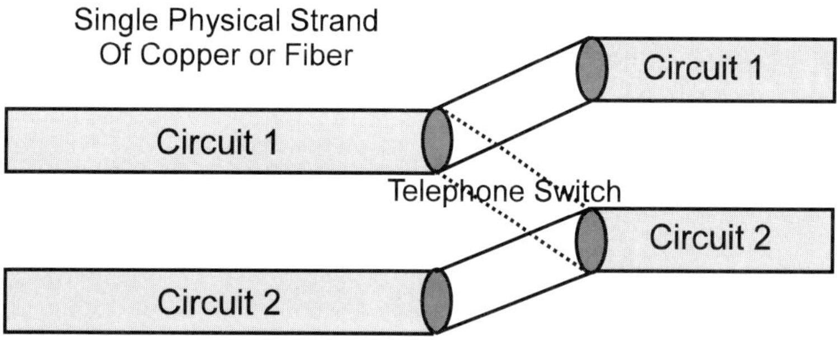

Figure 8.3 Two Circuits Connected to a Telephone Switch

communications. Packet switching uses a form of multiplexing that dynamically allocates bandwidth to the sending side when information is to be transmitted. Figure 8.4 illustrates packet switching between three workstations and the network.

With packet switching, routing may be via store-and-forward or broadcast methods. Store and forward methods store a packet at a node for later transmission to another node. In the broadcast method, such as with some types of local area networks, each node must

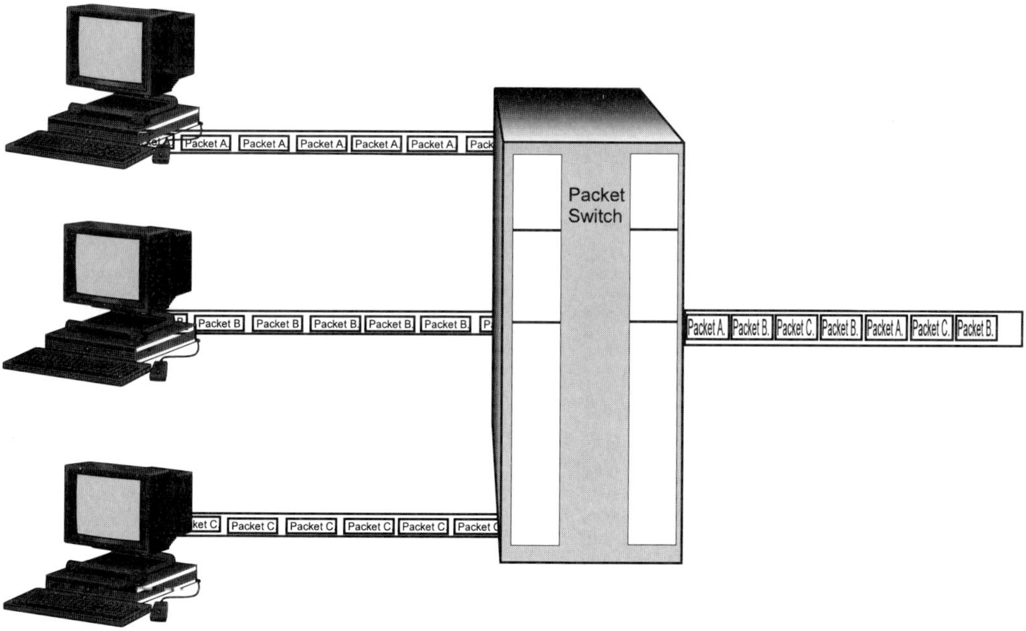

Figure 8.4 Packet Switching Between Three Networked Workstations

check the packet as it passes by to determine if it belongs to them before re-transmission can occur. Once the packet safely reaches the next node, the communication path between nodes is freed for transmission of another packet.

CONNECTION-ORIENTED, CONNECTIONLESS, AND VIRTUAL CIRCUITS

While packet switching services were a step in the right direction for the occasional user or for transaction-based activities such as interbank communications, another type of service was required for the dedicated or interactive user. This service is referred to as virtual circuit service. It is designed to provide end-to-end sequenced delivery, error detection, and correction and delivery confirmation of the data packets across the network.

A virtual circuit requires permanent or temporary dedicated links to all points in the network. A virtual circuit is a temporary channel between the sender and the receiver for a specific amount of time (i.e., the duration of the communications activity). This is similar to the type of circuit established for a telephone call. Virtual circuits guarantee sequenced, acknowledged delivery of each packet transmitted from the sender to the receiver. Virtual circuit services follow international specifications that are provided in vendor-specific software.

Connection-Oriented Services

A connection-oriented service is one in which a virtual circuit is established using call setup procedures that create an end-to-end circuit much like a telephone call connection. The sending node establishes the connection-oriented end-to-end circuit during call setup time. This circuit remains in place until a tear-down or disconnect request occurs. Data packets from both endpoints travel across the network using the virtual circuit. This provides an efficient method for guaranteeing acknowledged sequenced delivery of the data in an efficient manner as the packets travel over the same route for the duration of the connection.

Logical connections can either be permanent or temporary. Temporary virtual circuit transmission, also known as switched-virtual circuits (SVC), is designated a virtual call and is set up like a telephone call. The logical communication path does require call setup procedures and is disconnected upon completion of the transmission session. The permanent virtual circuit (PVC) is always (logically) permanently connected and no call setup or disconnection is required before data is transmitted.

Connectionless-Oriented Services

Connectionless-oriented services do not establish a virtual circuit from end-to-end. This type of connection uses the address contained within the packet for routing the data to its destination. An example of a connectionless-oriented circuit is the Internet. The network consists of many nodes primarily connected via dial-up or permanent circuits. Nodes send their packets to the next node in the network, and it is the responsibility of that node to determine the best routing to the next node in the network.

The X.25 packet switching networks developed in the mid 1970s were designed around a copper wire infrastructure that required extensive error checking because of the high error rates associated with copper. Basically every reliable packet that was sent required a response back to the sender to acknowledge its receipt. Since the early 1980s, optical fiber has replaced copper in most locations providing a higher degree of reliability therefore eliminating the need to check every packet. One would think that because of the reduced error rates associated with fiber, X.25 networking hardware and software would have been modified to take advantage of the improvements in the transmission media, however, this has not been the case. Today X.25 continues to use the same error processing methodologies used in 1974. Because of this, as well as other inefficiencies associated with X.25, a newer methodology was needed to support fiber, new micro-electronic technologies, and advances in software engineering. Two new technologies were developed and are called: Frame Relay and Cell Relay.

Frame Relay

Frame Relay is a network packet transmission technique referred to as one of the two fast packet technologies. While X.25 supports fixed length packets 4,098 bytes in length, frame relay supports variable length data packets up to 9,000 bytes with a minimum of error checking at very high speeds. It also takes advantage of the higher reliability of fiber so extensive error checking is not needed.

Frame Relay is principally supported by communications equipment manufacturers as an alternative to X.25 packet switching networks. However, it also has been embraced by most of the telephone companies of the world as a method for moving data at very high speeds. Despite being categorized with X.25 as a packet-mode service, it is much faster than X.25, supporting speeds up to 2.048 Mb/s.

Frame Relay supports two modes of operation: Mode 1 for private networks, and Mode 2 for public data networks such as those offered by telephone companies. Private networks can take advantage of Frame Relay capabilities using customer premise equipment. Greater throughput will be realized because no internetwork communications is required. With internetwork communications, overhead is needed for routing packets to their destination nodes and for network error detection that reduces throughput rates.

Frame Relay services are available globally and are offered by most public and private carriers. Frame Relay is supported by most of the worlds transmission service carriers including AT&T, MCI, Sprint, and most of the regional Bells. Most phone companies or owners of private networks can implement Frame Relay over existing networks without needing additional hardware or software other than at the end nodes. Frame Relay networks can also transport other types of packets and Frame Relay packets can travel over non-frame relay networks. Figure 8.5 shows the format of a Frame Relay packet.

Cell Relay—Asynchronous Transfer Mode (ATM)

ATM, the second fast packet technology mentioned in the beginning of this section (also called cell relay) will be the foundation packet handling methodology for most emerging communications technologies well into the future. ATM formats are designed to be com-

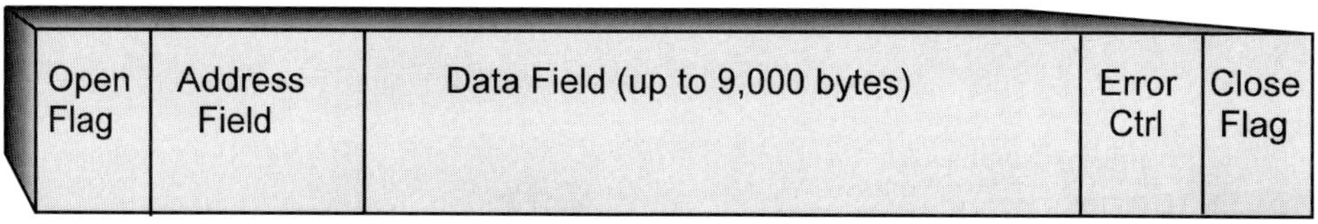

Figure 8.5 Frame Relay Packet

patible with a number of existing and emerging technologies and provides a methodology for the transparent transport of data to many destinations from a single transmission source.

ATM can be viewed as a transmission service provided by telephone companies and bypass carriers that employs optical fiber as its foundation network transmission technology. Principle use of ATM is for packet-based broadband integrated services digital networks (ISDN) and metropolitan area networks (MAN) operating at high speeds. It is designed to handle packet transport at speeds above 1.544 Mbps and will be capable of gigabit per second transmission rates. It also is designed to support both constant and bursty traffic conditions.

Most telephone companies in North America are currently implementing and testing ATM technology, and it should become widely available in the near future. Additionally, many hardware manufacturers are implementing ATM technology in equipment they provide for private networks. As a switching technology, ATM provides point-to-point connections between network nodes, giving each pair of communicating stations full network bandwidth.

ATM is designed around a small fixed length packet and operates by breaking the incoming data into short fixed length blocks of 53 bytes called cells. Each cell contains 48 bytes of data and 5 bytes of control information. A small fixed length packet size was decided upon in order to facilitate efficient switching by routing hardware. Because of its small size, it is ideal for higher bit speeds.

ATM is connection-oriented in that a virtual connection is established for each transmission, each cell travels the same path. The small cell size ensures little delay in cell switching. This allows ATM to handle both real-time and non-real-time data. ATM services also will support packets produced by other transmission technologies such as X.25 and Frame Relay. Figure 8.6 shows the format of an ATM-Cell Relay Packet.

Synchronous Transfer Mode (STM)

Unlike ATM, STM is an extension of traditional circuit switching and provides a fixed amount of bandwidth for packet transport. STM bandwidth can go unused if no data requires transport for a given channel. While STM is simpler to implement and is more cost efficient, it does not provide the throughput capacity of ATM. It is envisioned that most of the emerging technologies will use both ATM and STM for data transport.

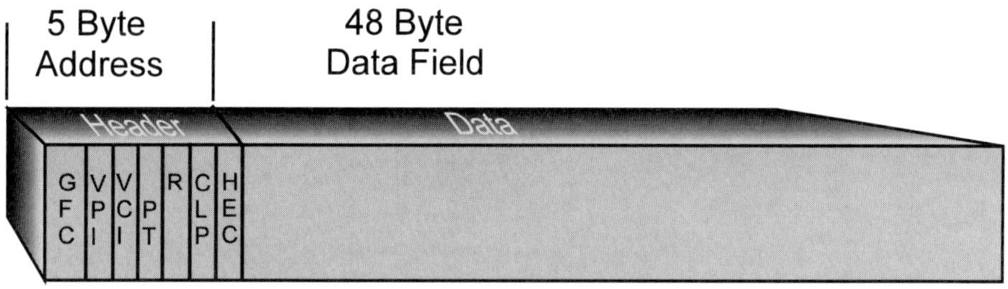

GFC Generic Flow Control
VCI Virtual Channel Ident.
R Reserved
HEC Header Error Control
VPI Virtual Path Identifier
PT Payload Type
CLP Cell Lose Priority
Data

Figure 8.6 ATM Cell Relay Packet

REVIEW QUESTIONS

1. What is a datagram?
2. How many types of datagrams exist?
3. What is a connectionless service?
4. What is a virtual circuit?
5. Name the two types of switching techniques discussed in this chapter?
6. What is the difference between a datagram and a packet?
7. What is a Cell Relay packet?
8. What is STM?
9. What is Frame Relay?
10. What is the difference between Frame Relay mode 1 and mode 2?

CHAPTER 9

Communications Services

INTRODUCTION

Communication services provide the method and facilities by which data is transported from one point to another. These services, typically contracted from telephone companies or independent link providers, offer a variety of options that you can choose from and implement depending upon your needs and the provider you select. Most telephone and communications service companies that provide data communications around the world have a wide selection of methods available for transporting your data. Communications services are those provided by vendors that are used to connect all points of your network together.

When planning and installing a network, it is very important to have a thorough understanding of all of the issues that need to be considered. The major reason behind this is the very high costs associated with the services. Remember: these costs are a monthly reoccurring expense that will be on-going, they will represent a major portion of an organizations' data communications budget. If you are not careful in the planning phase, it is quite easy to contract for services that are more expensive and often not required.

Many different types of connection services exist and include dedicated links, dial-up links, wireless communications, and high-speed connections. Also, each may have multiple price structures based on usage, distance, and contract duration. High-speed dedicated links broken down into a number of smaller connections would typically be used by large organization's because of large capacity requirements, whereas smaller companies will often use dial-up services to control expenses.

This chapter will present the issues relating to selecting communications services. We will also discuss the different types of connection services and explore the reasons for selecting one over another. Table 9.1 lists types of carrier services.

Table 9.1 Types of Carrier and Transmission Services

Type of Service	Bit Speed	Standard
Dial-up	to 33.3 Kbps	N. America
Leased Lines	to 64 Kbps	N. America
DDS	to 64 Kbps	N. America
Switched 56	56 Kbps	N. America
DS-X	to 800 Mbps	International
SMDS	to 655 Mbps	International
ISDN	to 2,048 Mbps	International
	Carrier Services	
SONET	to 13 Gbps	International
FDDI	100 Mbps	International
	Transmission Services	

BANDWIDTH CONSIDERATIONS

Before we can examine the different types of services available, we need to consider bandwidth a component of the selection process. Different types of data (audio, video, multimedia) require different amounts of bandwidth. Bandwidth capacity is a critical area that must be addressed when planning a network. It is an absolute requirement that you be able to provide sufficient capabilities to satisfy all of the organization's users without having a large amount of unused line capacity going to waste. To understand this problem fully and provide an acceptable solution, we need to look at how capacity has been provided in the past.

In the past, if an organization wished to implement a data communications network, the first step was to determine the required number of lines and the desired capacity. The next step was to install the lines and bring the network online. Planners needed to ensure that enough capacity was available to satisfy peak-period time demands; yet, this often translated into excess capacity in low-load situations. This is a problem of today as well.

Excess line capacity is wasteful in terms of the costs associated with having unused resources. It is important to note, however, that as line costs decrease, many companies do not consider this as great a problem. Also, in many locations around the world, waste is not an issue because the only link capability available is obtained from the government's telecommunications authority.

How then do you ensure that the links you are installing are being used to maximum availability and provide the highest return on investment? One way is for you to use different types of data communications services for different types of activities.

Your organizational environment is another point that must be considered. Not all data communications involves the use of contracted vendor-based services. More likely, it is a combination of dedicated hard-wired links and services installed within your organization, as well as commercial services leased from a communications service provider. This presents a situation where compatibility could become an issue because you could implement an internal methodology that may not interface to the outside world.

PUBLIC AND PRIVATE CARRIERS

Data processing and data communications services are provided by two types of carriers: common (or public) carriers and private carriers. Common or public carriers are companies that provide voice or data communications services to the public. They include organizations such as telephone companies, computer hardware vendors, time-sharing service vendors, and data carriers. Many of these companies offer similar services that will be discussed later in this chapter.

Common Carriers

While there are more than 3,000 telephone companies in North America, over 95 percent of all revenues are generated by the regional Bell operating companies (RBOCs), AT&T, MCI Communications Corp., Sprint, and General Telephone (GTE). These, as well as most other telephone companies, offer a full range of services from dial-up voice networks to value-added public data networks. Figure 9.1 illustrates the public switched telephone network carriers and the areas in which they operate. Outside of North America, many countries have a single provider that is usually a government agency.

Other Important World Wide Carriers

LDDS WorldCom is a full-service long distance carrier, providing domestic and international voice and data telecommunications services over a digital fiber optic nationwide network. Founded in the early 1980s, LDDS WorldCom has become one of the nation's four largest long distance carriers. LDDS WorldCom services include support for an array of digital services that include: virtual networks, frame relay service, and ISDN service.

Cable and Wireless is a major provider of data communications services in: Britain, Europe, the Middle East, Africa, Russia, Hong Kong, China, Japan, the Pacific Rim, and Australia; throughout the Caribbean; and on islands around the world. The company is the world's largest user of the Intelsat satellite network and owns one-third of AsiaSat, the region's first privately financed telecommunications satellite, the footprint of which covers half the world's population. It also has the largest commercial fleet of cable laying and repair vessels in the world. Cable & Wireless has established advanced digital networks in countries such as Hong Kong, the United States of America, the United Kingdom, and in the Caribbean, linked by fiber optic submarine cables.

France Telecom is the second largest telecommunications operator in Europe and the third largest worldwide in terms of international calls. Its network is highly digitized,

Figure 9.1 Regional Bell Operating Companies

permitting simultaneous availability nationwide of high-speed voice, data, text, and video through its commercial ISDN. France Telecom was the first company in the world to offer ISDN. France Telecom also offers Transpac, a packet-switched data network, one of its role is to transport data for the Minitel online information system.

Perhaps the best known symbol of France Telecom is its Minitel online information system. Available in homes and businesses for more than a decade using a small Minitel terminal, the service is also available for the PC. Minitel provides over 26,000 services, including an electronic phone directory, bill paying and e-mail, and is connected to the Internet. It has in excess of 6.5 million terminals installed with over 11 million connection hours logged annually.

Satellite Services

Over the past few years, satellite communications have become more popular for certain transmission activities because of the reduction in costs and greater circuit availability. This form of transmission service is excellent for long-distance and high-speed communications. Several telephone companies offer satellite service either by connecting subscribers through their land line (surface) facilities or, in some cases, through the placement of on-site uplink and downlink equipment installed at the subscriber's various locations.

Hughes Network Systems (HNS) is an example of an organization that provides wide area networking products to organizations around the globe. Hughes STRATEGY product is an enterprise network solution that encompasses a range of frame, cell, and packet services. They provide redundancy and source dynamic routing, along with accounting, performance monitoring and statistics information. STRATEGY also includes gateway, router, and bridging products that provide interconnectivity with other networks.

Value-Added Carriers

Value-Added Carriers provide services using transmission facilities leased from common carriers such as AT&T and MCI. Value-added carriers such as Tymnet and the McDonnell Douglas Network Systems Company provide network facilities that are connected through the use of computers and other specialized equipment designed to enhance the ability to transfer data. These enhancements include extensive error detection, code conversion, and encryption using data packets that are switched (transferred) from one computer to another until they reach their final destination. This is referred to as a public packet-switching network. Packet switching is a cost-effective method for transmission of data by small computer users because packets usually consist of a number of pieces of data.

International Record Carriers

International voice and data communications between the U.S. and other countries have been provided within the U.S. by such common carriers as AT&T, MCI, RCA Global Communications, ITT World Communications, TRT Communications, and FTC Communications. These organizations follow the guidelines issued by the ITU-T for interconnection. While they cannot directly have point-to-point international communications links in many locations, companies have formulated agreements with their international counterparts to provide end-to-end service to their subscribers.

Public Data Network Services

In many cases, using a value-added carrier service is less expensive than the services of the various common carriers; however, both provide public data network (PDN) services. Use of a public data network provides an advantage in that error control and routing are usually features of the PDN. Typically, users of PDN services must provide their own packet assembly/disassembly devices (PAD) to convert their data into synchronous packets. The PAD can be viewed as a minicomputer that creates high-level data link control frames of data that originate from user terminals and devices.

PDN providers usually have large support organizations and provide a variety of additional services. One such service is education in the use of network services and available systems. The PDN vendor also may provide communications networks for the user's offices. PDNs are generally implemented according to CCITT standards such as those described in X.21, X.25, and X.29. (Refer to Appendix A for a list of standards.) The carriers are regulated by the U.S. Federal Communications Commission and follow guidelines established by the International Standards Organization (ISO). Expenses associated with use depend on the types of lines and services required.

Time-Sharing Service Providers

Time-sharing services are provided by independent data communications and data processing companies to users who want to control their own processing or communications activities. These companies usually are not large enough to have their own data processing operation. They also may wish to use an outside service for security or convenience. Some time-sharing vendors have value-added services such as packet switching or protocol conversion that facilitate communications from one computer format to another. This feature allows the user to take advantage of software available on other manufacturers' hardware.

With the arrival of small, inexpensive, powerful personal computers and minicomputers, typical users of time-sharing services include organizations requiring services such as supercomputer activities. Service providers normally will include terminals and lines to a subscriber's location. Additionally, service providers make available certain types of application programs and, for a fee, will provide consulting and design services. Fees usually include a monthly service charge, as well as additional charges for connect time, number of lines printed, and number of bytes of disk storage used.

Time-sharing service vendors generally implement their systems according to international standards such as X.21, X.25 and X.29, which are identified in Appendix A, and use software provided by the hardware supplier. Time-sharing service vendors are regulated by the U.S. FCC and generally follow guidelines established by the ISO.

Packet Switching System Services

Many common carriers and PDN vendors provide an inexpensive, efficient method of transferring data using packet switching. Companies use this type of service to transfer data nationally and internationally without having to lease lines or pay for telephone company services they do not need. As an example, CompuServe, owned by H&R Block, provides a packet-based mail service to which many organizations subscribe and use. The system allows employees to access a secure mail system from any dial-up telephone by simply dialing an 800 number and logging onto the system.

Packet-switching system vendors normally provide value-added services to users such as code set conversion, data encryption services, communications format conversion, and error detection. The PDNs provide end-to-end data transmission regardless of format or hardware. The user need only send the data in the format it was created and the packet switching software or hardware assembles the data into packets and disassembles the packets back into the user's format. Companies providing packet switching systems generally provide the end links to subscriber sites. This is an excellent alternative for transferring large amounts of data not required for immediate processing.

TYPES OF CARRIER SERVICES

Most worldwide carriers offer several basic types of services as well as options to ensure compatibility between foreign networks. The major basic services are:

- Dial-up service (public-switched telephone network)
- Dedicated or private-line service, with or without conditioning to improve performance
- All digital service (DDS)
- Switched 56 Service
- T-level and DS-level High-Speed Wideband Service
- Switched Multimegabit Digital Services (SMDS)
- ISDN (Integrated Digital Services Network), an all-digital network service

In addition to the basic types of services, each carrier offers a variety of other services. Two examples are AT&T's 800 data transmission service and MCI's VNET service.

PUBLIC SWITCHED TELEPHONE NETWORK (PSTN)

Let us now explore the backbone of almost all data or voice communications networks, the Public Switched Telephone Network (PSTN). The PSTN consists of an infrastructure that provides central office switching, trunk lines and the transmission lines—including most of the local loop telephone circuits, that make up the telephone network. Figure 9.2 illustrates an example of a local loop, a trunk line, and the central office.

Central Offices

The central office is where your telephone line connects to the telephone companies switching equipment. This is the first level up the network from individual residences or office buildings. When making a connection to another computer, the circuit would typically go from your computer to the central office where it would be switched to a line attached to the other computer.

The Local Loop

The local loop is the wire that connects your telephone line to the central office. If your connection is a residence, you will have an analog line that uses two wires. If your connection is a business, you will have either an analog or digital line that uses four or more wires (for more information on wire and cable, see Chapter 7). Because of a recent change in the law (1996), cable television providers will also soon offer local telephone service. Their local loop will be the coaxial cable currently used for television. As the world's telephone companies switch to optical fiber, local loops also will migrate from copper

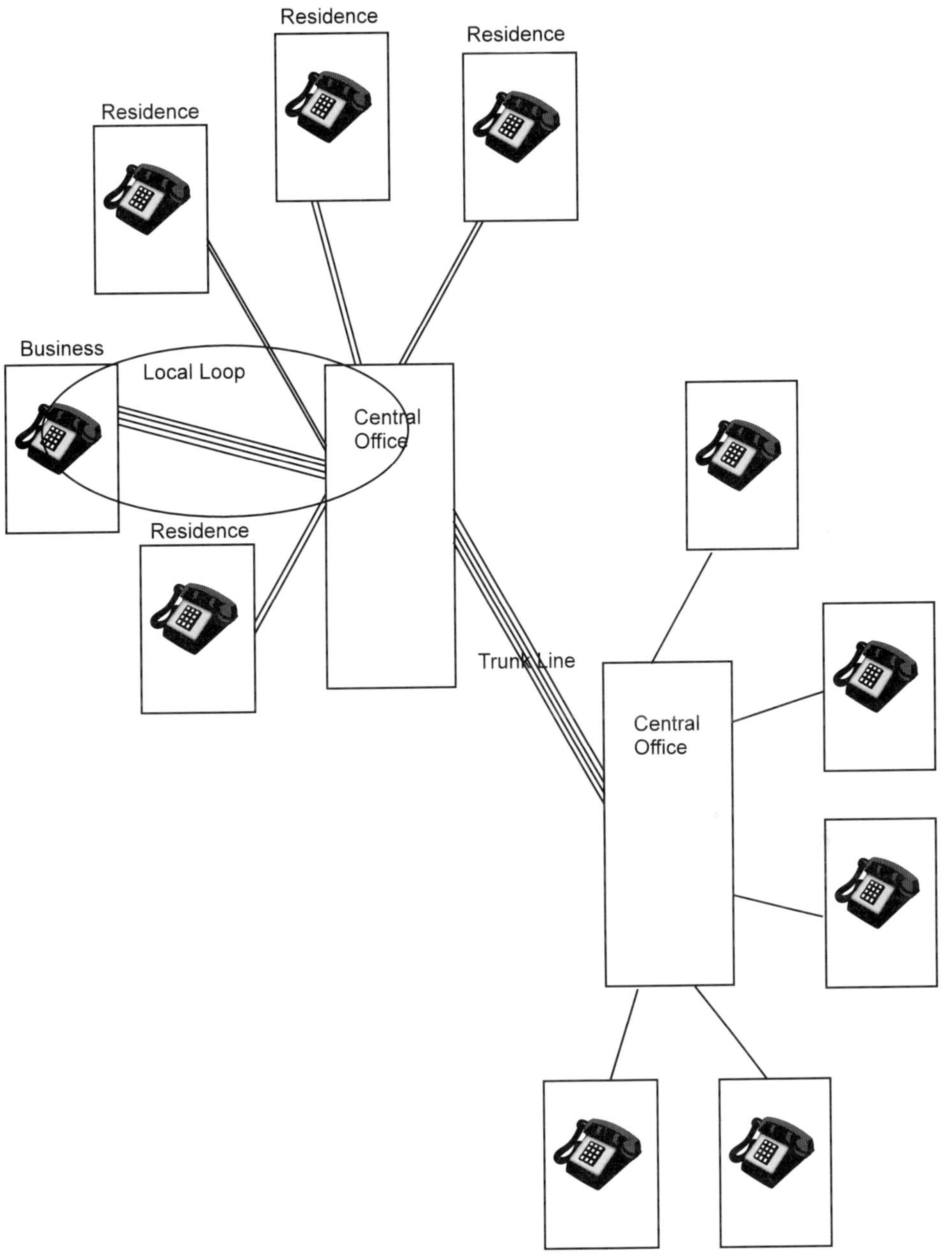

Figure 9.2 Example of a local loop, a trunk line, and the central office

wire to optical fiber. We will probably see a hybrid connection where coaxial cable will be used between a curbside box and your home, and fiber between the curbside box and the central office. The use of coaxial cable rather than fiber is based on the need to provide power to your telephone.

Central Office Trunk Lines

Connections between central offices use trunk lines. A trunk line can be an optical fiber, a coaxial cable, or a high-speed microwave circuit. Central offices between different carriers also are connected by trunk lines. An example would be an AT&T central office connected (trunked) to your local carriers central office.

Local Access and Transport Areas

A Local Access and Transport Area (LATA) is the geographic region set up to differentiate between local and long-distance services. If services are intra-LATA (typically the same area code), the local phone company is responsible for the circuit. If service is inter-LATA (usually different area codes), long-distance carriers are responsible. Of course, this is rapidly changing because the restrictions preventing local providers from offering long distance services, and long distance companies offering local services have been removed.

Until recently, most local loops employed for business used analog transmission. However, with the wide availability of digital technologies, a rapid conversion has begun to take place and digital circuits are now encountered more frequently. This is mainly due to the availability of fiber optics and its superior transmission characteristics. Additionally, the digital conversion and switching equipment required to support this type of transmission is more readily available. Bear in mind that analog transmission is still used and will be for some time to come because of its lower costs, wide availability and adequate capabilities for many activities.

Another reason that analog transmission is still widely used is that digital signals do not travel very far over standard telephone wires (about 2,000 feet) without expensive pieces of hardware called digital repeaters. These are needed to reamplify the digital signals at periodic intervals. However, digital signals can travel extensive distances without reamplification, when converted to the analog format and modulated to emulate a digital format. A modem is needed to convert the digital signal from a computer or a terminal into an analog signal.

The public-switched telephone network is also referred to as the direct distance dialing network (DDN). This network uses a variety of cabling methods to interconnect the various independent telephone company central offices with the subscriber sites. Figure 9.3 shows an example of the public-switched telephone dial-up network.

Dial-up Telephone Service

The dial-up telephone network is also known as POTS or plain old telephone service. It provides voice grade 3,000 Hz circuits that can be used for voice or data. Dial-up service

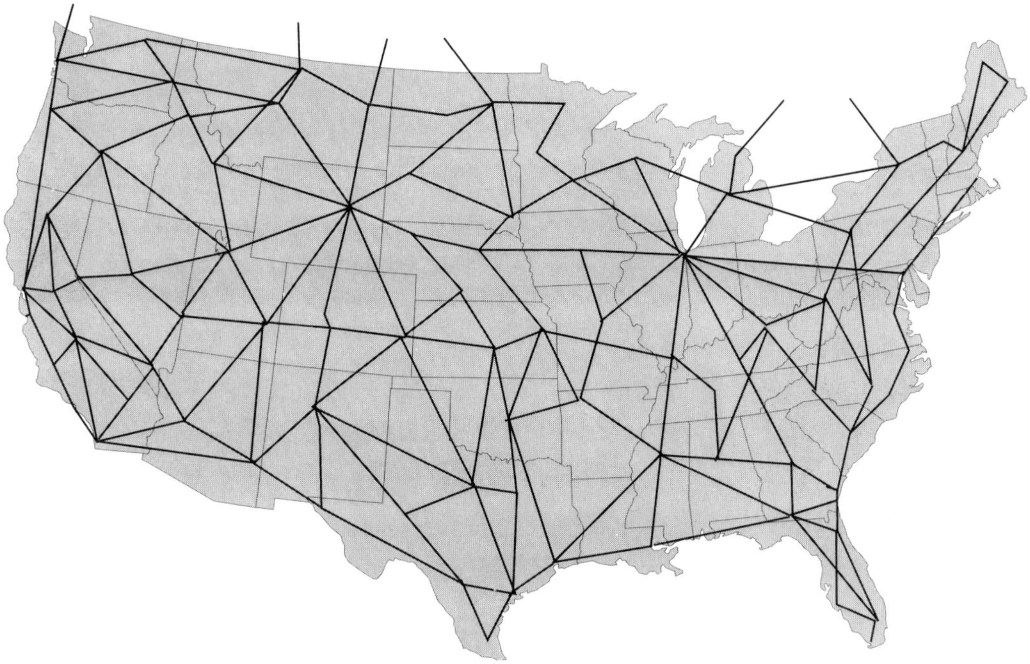

Figure 9.3 Public Switched Telephone Network

has been around since the turn of the century. While it is more than acceptable for voice, speed restrictions exist for data. A typical dial-up circuit would use two-wire or four-wire twisted-pairs for end point connections and limit data speeds to 33,300 bits per second or less. Because of variations in-line quality around the world, no minimum level of quality is guaranteed which means bit rates will vary depending upon your provider.

Why is the bandwidth limited on dial-up lines?

Twisted-pair wire is not the limiting factor for obtaining higher data communications bit rates over dial-up lines. In fact, twisted-pair wire supports bit rates in excess of 100 Mbps. Why then are we limited to 33 Kbps?

When Bell developed the telephone, 3,000 Hz of bandwidth over twisted-pair wire was selected. This decision was based on the fact that the current technology level, the expense to build switches, higher bandwidths were not required, and voice was the only service the telephone network was designed to support.

The bandwidth limitation described above is the major roadblock to faster dial-up data communications. In order for the telephone network to support higher bit rates, more bandwidth is needed. This can be accomplished in two ways: 1) By upgrading the analog switches around the world to support a higher bandwidth telephone network that would be a step backward because digital is far superior; 2) Convert the world-wide telephone network to an all digital format. An all digital format is the ideal solution because a standard already exists called ISDN that is rapidly being deployed around the world. This is a slow process, however, because telephone handsets need to be changed and central office switches need to be upgraded to support digital as well as analog transmission.

Dedicated or Leased Analog Line Service

Many organizations require the full-time services of a communications link that operates at higher speeds and is more reliable than dial-up lines. Communications service providers have developed a solution to this problem by offering hard-wired dedicated links between customers' locations. These dedicated or leased telephone lines are also known as private facilities; a leased circuit is permanently in place between two end points or shared among multiple points. Leased lines are usually configured to support point-to-point or multipoint line configurations. Leased circuits support two configurations, voiceband and wideband. Voiceband has a 3 Khz bandwidth and supports bit rates exceeding 100 Kbps. Wideband supports a higher bandwidth and bit rates into the millions. This type of line normally consists of four-wire twisted-pair full-duplex circuits, however at newer locations or office buildings it may also be coaxial cable, or optical fiber. Figure 9.4 shows an example of leased-line service.

The subscriber has sole use of the link, which is a permanent circuit. Line availability is guaranteed, along with some base level of service. The base level of service can include maximum acceptable values for delay distortion, impulse noise bits, and bit and frequency error rates. Data transmission is normally accomplished using one of the synchronous transmission methods.

Conditioning

Standard four-wire dedicated links often do not provide sufficient noise and transmission control for high-speed data communications. Lines can be tailored for better performance using a feature called conditioning. MCI, Sprint, and the various Bell operating telephone companies provide leased analog lines suitable for data transmission based on specifications established by the FCC that defines signal-to-noise ratios and permissible signal strength gain from the transmitter. Table 9.2 lists parameters associated with conditioning.

FCC Tariff No.260 defines additional capabilities for a basic voice channel that is intended for use with data applications. These additional capabilities, referred to as line conditioning, include five levels of C conditioning and two levels of D conditioning. C-level conditioning controls attenuation and envelopes delay distortion by providing

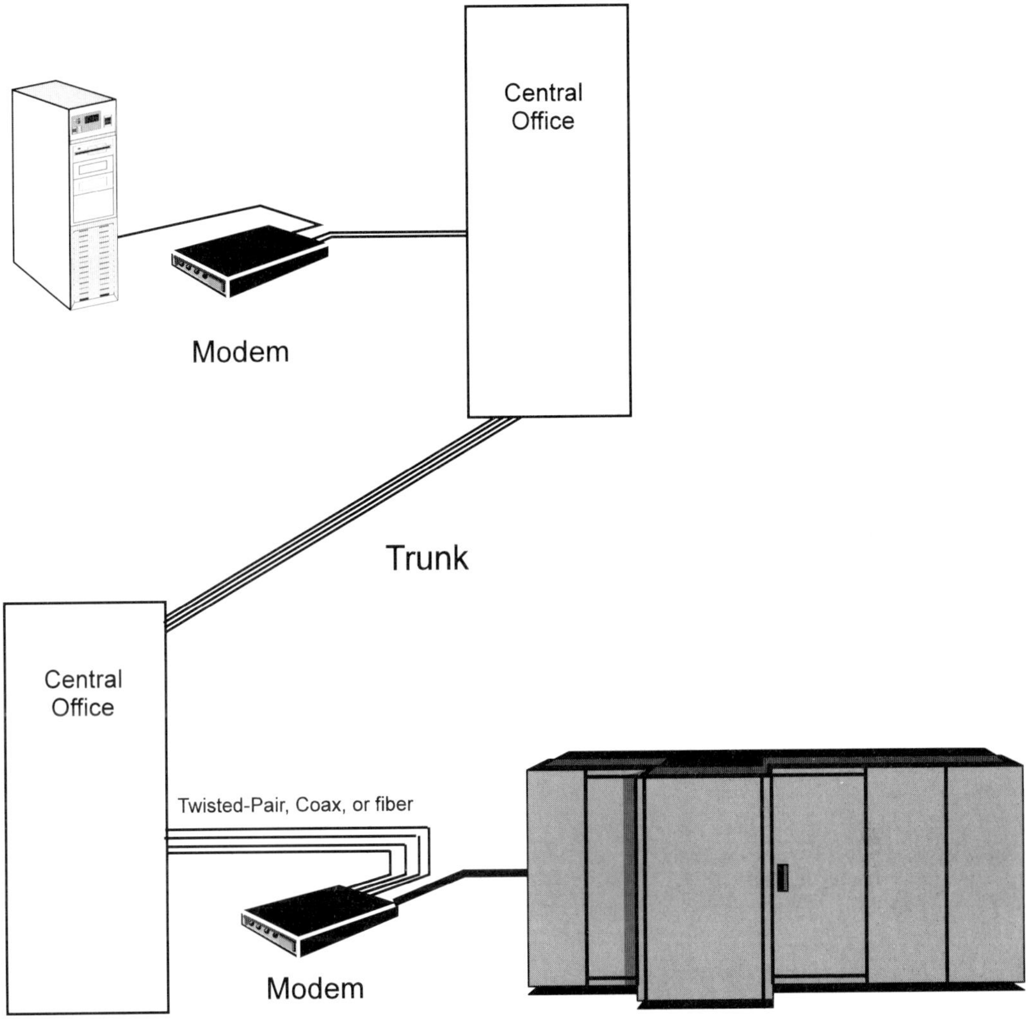

Figure 9.4 Example of Leased Circuit

equalization of the circuit. (See Chapter 7 for more information on attenuation distortion and envelope delay.) Equalization capabilities allow both ends of the link to be balanced to each other in terms of noise and signal strength. This is accomplished by inserting equalizers on the line between the customer's site and the central office. D-level conditioning controls nonlinear distortion and signal-to-noise ratios. D conditioning is accomplished by matching the modem's signal strength, noise levels, and distortion parameters to the line specifications.

Table 9.2 Levels of Conditioning

	Nonconditioned 3002 Channel	With C1 Conditioning	With C2 Conditioning	With C4 Conditioning	With D Conditioning
Frequency Range in Hertz (Hz)	300–3,000	300–3,000	300–3,000	300–3,200	
Attenuation Distortion (Net Loss at 1,000 Hz)	Frequency Range / Decibel Variation 300–3,000 / −3 to +12 500–2,500 / −2 to +6	Frequency Response / Decibel Variation 300–2,700 / −2 to +6 1,000–2,400 / −1 to +3	Frequency Response / Decibel Variation 300–3,000 / −2 to +6 500–2,800 / −1 to +3	Frequency Response / Decibel Variation 300–3,200 / −2 to +6 500–3,000 / −2 to +3	
Delay Distortion in Microseconds (µs)	Less than 177 µs from 800 to 2,600 Hz.	Less than 1000 ms from 1,000 to 2,400 Hz. Less than 1,750, us from 800 to 2,600 Hz.	Less than 500 µs frcm 1,000 to 2,600 Hz. Less than 1,500 µs from 600 to 2,600 Hz. Less than 3,000 µs from 500 to 2,800 Hz.	Less than 300 µs from 1,000 to 2,600 Hz. Less than 500 µs from 800 to 2,800 HZ. Less than 1,500 µs from 600 to 3,000 Hz. Less than 3,000 µs from 500 to 3,000 Hz.	
Signal to Noise (db)	24	24	24	24	28
Nonlinear Distortion Signal to 2nd Harmonic (db)	25	25	25	25	35
Signal to 3rd Harmonic (db)	30	30	30	30	40

Digital Data Services (DDS)

Digital Data Service (DDS) was originally an AT&T service, but it now more generally describes a digital service provided by most major carriers and regional telephone operating companies (see Figures 9.5 and 9.6). DDS services consist of leased digital line circuits that provide direct-connection synchronous transmission of pure digital sources. DDS provides network timing and synchronization using a master clock and subordinate clocks, which keeps all clocks on the various network nodes operating at the same rate. Timing signals also are provided to subscribers for synchronization within the network.

AT&T's DDS network access is provided to end users at over 120 local access and transport areas (LATA) serving more than 400 cities in the U.S. It can be connected to Canada's Data Route service at four border crossings and connects to more than 70 locations throughout Canada.

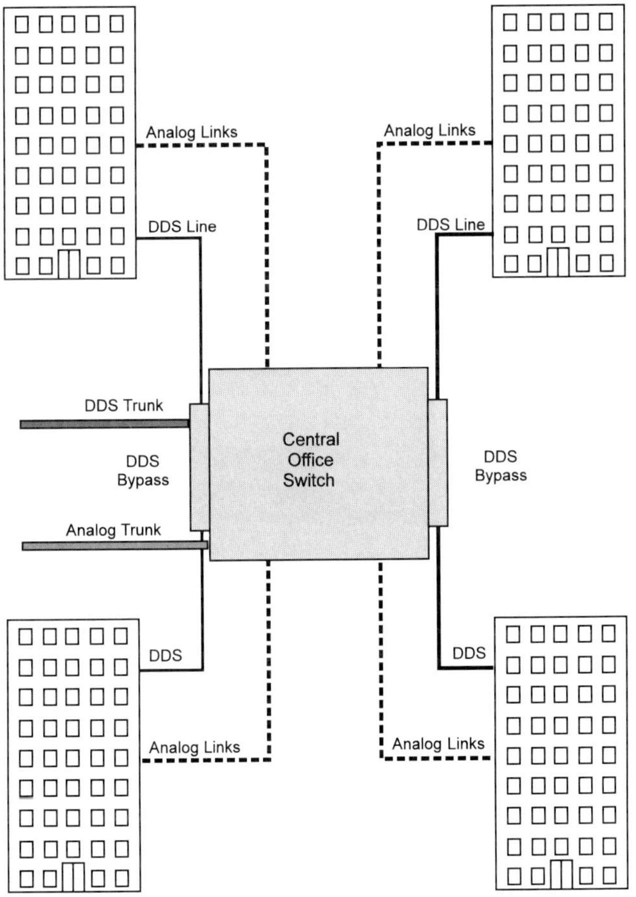

Figure 9.5 DDS Digital Service

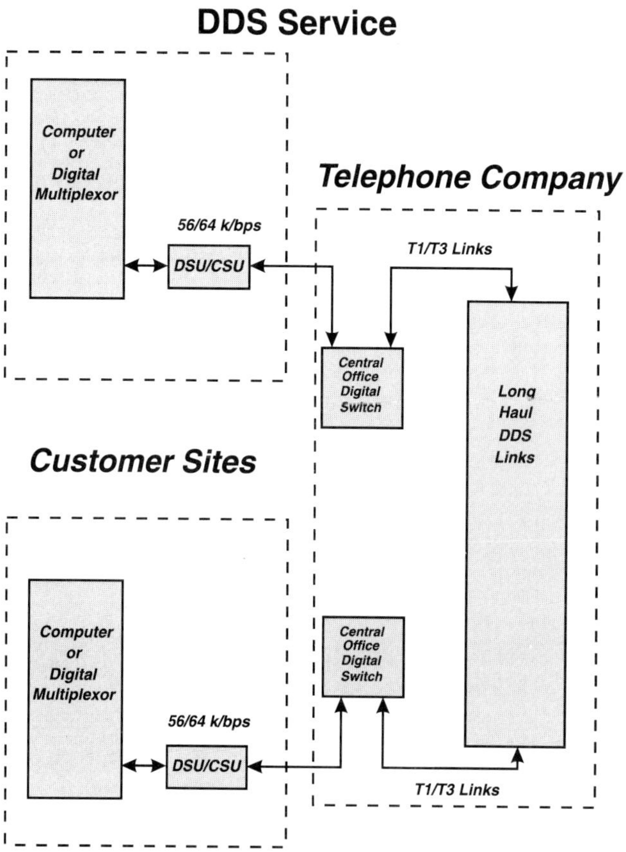

Figure 9.6 DDS Connections

DDS operates at rates from 4,800 bps to 64 Kbps. DDS also requires a Digital Services Unit/Channel Services Unit (DSU/CSU) device, which is similar to a modem except the DSU/CSU operates digitally. DDS is primarily a full-duplex service, but can be operated in half-duplex or simplex mode if necessary. The transmission rate is set at the telephone company's central office.

A DDS centralized testing facility provides dynamic testing and automatic rerouting of the data upon circuit outage or network degradation. DDS is usually guaranteed to provide 99.5 percent error-free transmission seconds at 56 Kbps, with 99 percent transmission availability. This type of service is implemented using a hierarchy of multiplexers within the various telephone companies systems. Typically, DDS supports high-speed links such as intermediate length regional lines (T1, T2 carrier [64 Kbps]) and long-haul microwave facilities.

DDS service requires the removal of loading coils from the local loops (the computer site and the central office), thus providing an unloaded leased line capable of carrying digital pulses to and from the customer's location. DDS also requires the use of regenerative repeaters every few thousand feet. Repeaters are needed to take the incoming digital signal, refresh it and pass it along to the next repeater or switch. DDS service is somewhat expensive, but the cost is rapidly decreasing in response to competitive service offerings. (DDS must compete against the newer high-speed digital services and fractional T1 digital services.) An important point to remember is that DDS can operate over a twisted-pair configuration.

Switched 56 Service

Within the past couple of years a service called Switched 56 digital service, has been implemented by most of the major telephone companies. Switched 56 service provides digital transmission at 56 Kbps over a leased line on a dial-up basis. It is much less expensive than a dedicated DDS 56 Kbps service. Charges for Switched 56 services are similar to the way charges for dial-up voice are incurred. The user pays a one-time installation charge, a monthly service charge, and a charge for the amount of time the line is used. The dial-up nature of Switched 56 makes it ideal for high-bandwidth, intermittent data applications.

Connection to the central office can be made using either two-wire or four-wire circuits. This is usually dictated by the type of CO switch in use. Four primary connection arrangements are available. AT&T 4ESS/5ESS and Northern Telecom DMS 100 CO switched access. This is similar to the four-wire 56 Kbps DDS service presented in the previous section. Datapath, based on Northern Telecom two-wire technology, Switched Digital Capability (SDC), and Integrated Services Digital Network (ISDN) 2B + D (two-wire).

Some switches support both types of technologies, however, range differs. Four-wire service provides greater distances. Two-wire service supports lower rates (2.4, 4.8, 9.6 Kbps). Two-wire and four-wire services are compatible. The Switched 56 services offered by local exchange carriers (LECs) and interexchange carriers (IXCs) are fully interoperable, calls can be made from local to long distance to international destinations similar to voice service.

Depending upon needs, there are different ways to access Switched 56 service. Intralata (originating and ending within the LEC's local area) or interlata (originating in one LEC's area and ending in another's). Gaining access is similar to regular telephone service. The local telephone company handles Intralata calls (local service). Long distance (interlata) calls are handled by an IXC, like MCI or AT&T. For long distance calls, the local carrier interfaces the call to the inter-exchange carrier (IXC) point of presence (POP). The IXC carries it to their end point POP where it is carried by the local telco to its end destination.

A switched CSU/DSU is used to connect your computer to the switched 56 network. A switched CSU/DSU is usually more expensive than a dedicated CSU/DSU because it can answer and dial.

Switched 56 supports bandwidth-on-demand allowing Switched 56 circuits to be combined for additional bandwidth. It is designed to support videoconferencing, group IV (high resolution) fax, large file transfers, remote backup and LAN internetworking.

HIGH SPEED WIDEBAND SERVICES

A variety of wideband digital and analog services may be offered by public carriers depending on location. An analog service is usually incorporated for speeds up to 1.544 Mbps because analog circuits can manage reliable data transfer into the millions of bits per second. One problem with analog circuits, though, is that they experience a higher bit error rate than digital circuits. High-speed analog circuits come in various bandwidths suitable for various equipment and requirements.

Carriers also offer a variety of digital circuit services. These are more expensive than analog circuits, however, they do provide major benefits, including: better noise control, higher bit rates, and extensive management features.

Of major importance are the T type carrier services. These are also referred to as DS-x services. Two categories of T services are DS service and fractional T1 service. Both of these services can employ either analog or digital facilities.

High Speed Digital Services

In addition to DDS and Switched 56, many other high-speed wideband voice and data transmission services operate digitally. To provide digital transmission, the carriers had to develop a method to manage and control the digitized channels. Channel information (data or voice) is carried within data frames, also known as DS-1 signal frames, that can contain 192 user information bits and one control frame bit. Twelve DS-1 signal frames are grouped together to create a D4 superframe (see Figure 9.7).

Each superframe contains information for signaling and synchronization (in addition to the 12 DS-1 signal frames). DS superframes are designed to be carried across all types of DS links. All U.S. domestic and some international carriers follow the DS specifications, but in some cases the available bit rates vary from carrier to carrier. This is mainly due to the carrier's need to maintain timing and control on the link.

DS links usually consist of customers' time-division multiplexed traffic channels. Time-division multiplexing (TDM) is the sharing of a single line between multiple devices by allocating time on the line to each device. (Figure 9.8 shows DS links and multiplexer connectivity.) Digital service is divided into categories based on the speed of the link. A DS-0 link is equivalent to one voice circuit of 64 Kbps. A traffic channel (DS-0) may carry either voice or data, and will contain signaling information to identify the data carried.

DS-1 is a 1.544 Mbps link that is usually multiplexed into 24 64 Kbps channels. DS-1 is generally referred to as a T1 carrier. While DS-1 is rated as a 1.544 Mbps link, unless the carrier offers clear-channel capability, only 1.344 Mbps are available to the end user. This is primarily due to the various carriers' need to maintain timing on the network using what is referred to as one's density. One's density uses 192 Kbps to maintain proper timing between nodes on the network.

DS-1C operates at 3.152 Mbps and consists of two T1 TDM lines. It can be broken into 96 traffic channels of 64 Kbps each. This is referred to as a T1C carrier.

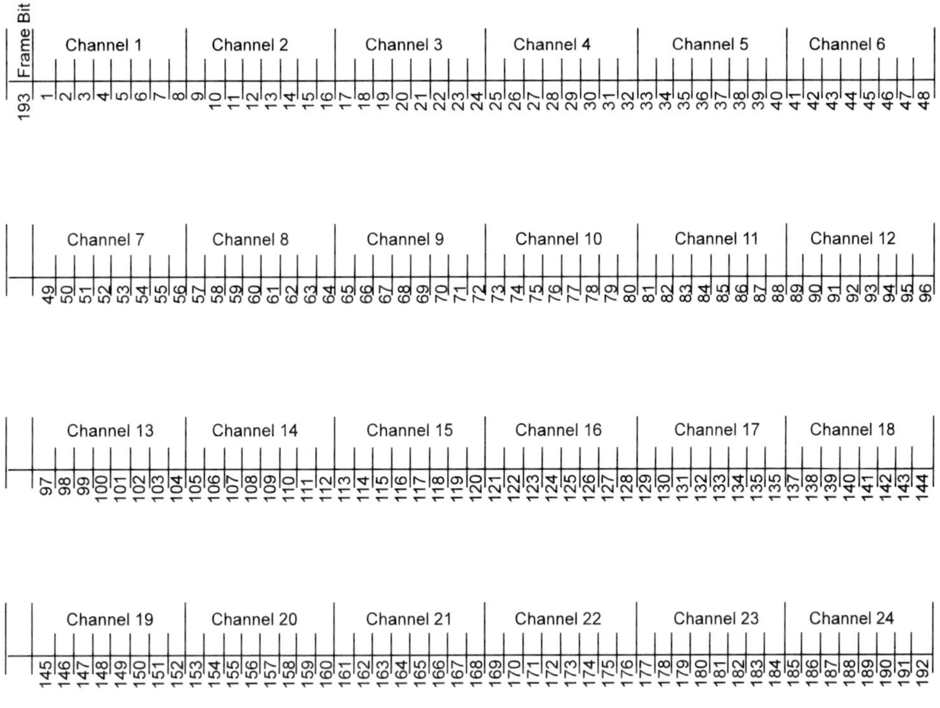

Figure 9.7 Example of Multiplexing and DS Framing

DS-2 is referred to as T2 carrier service (6.312 Mbps) and consists of four T1 TDM links providing 96 DM-0 channels (64 Kbps).

DS-3 service (T3 carrier) operates at 44.736 Mbps and provides the capacity for 28 T1 channels supporting 672 DS-0 voice channels.

DS-4 service (T4 carrier) provides 168 TDM T1 lines for a total capacity of 4032 DS-0 channels of 64 Kbps. The total speed of DS-4 service is 274.176 Mbps.

Many types of service support the links described above, such as fractional T1, insert, drop, and bypass services. Fractional T1 allows customers to use a portion of a T1 link for end-to-end dedicated services; this can be provided by the carrier company. Insert and drop capabilities allow the user to specify insertion and drop-off points for parts of a leased T1 circuit.

DS services are offered worldwide. However, in Europe, T-span services use a different channel allocation that creates some bit rate incompatibility between global networks. In North America T1 rates are 1.544 Mbps spanning 24 channels and in Europe, they are known as E1 lines and support a 2.048 Mbps rate spanning 30 channels.

Today, many carriers are offering clear-channel signaling, which allows the user to bypass the framing conventions normally found with DS-0 and DS links. Clear-channel

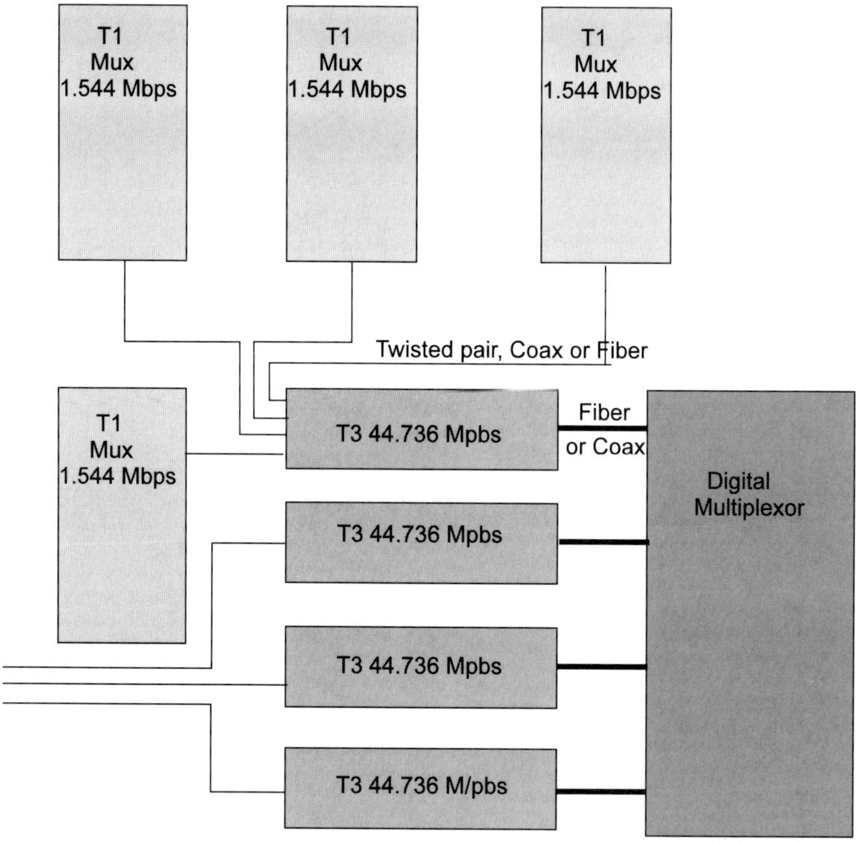

Figure 9.8 Example of a High-Speed Digital Transmission Hierarchy

capabilities provide the user with the entire available bandwidth minus a small amount of bit overhead needed for timing control.

E-SERVICES

DS-X services are offered worldwide. In Europe however, T-span services use different bit rates and the circuits used are referred to as E-services. An E1 circuit has a capacity of 2.048 Mbps, normally divided into 30 channels. E2, E3 and E4 lines are also available in some areas, providing proportionately larger capacities. And as with the T-carrier lines, they can be used as either a single clear channel or multiple channels. Interconnecting T-services and E-services on the same network is seldom a problem, because many multiplexers provide compatibility between the two. Exceptions may be found in Japan, Germany, or France. These countries vary slightly in their E1 formats, that may create compatibility problems.

FRACTIONAL T1 CARRIER SERVICES

Because at the present time an entire T1 offers more bandwidth than most companies need, fractional T1 was developed. Fractional T1 lets a business select only the amount of bandwidth required. Supported rates include 56, 64, 128, 256, 384, 512, and 768 Kpbs. Or full T1 speeds of 1.544 Mbps. (Figure 9.9 shows an example of a fractional T1 link.)

Installation of a fractional T1 circuit depends on the connection points and whether they are in the same LATA.

If connection points are intra-LATA, the local telephone carrier is responsible for the circuit. If the connection is inter-LATA, long-distance carriers are responsible.

A full T1 line must be installed in the building where service is to be provided. The line is connected to a Multiplexer and DSU/CSU.

Price of a fractional T1 varies with the amount of bandwidth requested and the distance between the connections. The most popular fractional T1 service is ACCUNET Spectrum of Digital Services (ASDS) offered by AT&T. It currently features analog multipoint service, digital multipoint service, end-to-end, 9.6 Kbps digital service, subrate data multiplexing, and diversity options. There are four access options to these services with ASDS:

- Analog Voice Grade Private Line (AVGPL)—ideal for applications such as voice conferencing and broadcast polling.
- Digital Data Service (DDS) at 56 Kbps and 9.6 Kbps—a high-quality, economical transport option ideal for many applications.

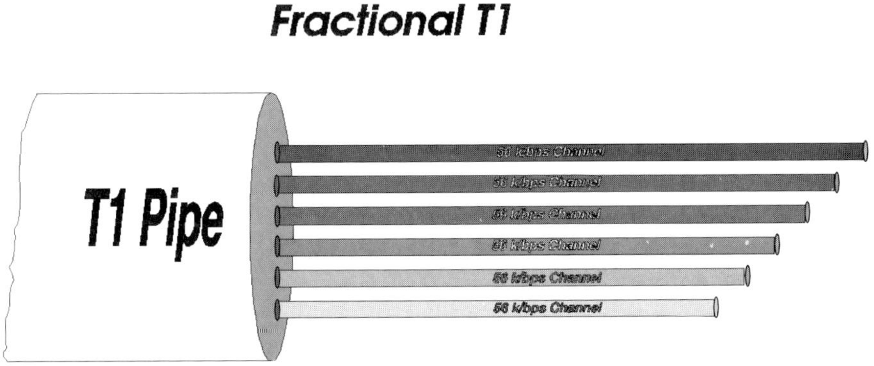

Figure 9.9 Example of a Fractional T1 Link

- ACCUNET T1.5 Service (DS1) at 1.544 Mbps only based on multiplexing of up to five 9.6 Kbps circuits into a 56 Kbps data stream.
- Generic Digital (DS0) at 64 Kbps.

SWITCHED MULTIMEGABIT DIGITAL SERVICE

Switched Multimegabit Digital Service (SMDS) is a public high-speed data transmission service originally developed by AT&T and the regional Bell operating companies to support high-speed public local area networking.

Created in 1989 and based on Bellcore standards, SMDS was designed to provide a transport service for LANs, host computers, and WANs. The original Bellcore standard specified how customer premise equipment (CPE) can access a SMDS switch located at a central office using twisted-pair wiring at the T1 rate or optical fiber at the T3 rate. Figure 9.10 illustrates an example of a Metropolitan Area Network using SMDS Services.

The scope of SMDS has evolved since its original design to support WANs. Additionally the standard has been enhanced with the inclusion of certain specifications from the IEEE 802.6 MAN standard. It now includes higher rate definitions and in the future will operate at the OC-3 SONET rate.

SMDS primarily works with metropolitan area networks and supports the ability to interconnect geographically dispersed multiple local area networks operating at speeds up to 16 Mbps and is considered an on-demand high bandwidth service. It is a protocol-independent service that allows bandwidth to be dynamically allocated.

SMDS can be viewed as a high-speed protocol or data transfer methodology, and will use either ATM or frame relay type packets. For supporting frame relay, SMDS breaks the data stream into packets that can be as large as 9,188 bytes in length. ATM type cells will remain at 53 bytes. One problem using SMDS with frame relay is it does not fully support isochronous traffic (evenly distributed in time) such as voice and video. Frame Relay techniques send packets of differing lengths sporadically.

SMDS uses a connectionless-oriented service and cells are routed to their destinations down any available path. They do not necessarily follow the same route. At their destination, SMDS cells are reassembled in the proper order.

INTEGRATED SERVICES DIGITAL NETWORK (ISDN)

Overview

At the present time, telephone companies and independent carriers support a mixture of links around the world. These consist of two-wire twisted-pairs, four-wire twisted-pairs, coaxial cable, microwave, satellite links, and more recently, fiber optic cable. The different link types exist primarily due to millions of miles of installed cable, the evolution of technology over time, user requirements, and costs. Looking at the installed and existing capabilities in a traditional sense, three major problems become immediately apparent,

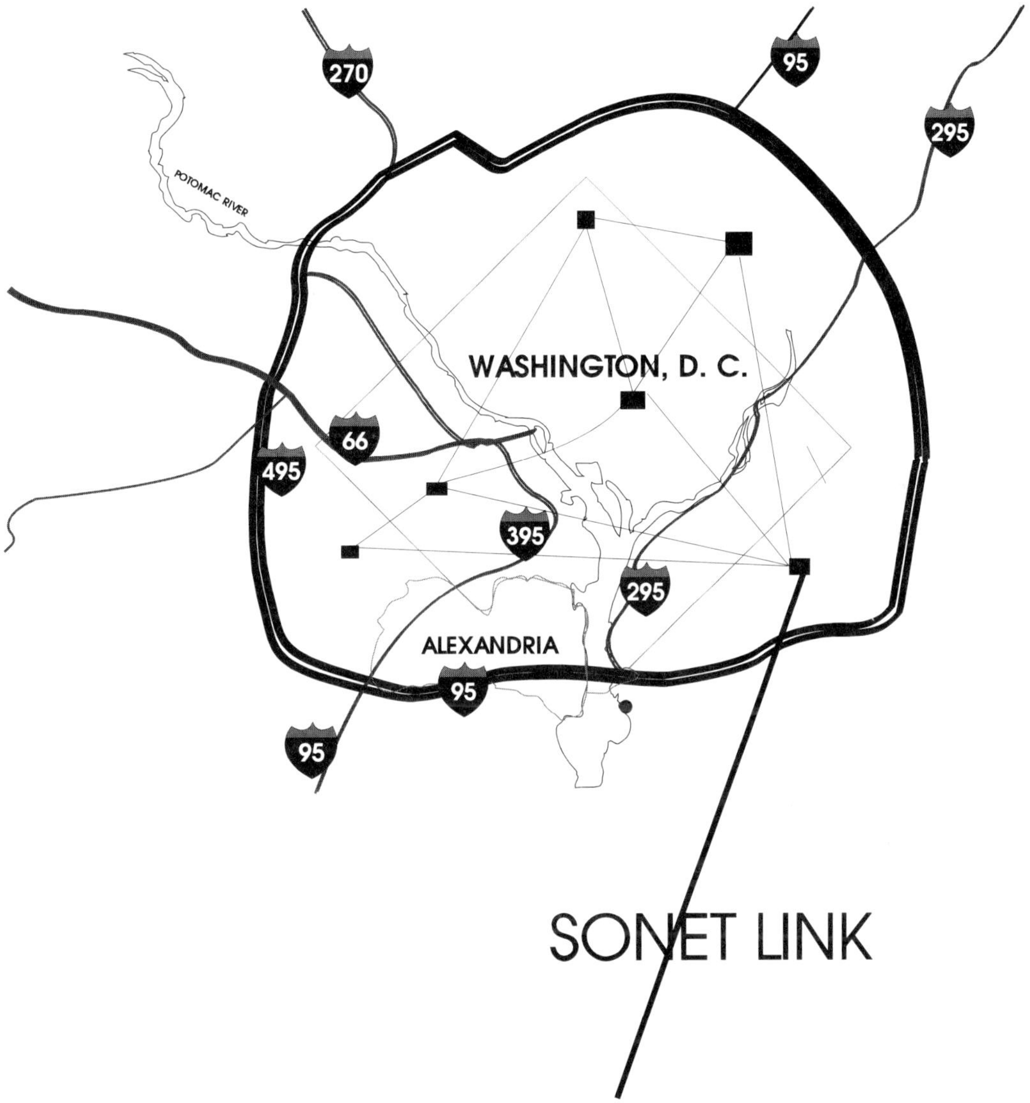

Figure 9.10 Example of a SMDS Metropolitan Area Network

different speed restrictions exist for each type of link, i.e., twisted-pair wire versus optical fiber; different error rates apply to the different types of medium used; multiple protocol conversions can often take place. Figure 9.11 shows an example of a mixed media network.

While this presents little problem for low-speed links, higher speed requirements place a great strain on existing facilities. As the world becomes more and more digitally

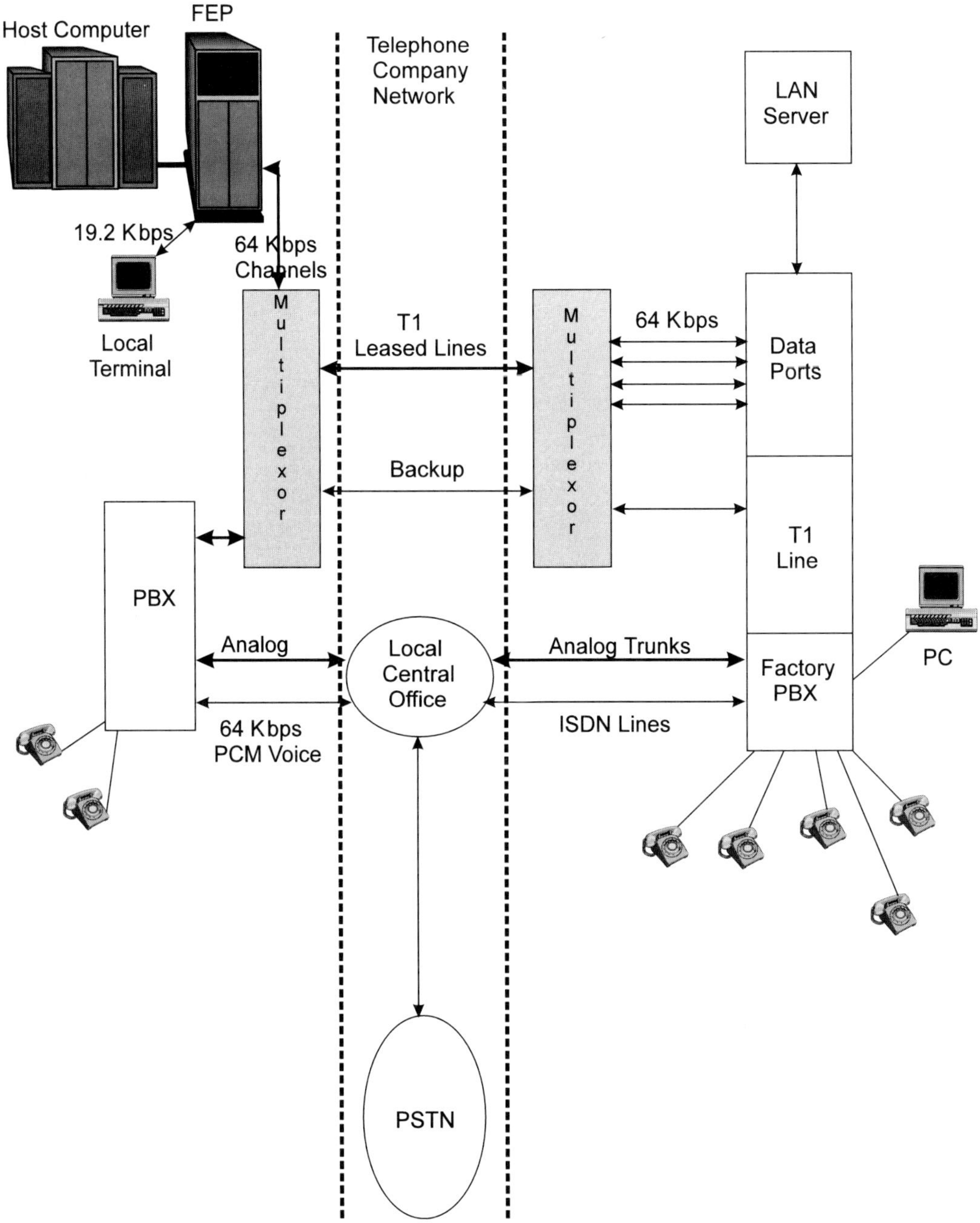

Figure 9.11 Mixed Media Network

oriented, existing techniques can no longer efficiently handle the demands required for digital voice, high speed data, video, PBX, local area networks, and the necessity for integration of these services. ISDN is a series of developed and pending network standards designed in 1980 to provide an all digital end-to-end pathway for the transmission of digital information. It consists of standards defined by the ITU-T. These are referred to as the I.400 series of standards.

ISDN is an all-digital network standard designed to provide a pathway where multiple conversions of the data along the transmission circuit will be avoided. This will reduce distortion and processor delays. ISDN will allow the concurrent transmission of voice, data, video, and text over the same digital pipeline. Intelligence, capacity and an end-to-end digital pipeline are three key features of the ISDN Network.

When ISDN was announced in 1985, it appeared to provide a solution for the problems associated with existing facilities. Today, most carriers have either implemented or are in the process of implementing ISDN, however, they exist in pockets around the world. In other words, due to minor incompatibilities, many of these networks cannot communicate with each other. Several problems must be resolved before a worldwide integrated ISDN network is a reality. Among the problems that are being resolved are different implementations of the standards. Implementation differences included limited availability, interoperability conflicts, high prices, non-standard technologies, speed differences between North America and the rest of the world, and differences in the number of channels a circuit can carry. Figure 9.12 illustrates an example of an ISDN environment.

North American National ISDN-1. To address these issues, Bellcore (Bell Communications Research Inc.), in an agreement with other major equipment vendors, has worked out a set of specifications called National ISDN-1 that is intended to provide a way to standardize ISDN equipment and services. A primary advantage of National ISDN-1 is that it supports standardization of network-to-network communications for different implementations of ISDN.

National ISDN-1 represents an industry wide effort to establish consensus on standard technical specifications and implementation agreements. It refers to an agreement to make ISDN switches from various manufacturers compatible. With implementation of National ISDN-1, subscribers will be able to call anywhere National ISDN service is available.

Another set of ISDN specifications recently completed are called National ISDN-2. They are designed to address standard billing and improved operations. ISDN-2 also includes administration and maintenance features.

Net3 European Standards. Net3, the pan-European ISDN standard has recently been completed, however, it still remains difficult to link up with the U.S. National ISDN-1/2 standard, because it still has major differences with Net3. Supplementary services also vary between many countries, and Europe lacks a common software interface. The European Telecommunications Standards Institute (ETSI) has developed a common European interface standard, which was introduced in 1993.

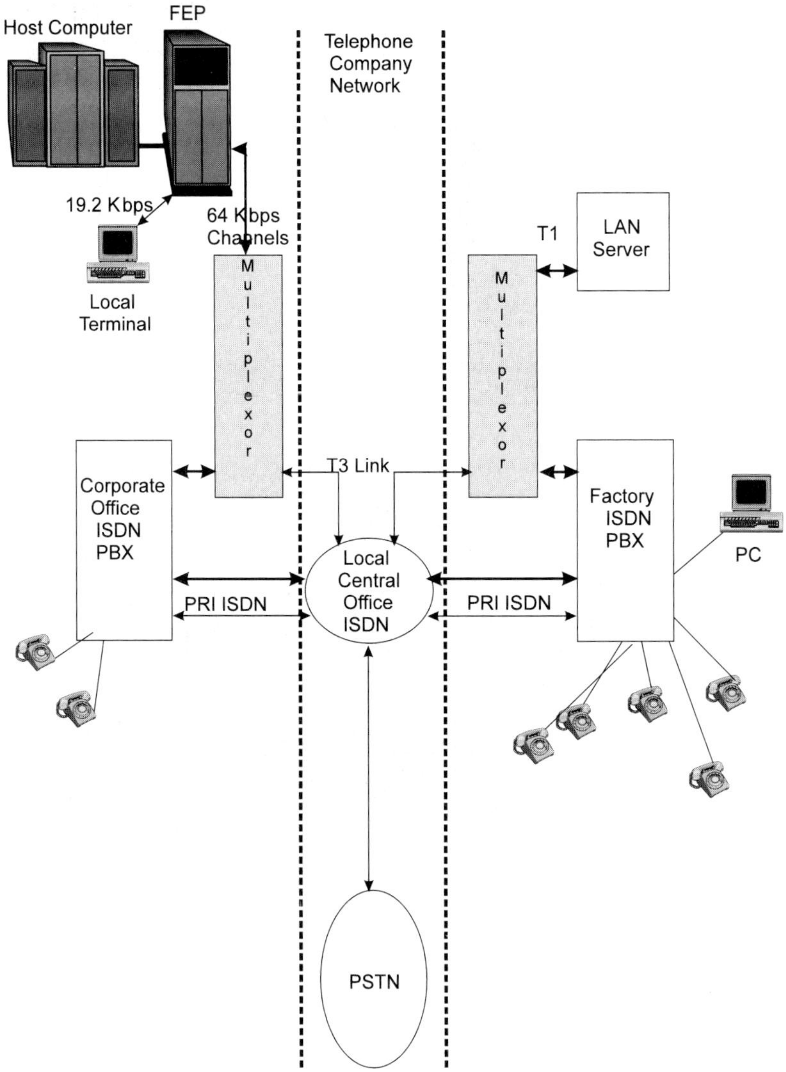

Figure 9.12 ISDN Based Network

ISDN Design Criteria

ISDN standards and networks are designed to provide the following benefits:

- An all digital transmission methodology regardless of the source, data, video, voice, or any mixture. ISDN will provide an all digital pathway from end to end.

- Lower error rates. Error rates will be reduced over conventional analog data transmission methodologies due to end-to-end control using ISDN network specifications, not link-to-link error specifications.
- ISDN will provide greater capacity because it is compatible with and will use the capabilities of SONET and SMDS. ISDN also supports standard bit rates for the various levels of service offered which makes it compatible with Frame Relay and ATM services.
- ISDN provides compatibility with the Seven Layer Open Systems Interconnect (OSI) model and uses the services provided at the upper layers.
- Reduced costs will be the results of a standard approach to network communications because hardware from different vendors will be compatible.
- Two types of ISDN have been defined, these can be classified as either narrowband ISDN or broadband ISDN (BISDN).

Narrowband ISDN

The creation of narrowband ISDN, the first ISDN specification to be developed, was the result of the emergence of ISDN standards defined in the mid-1980s by the CCITT. ISDN was created to operate over existing twisted-pair wiring that is used in the existing public telephone network.

ISDN circuits consist of multiple channels that are designated as "B" channels (used for voice or data), and "D" channels that can be used for passing control information or packets of data to central office equipment. The "D" channel provides the ability to communicate in and out of information band area so it will not disrupt the "B" channels. This allows them to be used exclusively for the transportation of information.

Narrowband ISDN was designed to support several basic classes of service, two of which form the foundation for ISDN. They are called the Basic Rate Interface (BRI) and referred to as 2B+D, and the Primary Rate Interface (PRI) called 23B+D.

BRI consists of two 64 Kbps "B" or bearer channels, and one 16 Kbps "D" channel. BRI was designed to provide connections from the home or small business using a twisted-pair circuit and provides a total bit capacity of 144 Kbps. The "B" channels can be used for either voice or data, or may be combined for higher bandwidth usage.

PRI consists of 23 64 Kbps "B" channels and one 64 Kbps "D" channel designed to operate over four-wire twisted-pairs for a total bit capacity of 1.544 Mbps. PRI capabilities were designed to accommodate larger businesses that require T1 capacity. The "B" channels can be combined into different configurations in order to satisfy bandwidth-on-demand needs or for special applications such as video conferencing. (30 channels in Europe).

Narrowband ISDN also provides additional capabilities for special needs. The 3HO+D designation provides three 384 Kbps channels plus one 64 Kbps channel for packet signalization. This interface was originally designed for video conferencing however because of efficiencies gained in converting video signals to digital formats that can use as little as 128 Kbps, 3HO+D has lost some of its importance.

Narrowband ISDN also defines H11 which is a nonchannelized 1.544 Mbps capability, H12, a nonchannelized 1,920 Mbps channel and multislotted multiples of 64 Kbps channels under customer control up to 1,544 Mbps.

Standard Interface Connections

In addition to standards for channels and signaling, interfaces to the network are also standardized. At the present time, ISDN consists of the following components:

Terminal Interface Types. There exists two basic ISDN terminal types because there are two different types of terminal equipment, devices that have built-in ISDN compatibility, and those who do not and must be interfaced. The two types are designated as Terminal Endpoint Type 1 devices (TE1) and Terminal Endpoint Type 2 devices (TE2). TE1 is an ISDN compatible-terminal such as an ISDN integrated voice data personal computer or an ISDN video codec used for video signal conversion and transmission. TE1 type devices can be plugged directly into an ISDN network. TE2 devices such as those that use EIA-232, V.24, Ethernet and Token Ring are incompatible with ISDN recommendations and require a Terminal Adapter (TA) for conversion and connection to a basic rate interface.

Terminal Adapters. The TA provides an interface such as that provided by a modem in today's analog world. TAs can be implemented as a standalone device or as a PC circuit card that may support both circuit switched and packet switched services. The TA provides the electrical/mechanical interface signal conversion and bit rate adaptation between a non-ISDN user device(TE2) and the ISDN network. The TA also provides synchronization and auto dial-answer functions, as well as manual call control functions for either voice or data circuit switched services.

The TAs major task is adaptation of the V series bit rates to the ISDN "B" channel rate. This function is performed by rate adapters (RA) located in the TA. The RA1 converts V series voiceband bit rates to intermediate bit rates of either 8, 16, or 32 Kbps. The RA2 takes the intermediate rate or the V series wideband modem rates and converts them into 64 Kbps rates.

Network Termination Devices. Two network termination devices exist for connectivity between the TA and the network. The NT1 is used for device connection of basic rate devices to BRI circuits. The NT2 is used for interconnection of multiple line devices such as PBXs, LANs, etc. to PRI circuits.

Additional Interfaces

R Interface—Rate adapts RS-232 to 64 Kbps (RS-232, V.35, X.21, X.25)

S Interface—Standard four-wire plug for basic rate

T Interface—Same as S interface

U Interface—Central Office connection to home using two-wire

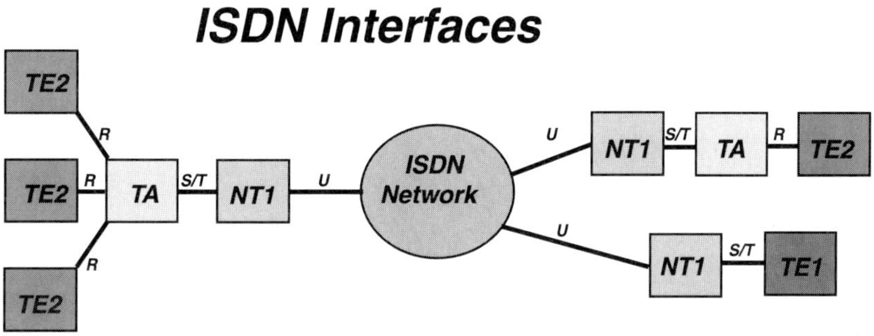

ISDN - Integrated Services Digital Network
NT - Network Termination
TA - Terminal Adaptor
TE - Terminal Equipment

Figure 9.13 ISDN Terminal Interfaces

V Interface—Proposed specification for the connection of ISDN PBX type devices

Z Interface—Analog Interface for Voice Grade Circuits

Figure 9.13 Shows an example of several ISDN interfaces.

OUT-OF-BAND SIGNALING

As you have already learned, network communications require many interrelated activities for the transfer of data. One of these activities is the actual movement of the data packets from one node to another. Another activity is the generation and movement of control information between the nodes in the transmission path. The control information describes routing requirements, carries billing data, and may provide special handling instructions for the data being transferred. The control information is usually carried on the same circuit as the data. Combining data and control information on the same circuit is known as in-band signaling. Figure 9.14 illustrates out-of-band signaling.

An example of in-band signaling can be illustrated in the existing telephone network. Routing and billing information coexist with the information signal along the voice pathway and are considered to be in-band. This works quite well for voice and low-speed data, however, it does create somewhat of a problem in that it adds overhead in the form of reduced capacity and throughput. The results of in-band signaling can be most readily noticed in the amount of time required to establish a connection (known as call-setup time).

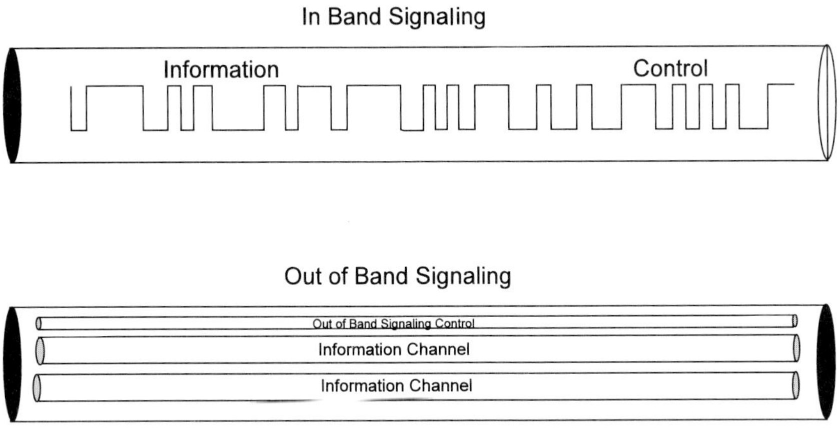

Figure 9.14 In-band and Out-of-Band Signaling

With ISDN, control and signaling information use out-of-band signaling. Out-of-band signaling provides a noticeable improvement in connection time as well as provides an improvement in overall throughput. All activities required to establish the routing, make the connection and generate billing information are placed on the "D" channel. At the present time, the technique and technology for handling this activity is known as Common Channel Signaling #7 (CCS#7).

CCS#7 is a common channel signaling protocol that is being adapted worldwide for support of packet switching which defines the basic communications support required for advanced telephony services. CCS#7 provides for the exchange of trunk information between switches via data links instead of on a per trunk basis. It includes the follow characteristics:

- Common channel signaling
- Part of the a seven layer protocol stack compatible to the OSI model.
- Provides a high degree of flexibility
- Internationally compatible
- Supports transmission rates up to 64 Kbps.

Broadband ISDN

When ISDN was first defined in the mid 1980s, widespread deployment of optical fiber by most carriers was just beginning to take place. Because ISDN was principally designed for a copper-based wire environment, no plans had been developed to take advantage of high-speed, high-bandwidth optical fiber for speeds greater than 1.544 Mbps. As ISDN started to evolve, several problems rapidly became apparent.

1. Different manufacturers of ISDN hardware and software used different versions of the specifications to create their products. This created incompatibilities between carriers', vendors', and users' equipment.
2. With the development, implementation, and widespread use of SONET technology, bandwidth capacity will become available well into the gigabit per second range, however, ISDN as defined could not support future application requirements such as digital video requiring bit rates above the T1 level.

At the beginning of the 1990s, the CCITT began the process of defining a strategy for broadband ISDN or BISDN to address those new application needs. Broadband ISDN using SONET's networking foundation is being designed to take advantage of optical fiber and the gigabit per second speeds it offers. Initial BISDN specifications call for bit rates of 155 Mbps and greater using either the ATM employed by cell relay technology or a STM mode of operation.

BISDN will provide users with higher throughput, better performance, and standard interfaces, because of its relationship to SONET and other emerging technologies designed for the high-speed movement of voice and data. BISDN standards are also being developed to address trouble indicators, fault locators, and performance monitors. These will be designed to interface to a variety of customer equipment and to BISDN central office switches to provide network operational viability.

BISDN will additionally provide a number of new services to its user base. The CCITT has defined two classes of service called interactive services and distribution services. Interactive services are designed to handle two way information exchange such as conversational services, messaging services, and retrieval services. Distribution services primarily handle information movement in one direction from the service provider to the BISDN subscriber. Distribution services include broadcast services and cyclical services.

Interactive Services. Conversation services will encompass the two way movement of information in real-time between two end users. Applications that will fall into this category include video conversational services such as video telephone and video conferencing.

Messaging services will be another feature of interactive services and will provide E-mail, video mail, store-and-forward message handling, and format conversion services.

Retrieval services will provide users with the ability to retrieve information stored in public information databases.

Distribution Services. One of the main areas addressed by distribution services is broadcast services. This service will provide a continuous flow of information from a central source to multiple subscribers. An example of this is broadcast television. With BISDN broadcast services, it is envisioned that by using SONET delivery capabilities, it will be possible to use a single source delivery mechanism such as fiber for data, video, imaging, broadcast television, and voice. Broadcast services will allow users to access the information but not provide any user control capabilities.

Cyclical services will provide information from a central source to multiple subscribers and permit some degree of user control for accessing specific portions of the data. A typical application for this type of service is video text services for the distribution of electronic print information. Using cyclical services will allow the user to access specific pages of an issue or a specific issue.

TYPES OF TRANSMISSION SERVICES

The difference between transmission services and carrier services is that the transmission service always represents the physical side of the network. Transmission services support the various carrier services previously described. Two different transmission services exist, they are:

- Synchronous Optical Network (SONET)
- Fiber Data Distributive Interface (FDDI)

SONET

In 1985, Bellcore proposed a new networking transmission approach called the Synchronous Optical Network (SONET). SONET was developed because of a demand by industry for a new technological approach that addressed high-speed digital transmission over optical fiber. SONET is a transport vehicle offered by carriers capable of delivering transmission rates well into the gigabit range. Until 1988, no standards for the use of optical fiber were defined that year, however, ANSI published the first of a series of synchronous optical network recommendations that has evolved into the SONET standards.

The SONET standards originated out of a cooperative effort between various standards bodies and private industry and has evolved into a series of international standards sanctioned by many national and international organizations. SONET standards not only define transmission rates and network management, but also define standard optical interfaces that allow a great amount of configuration flexibility. Phase 1 of the standard specifies a hierarchy of rates and formats for optical transmission. Phase 2 of the SONET standard when completed, will detail network maintenance, protection, and internetworking between different vendor's equipment.

With the advent of SONET, new applications impractical because of high bandwidth requirements will become a reality. Video is a prime market for use of high bandwidth for as our television industry matures, high definition television (HDTV) will require digital capabilities able to support gigabit transmission rates. This is because the new HDTV standards specify a digital methodology. Another television service called video-on-demand will allow users to program and schedule their television viewing based upon specific schedules or interest and have the programs delivered at specific times.

New applications such as virtual reality will begin to mature, spawning a new industry. Using SONET as a delivery vehicle, users will be allowed to access programs in a three dimensional interactive fashion. Virtual reality can be described as a technology that places the user in the middle of the program, and to accomplish this will require massive amounts of bandwidth to carry the program material. Figure 9.15 shows an example of distant networks connected via a SONET Link.

The basic building block of SONET is a DS-0 (64 Kbps) channel which is multiplexed into a Synchronous Transport Signal Level 1 (STS-1) transmission frame or packet. STS-1 uses a synchronous frame alignment (packets are transported using a clock) which enables DS-1 data frames to be extracted from the STS-1 frames without the frame being disassembled and reassembled. The majority of the STS-1 frame consists of the SONET payload envelope which is a large data field that allows STS-1 to carry a clear channel DS-3 signal or a number of lower rate signals.

The basic STS-1 transmission rate is 51.8 Mbps, but SONET will be expanded in the future to take advantage of higher bit rates. SONET frame handling capabilities makes it comparable with other existing and emerging technologies such as: T1, T3, ISDN, cell relay (SMDS) and Frame Relay services. It additionally supports ATM and STM operations.

SONET specifications include an electrical signal interface designation and optical equivalent for various speeds. The specifications include: Electrical signal STS-1, Optical Equivalent OC-1 with a data rate of 51.840 Mbps; Electrical Signal STS-3, Optical Equivalent OC-3 with a data rate of 155.520 Mbps; OC-9 with a data rate of 466.650 Mbps; OC-12 with a rate of 622.080 Mbps; OC-18 with a rate of 933.120 Mbps; OC-24 with a rate of 1.244 G/bps; OC-36 with a rate of 1.866 Gbps; and OC-48 with a rate of 2.488 Gbps.

At the present time, STS specifications have not been defined for OC-9 through OC-48. Provisions have also been made to take SONET rates up to 13 Gbps in the future.

Fiber Data Distributed Interface (FDDI)

FDDI emerged in 1990 as a set of standards from the ANSI and is comparable with the OSI models seven layer architecture for open systems networking. FDDI is a high-speed cabling methodology designed for use on optical fiber or copper twisted-pairs wire. It is primarily targeted for private networks that require a high-speed interface for connecting LANs. FDDI can be considered an extension to Token Passing Ring and Ethernet LAN technology that links LAN islands onto a broadband backbone. FDDI operates at 100 Mbps and employs a dual cable counter rotating token passing methodology that allows devices to access the ring when they are in possession of the token. This token passing technology is similar to the 4 and 16 Mbps LAN capabilities currently in use today. Two FDDI technologies exists and are called FDDI-I and FDDI-II. Figure 9.16 shows an FDDI Backbone with several DAS and SAS stations.

Architecture

The FDDI architecture consists of four operational layers that map into the bottom two layers of the International Standards Organizations (ISO) Open Systems Interconnect (OSI) model discussed in another module. The OSI architecture defines a series of standards for communicating between nodes of a network, two different networks and the two

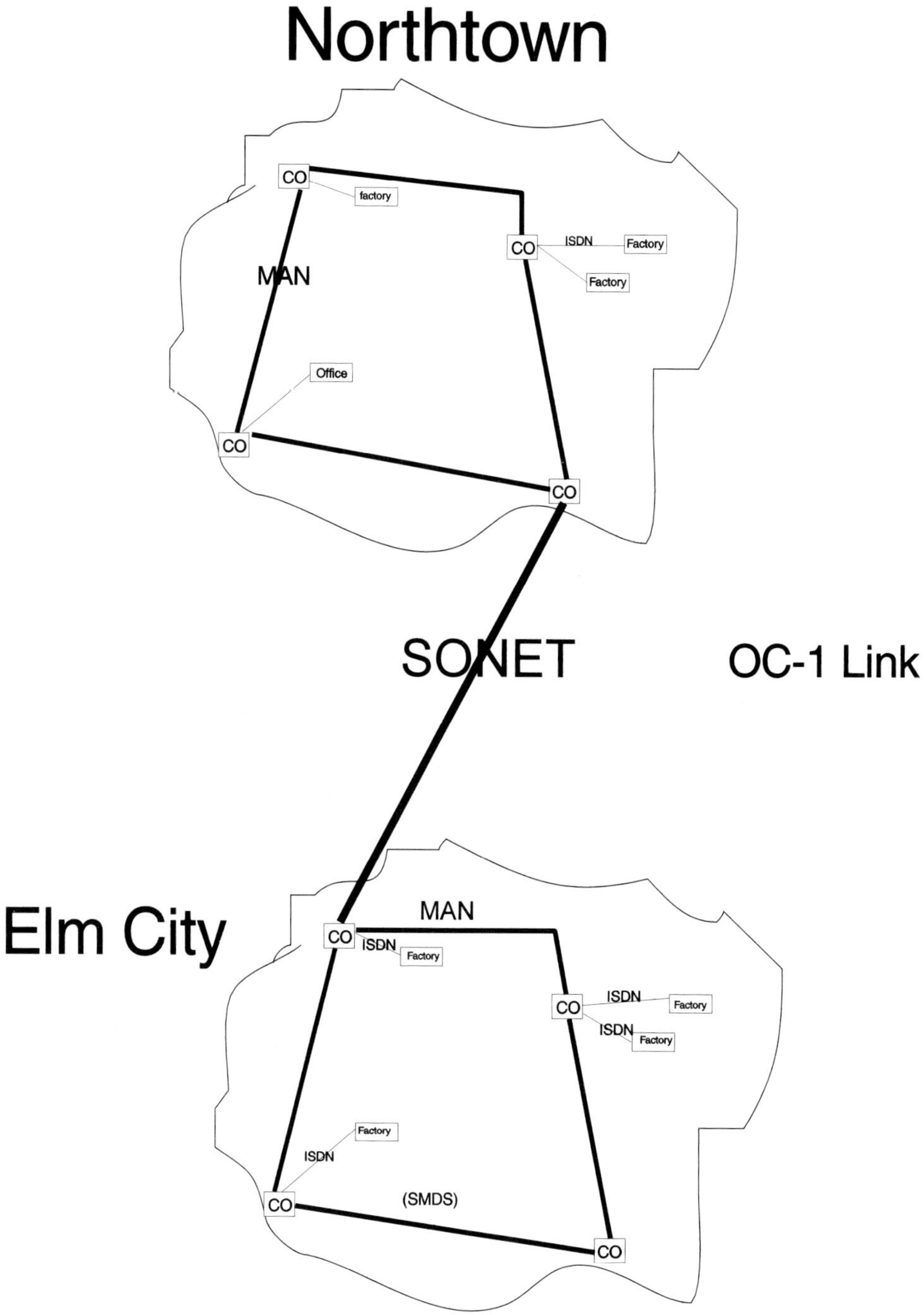

Figure 9.15 Example of a SONET Link

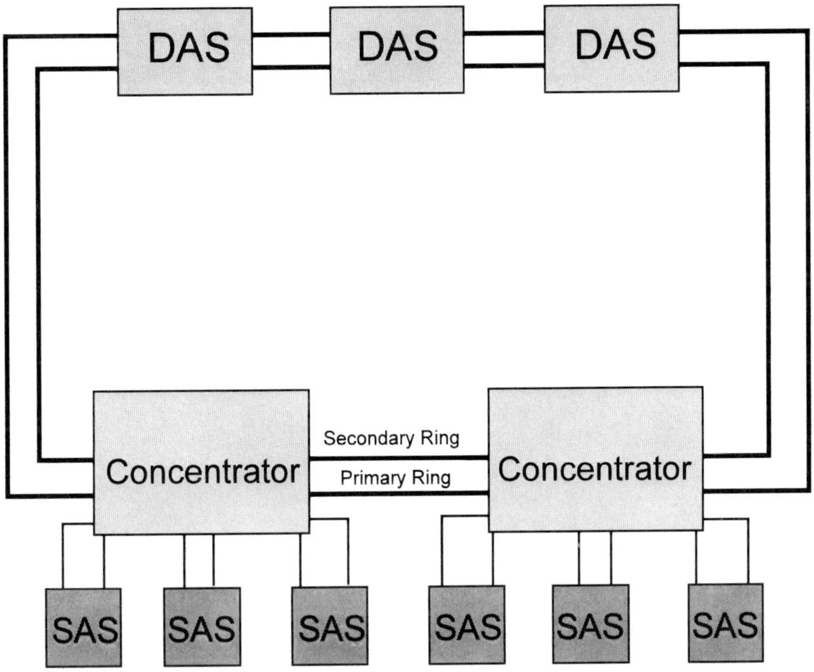

Figure 9.16 FDDI Backbone Ring

end points of the communications link. Layer one is considered the physical layer and consists of two sub-layers.

Sub-layer one is called the Physical Layer Medium (PMD) that defines the fiber optic interface. It specifies a 62.5/125 micron multimode graded index fiber operating in the 1,300 nanometer wavelength at 100 Mbps.

Sub-layer two is called the Physical Layer Protocol (PHY) and specifies the type of data encoding used, how control information is generated and detected, and specifies receiver characteristics.

The second layer is called the Media Access Control layer (MAC) and defines the data frame format and the token passing protocol, describing how access is gained to the backbone ring.

The last layer addresses Station Management (SMT) and handles connection management and fault detection functions of the FDDI ring. Additional responsibilities include ring initialization, ring monitoring, and fault recovery. Figure 9.17 shows workstations attached to an FDDI ring.

FDDI-I

FDDI-I is a fiber-based digitally-oriented technology that uses multimode fiber cable for the transport of data. It can support up to 1,000 connections over a 200 Km distance. Packet sizes are limited to 4,500 bytes or less, and a token passing access protocol is used

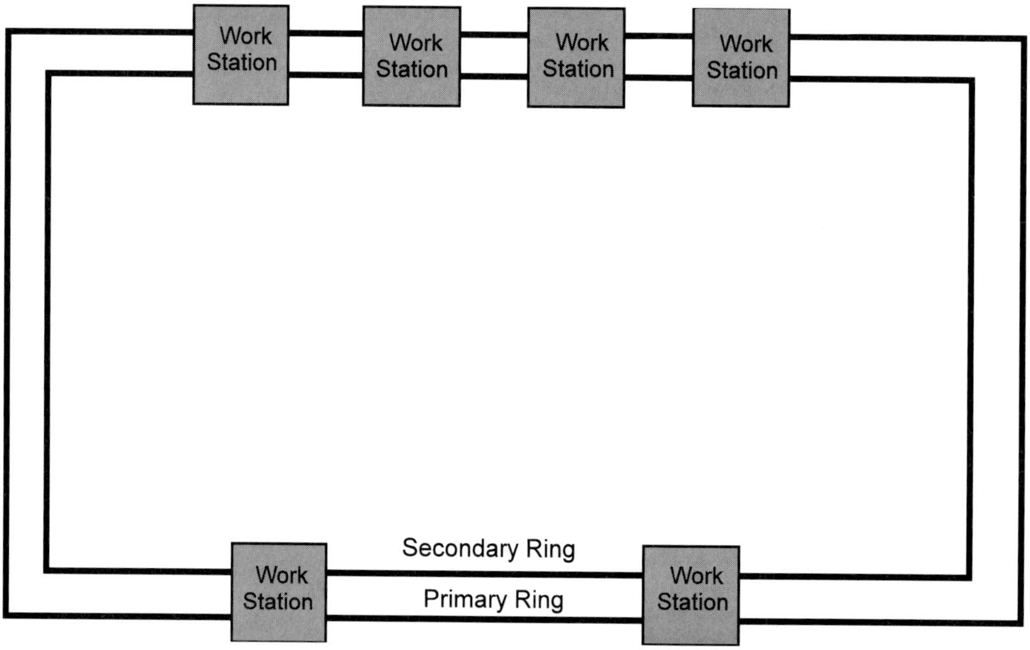

Figure 9.17 Workstations on an FDDI Ring

for management and control. While FDDI-I utilizes synchronous timing for network control, packets are sent in an asynchronous manner (one at a time from the destination node).

FDDI-II

Because of the restrictions of FDDI-I, a new specification is under development that addresses those limitations. This new specification defines extensions to FDDI-I and is known as FDDI-II or TPFDDI. TPFDDI, or twisted-pairs FDDI, allows data to be transmitted at the FDDI rate of 100 Mbps over copper twisted pair wires. It uses a modified token passing technique for accessing the ring and will additionally support voice traffic. FDDI-II is designed to support voice and video; however, the specification is still under development and is incompatible with FDDI-I.

FDDI Hardware

Several hardware components are used with a FDDI backbone and include the duel attachment interface that provides two fiber rings. One is used for primary transmission and the other for backup in the event of a primary ring failure; a single attachment hub oriented fiber optic star; and a FDDI concentrator that allows attachment of individual LANs to the FDDI ring.

FUTURE TRANSMISSION TECHNOLOGIES

Several new clases of services called digital subscriber line (DSL) services are in the process of being developed by the various telephone companies of the world to provide high-speed connectivity over standard telephone lines. These include asymmetrical digital subscriber line (ADSL), symmetrical digital subscriber line (SDSL), high-speed digital subscriber line (HDSL), and very high-speed digital subscriber line (VDSL) services. ADSL and SDSL appears to be the first products to be offered.

ADSL/SDSL Service

ADSL is a high speed wideband service developed by AT&T and is designed to deliver high bandwidth to homes or small businesses over a single twisted copper pair. ADSL will provide high downstream bandwidths but relatively low upstream bandwidths laying the foundation for a new type of service geared to remote and Internet users. A number of carriers including GTE, NYNEX, Pac Bell, and U.S. West plan to implement support in 1996 for T1 speeds in one direction and 64 Kbps in the other. This could be an ideal solution for remote users as this service could satisfy high speed data requirements between remote LANs as well as Internet users.

A typical application might involve a remote user requesting a large file. The request could be sent at 64 Kbps and the file could then be returned at T1 speeds. Connecting a branch location to a main office using ADSL would require that the main office have a T1 line connected to its central office and the remote location, and an ADSL link to its local CO.

ADSL communications between local and remote sites would use the networks of the local exchange carriers. At the uplink CO, an ADSL termination unit called an ATU-C would take the T1 signal and convert it to an ADSL signal for transmission to the remote location. At the remote location, a second unit, the ATU-R, could separate analog voice signals sending them to the telephone, and high-speed digital signals to an ADSL modem.

SDSL is an extension of the ADSL technology. SDSL takes that concept but adjusts the data flow to provide the equivalent of four-wire T1 lines—1.544 Mbps in both directions. When ADSL and SDSL services are rolled out and devices are manufactured, costs will be in the range of $500 to $1,400 per line. SDSL in combination with competition for local exchange service could bring down rates charged by carriers for what is the equivalent of a T1 pipe. Rules being considered in federal telecommunications legislation would give competitive access providers (CAP) access to the installed copper wires now used by the local exchange carriers. CAPs could offer T1-type service using SDSL, and at a lower price than traditional T1. SDSL is a firmware upgrade that supports ADSL with speeds up

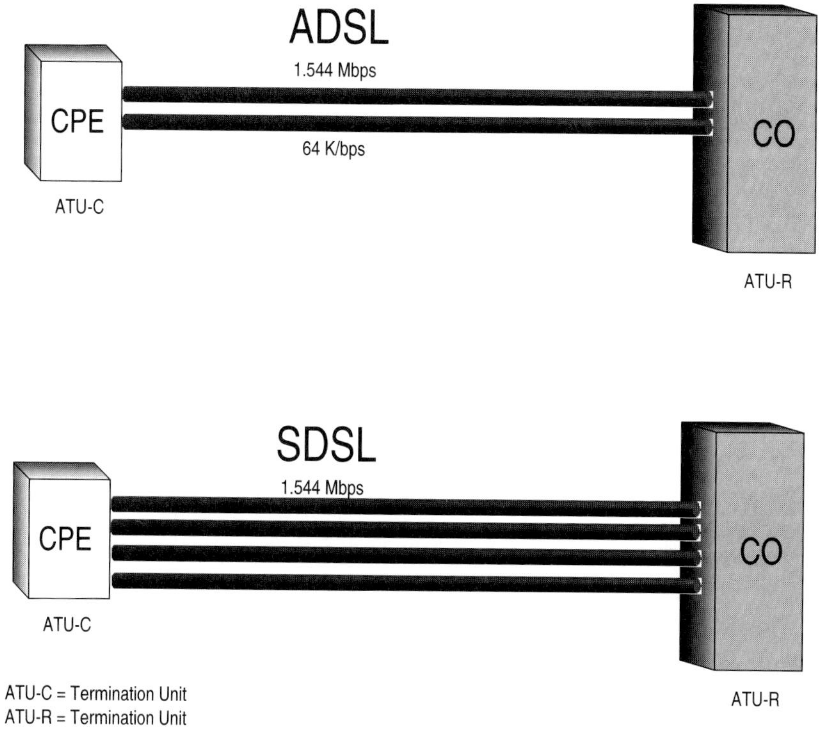

Figure 9.18 Example of ADSL and SDSL

to 6 Mbps in one direction and 64 Kbps in the other. Figure 9.18 illustrates an example of ADSL and SDSL Links.

Cable (CATV) Transmission Services

Cable Access Television (CATV) transmission services will allow high speed data access (such as to the Internet) via a cable TV network. This type of access will require a cable modem having two connections, one to the cable outlet and the other to the computer. A LAN technology called Ethernet has been mentioned as the possible vehicle to accomplish this activity. Figure 9.19 illustrates an example of CATV service.

During initial CATV modem service, an unbalanced data stream setup will probably be more common than a balanced setup. In the unbalanced model, the downstream channel has a much higher bandwidth allocation than the upstream. One reason is that current Internet traffic tends to be unbalanced in nature. Activities such as access to the

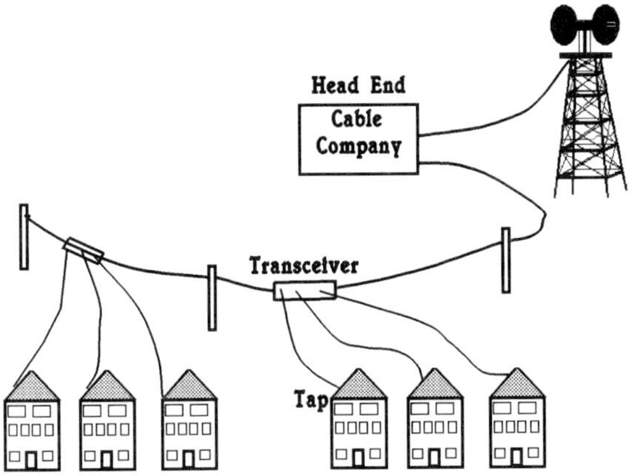

Figure 9.19 Example of CATV Service

World Wide Web and newsgroups have more data flowing to the computer than to the network. Web access and e-mail messages do not require large amounts of bandwidth from the computer to the network.

Initially, CATV speeds will have a wide variance depending upon the CATV provider. In the downstream direction, speeds can be as fast as 36 Mbps. In the upstream direction, speeds can be up to 784 Kbps.

As stated above, CATV will use a cable modem, however, it will be quite different from the standard telephone-based modem. They both modulate and demodulate signals, however, the similarity ends there. This is because cable modems are substantially more complicated than their telephone counterparts. Cable modems are designed to provide a variety of functions, including encryption/decryption capabilities, error control, and router functions.

CATV will send and receive data in two slightly different fashions. From the network to the computer, the digital signal is modulated and then placed on a standard 6 MHz television carrier occupying one of the assigned channels. The upstream data signal presents a more difficult environment because in bi-directional cable systems, the data transmitted from the computer across the CATV network uses a much smaller bandwidth. This provides an environment in which interference from short wave radio and impulse noise can create data errors.

CASE STUDY (CONTINUED)

The plans for implementing a data communications network at the Bank of Scotts Valley are progressing as planned. To date, the group responsible for planning the service centers' hardware and software and defining the network has had considerable success. It has completed the definition and selection of the computer centers' mainframe computers,

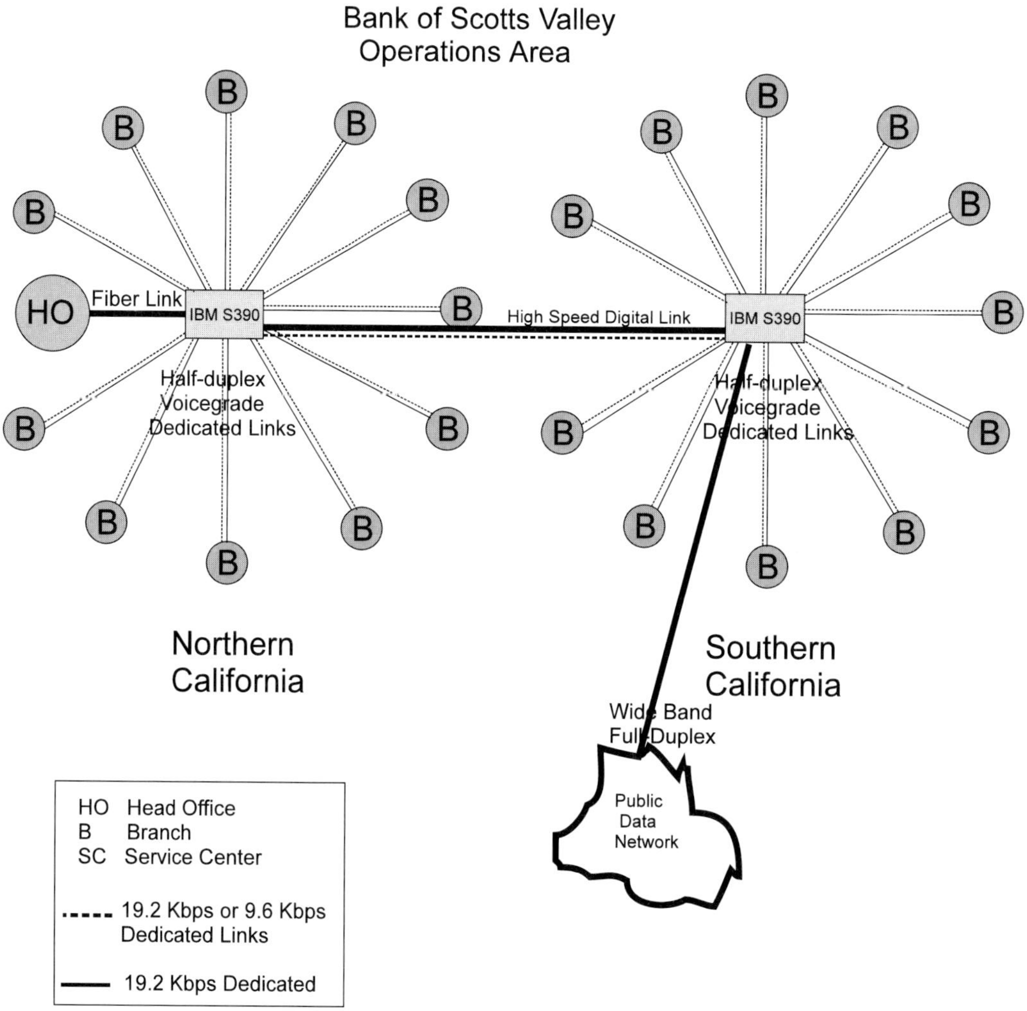

Figure 9.20 ISDN Terminal Interfaces

which will be IBM S/390 processors. It has defined and identified the required operations, applications, and communications software needed by the bank. Also completed is a general plan for the operational mode and types of lines required by the service centers and branches. Figure 9.20 shows the Bank of Scotts Valley's progress to date.

The bank's planners are now ready to begin defining the different types of communications link services needed by each of the locations that make up the bank's network. Communications between the branches and the service center consist of teller, administrative, and ATM activities. Each of these activities requires different capabilities. At some future point, the planners also want to provide access to customers via the Internet, as well as implement an Intranet capability for employees, however this is not part of the current implementation plans and will be addressed at some future point in time.

Teller and administration activities involve continuous communications to and from the service center in support of teller functions such as check cashing and savings transactions, and administrative activities associated with loans, stocks, and investments. Communications requirements include random-sized data packets from several terminals at the same time or in rapid succession. This dictates the need for lines that will provide enough throughput rate to ensure acceptable response time. The bank's planners have decided to install two 19.2 Kbps analog dedicated conditioned links to provide this service. The bank will contract with the local telephone company for the links. The bit rate selected is based on the perceived requirements at the present time and for anticipated growth.

The ATM terminal needs a dedicated primary communications link, because it will support random activities 24 hours per day. The ATM also will require a backup link should the primary link fail. Data requirements between the terminal and the host computer are minimal because each transaction will carry only about 35 bytes of information. This is because many of the basic user interface functions, such as the PIN validation, are contained within the terminal. The planners have determined that a low-speed, 1,200 bps dedicated and conditioned analog link between each ATM terminal and the service center will be required. This link will be provided by the local telephone company.

Because a backup link is required, an internal modem within the ATM will be used should the primary link go down. It will be programmed to call a number at the service center and re-establish the connection.

Each service center will require facilities to support multiple dedicated 19.2 Kbps links. They also must have high-speed dedicated links between each other to support interbranch banking activities. The planners initially considered incorporating ISDN service but decided not to go in that direction because ISDN is still an emerging technology and has not been fully implemented throughout the state. Instead, they have decided to use digital T1 links provided by one of the long-distance carriers because several different telephone companies operate throughout the state. Two T1 links will be installed to support both voice and data traffic.

REVIEW QUESTIONS

1. What three companies provide 95 percent of all carrier services in North America?
2. What services do international carriers provide?
3. What services do public data network vendors typically provide?
4. What types of packets are available for use in packet-switching networks?
5. What does PSTN refer to?
6. What is a type 3002 line?

7. Name the two types of conditioning available with leased lines.
8. What transmission service do most common carriers support?
9. The design criteria for ISDN were based on what?
10. What is 2B+D considered to be?
11. For what activity is the D channel used?
12. What is the bit rate for the ISDN 3H0+D channel?

CHAPTER 10

Data Communications Hardware

INTRODUCTION

A major part of planning and implementing a data communications network is evaluating and selecting software and hardware components. Myriad products with the same or similar capabilities are available, manufactured by a multitude of vendors. The process of reviewing hundreds of products is time-consuming, but absolutely necessary to produce an efficient, cost-effective system. Expensive mistakes can occur if the network components are incompatible or cannot be upgraded to communicate with newer equipment. It is important to determine if vendor claims of compatibility and field upgradability are valid. Vendor claims of performance and compatibility are usually correct, but the test cases they use are highly controlled, with only a limited number of configurations used to perform the validation.

Questions frequently asked when planning for network and hardware communications components include:

- Will they meet our basic communications and networking needs?
- Do we have to acquire additional equipment or convert to new equipment when our network expands or when we need to replace older equipment?
- Are we buying capabilities we do not need at the present time, and is it cost-effective to do so?

Choices often are limited because of an organization's need for specific hardware configurations. This also applies when equipment is to be placed in remote locations where the vendor's support policies may not cover it. Equipment to be installed in remote locations and operated in a self-attended mode (unattended operation) require higher mean time to failure parameters for a higher degree of reliability. The self-attendance requirement usually adds to the product's cost and generally limits selection to a few products. Another factor to consider is the vendor's minimum configuration product purchase requirements. Frequently, these policies force the buyer to purchase equipment that has more capabilities than required, thereby driving up the costs of network implementation and maintenance.

The term *data communications hardware* generally describes equipment designed specifically to support a computer data communications environment. It is important to recognize that the equipment also may be required to support a digitized voice environment. This chapter addresses the nonvoice digital environment that is used primarily for computer-to-computer or terminal-to-computer communications. However, it is unwise to assume that digital data and voice reside in separate worlds; such an assumption can create future problems because a combined environment might be a requirement further down the implementation path. Figure 10.1 illustrates various types of equipment in a typical IBM network.

As data communications evolved, the computer or central processor of the past was set up to manage communications software, multiple concurrent tasks, and the physical communications lines. In many cases, the same situation still exists today. In small networks where traffic is moderate, this arrangement still works well. However, there is a price to pay for this arrangement, because a processor with a somewhat high MIPS (millions of instructions per second) rate is required to support the combined data processing and communications activities. This high MIPS rate requirement is primarily due to the tremendous overhead needed for both memory-resident software resources and CPU processing cycles.

Terminal/LAN Controllers

Direct attachment of terminals to a host computer is not usually considered efficient because of the wide disparity in bit rates between the channel (Mbs) and the terminal (thousands of bits per second). Most computer manufacturers provide either device-level or board-level connectivity for terminals.

The terminal controller device or board is designed to support a variety of terminal types and printers in a local or remote environment. In the non-IBM world, the boards or terminal control devices may be referred to as asynchronous or synchronous terminal adapters. Terminal controllers associated with the IBM world are referred to as cluster controllers. The terminal controller can be viewed as an intelligent microprocessor-based device having specific dedicated responsibilities. Figure 10.2 shows the internals of a typical terminal controller.

The terminal controller performs a number of functions depending on the manufacturer and how it is to be configured. The controller is primarily responsible for establishing and maintaining connectivity to individual or groups of terminals. The role of the terminal controller also is evolving because many are being used for additional activities such as the control of LANs.

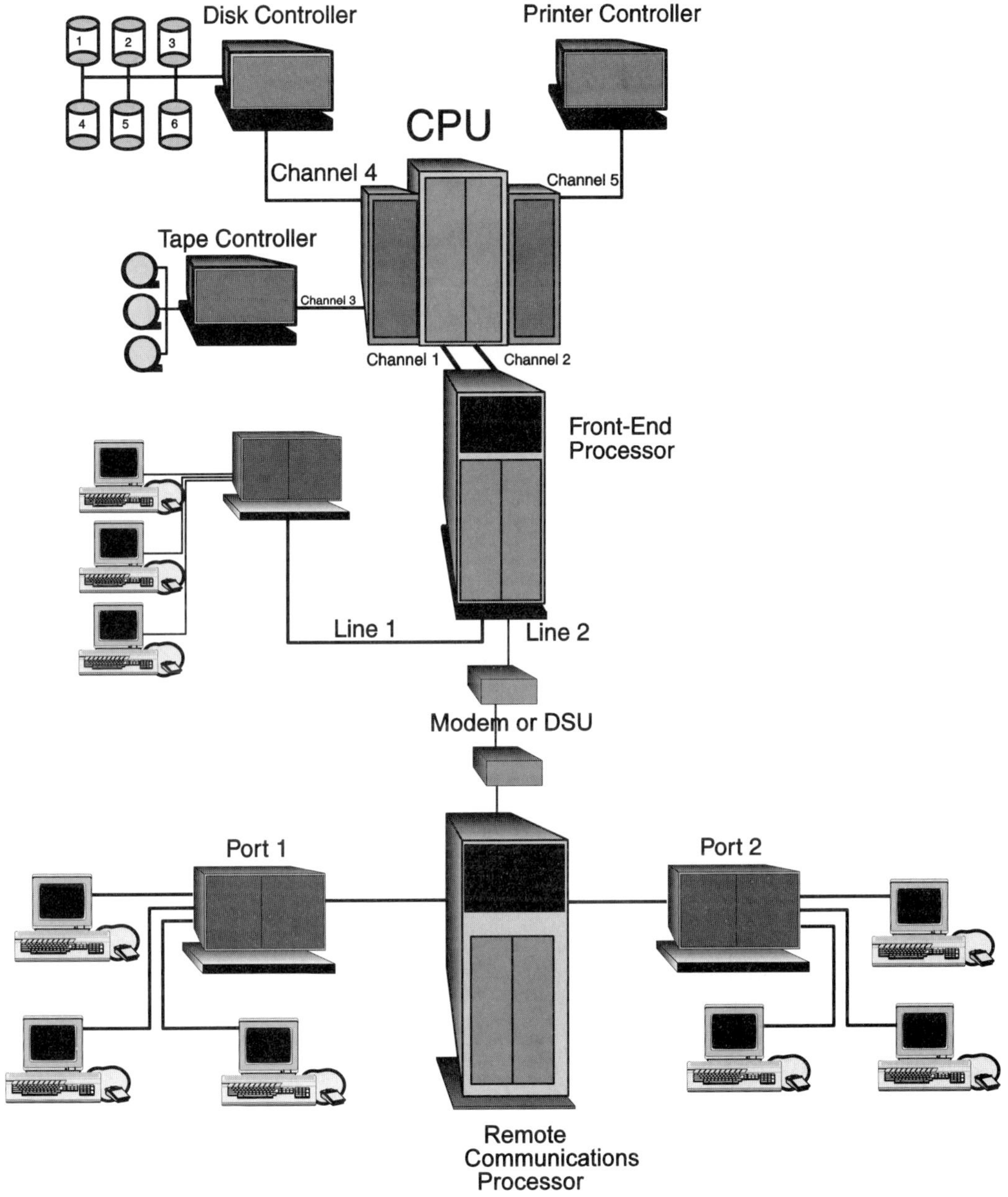

Figure 10.1 Typical IBM Communications Network

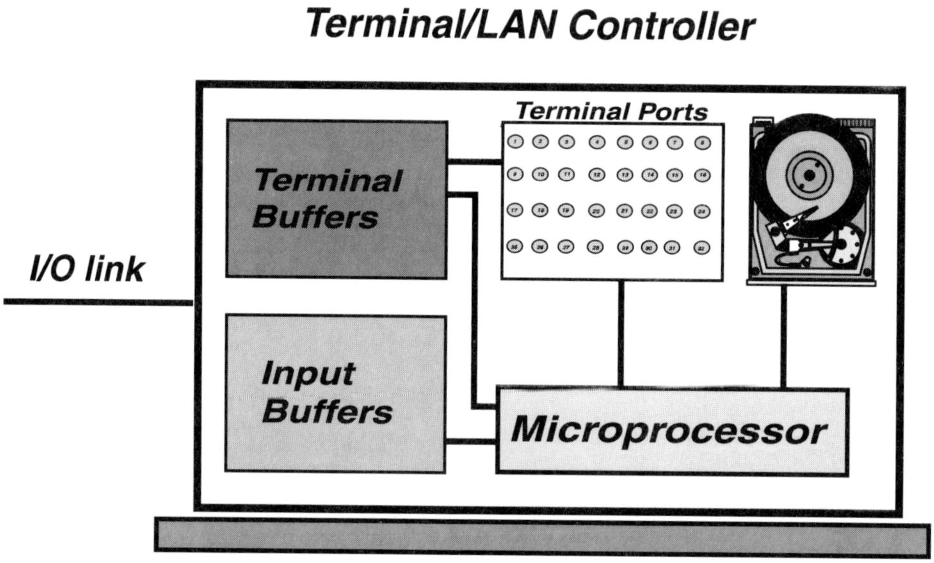

Figure 10.2 Example of LAN/Terminal Controller

One of the controller's basic functions is to maintain the physical addresses for all terminals it manages. Another is performing link control framing and deframing for data transmission between itself and the host computer or front-end processor. A third function is error detection. If an error is encountered, the controller is responsible for coordinating and requesting retransmission.

A terminal controller normally contains memory and allocates buffer space for blocking and deblocking of messages between the terminal and the host computer. It controls the transfer of data between the computer link and the terminals and ensures that all terminals have equal access to the host.

The controller is hard-wired to each terminal and manages polling and selection requests to and from each terminal after it has received the poll/selection request from the FEP or host computer. When the controller is installed or the terminal configuration changes, the controller needs to be configured (set up for the types and numbers of terminals it will control). This is normally accomplished using vendor-supplied software loaded from an installation disk or hard disk drive.

Identical in operation to the terminal controller is the value-added terminal controller manufactured by such companies as Memorex Telex Corp., Televideo Systems Inc., Telex Communications Inc., and Charles River Data Systems Inc. The value-added terminal controller offers features greater than those found on the original manufacturer's equipment. Usually, this device is designed for a specific configuration and is able to perform protocol conversion. It also may be designed to handle character set conversion (ASCII to EBCDIC) and to support a larger number of terminals than the original manufacturer's equipment.

Many of the newer controllers provide emulation for asynchronous and synchronous terminals at the same time, allowing the user to have a mix of low-cost and high-end terminals. Some controllers can link to multiple host computers and provide its terminals with a selection capability at the touch of a key.

Speeds up to 19.2 Kbps are standard, but several controllers operate at higher speeds (up to 56 Kbps) than the original equipment available with the system. Many value-added controllers contain additional intelligence and memory, and they support enhanced color displays and graphics.

TERMINALS AND PCS

Terminals enable users to interact with a computer. Excluding those designed for specific activities, terminals are considered to be input/output (I/O) devices that usually have a keyboard for input, a cathode ray tube (CRT) or video screen for output, and possibly a cable-connected printer for hardcopy output. Figure 10.3 illustrates an example of several typical terminals and personal computers.

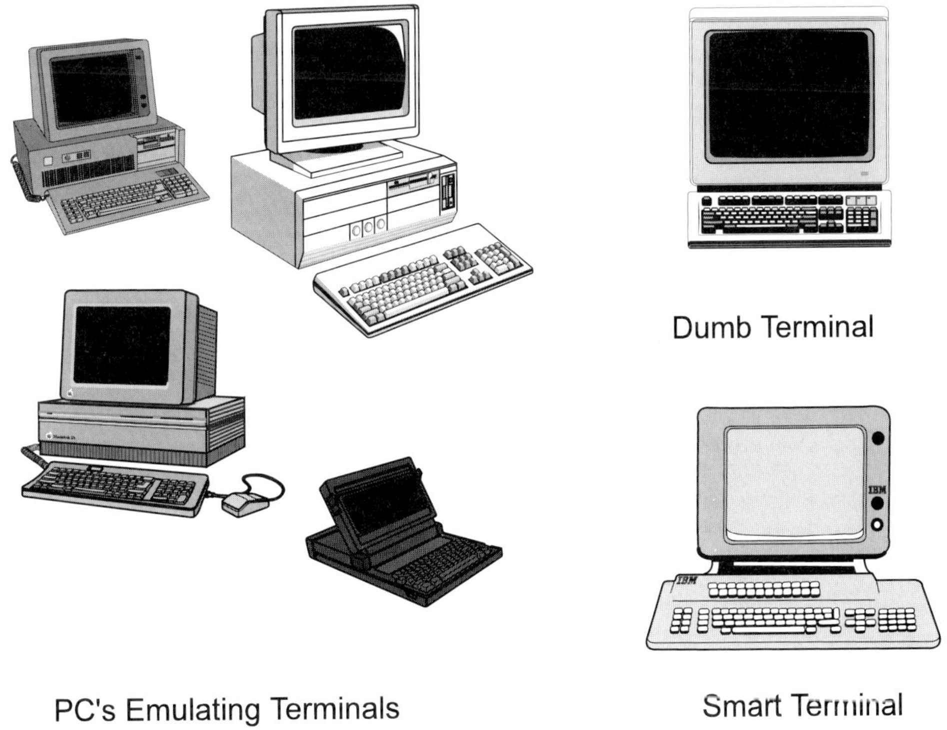

Figure 10.3 Examples of Purchase and Terminals

Several types of terminals in a variety of shapes and sizes are manufactured. Most CRTs are generally configured for 25 lines vertically and 80 characters horizontally, or 43 lines vertically and 133 characters horizontally. The 25th line is usually reserved for communicating status information between the computer system and the user and may be used to indicate that the system is busy, disconnected, or available. Some terminals provide a greater number of resolution widths and lines, but these must have enough intelligence to execute a screen image sizing program.

Terminal functions also can be performed by personal computers (PC) equipped with terminal emulation hardware, software, or both.

Dumb Terminals

A terminal described as "dumb" (as opposed to "smart") is a terminal that has no microprocessor or memory. The internal equipment of this type of terminal is usually nothing more than a video display tube with several wires. Almost all terminal configuration is performed by the software in the attached computer. Most dumb terminals function asynchronously, sending one character at a time, and are considered to be low-speed devices. Because they have few components, they generally cost less than terminals with some intelligence. This type of terminal is also referred to as a scroll-mode terminal because each byte is transferred immediately on being entered at the terminal. Dumb terminals are generally used as low-cost inquiry and data entry devices. They usually do not provide enough capability for the interactive end user.

Several terminal manufacturers produce dumb terminals with limited intelligence. These terminals contain microprocessors and memory, and operate identically to those described above, except that they also provide buffering and line control management. Many dumb terminals allow some form of basic configuration and have micro switches you can use to change the display setup. This type of terminal also may have some error detection capability, but usually will not have control capabilities for retry. It might also have serial printer support to dump the screen to a printer for a hard copy.

Smart Terminals

Smart terminals contain intelligence in the form of a microprocessor and memory. The terminal design can be general or specific, depending on the application it is intended to handle. Examples of specific designs include airline terminals and automatic teller machines. Smart terminals are sometimes referred to as page-mode terminals, because they hold pages in memory and transfer the complete page rather than send bytes one at a time to the terminal controller lines.

Probably the most popular smart terminals in use today are the VT family of terminals from Digital Equipment and the IBM 3170 series of terminals. The standard smart terminal can be either asynchronous or synchronous. Digital terminals operate asynchronously, while the IBM 3170s employ synchronous operation.

Almost all smart terminals are microprocessor-based and have extensive programming in read-only memory (ROM). This code contains diagnostic programs that monitor and test activities such as character generation, memory, framing and deframing of the transport data, and extensive error detection.

Smart terminals provide pixel addressability, which is the ability to move a text or graphics character to a specific location on the display. They also provide a protected field and a field tab capability. A protected field is an area of the display that cannot be modified by the user. Field tab capability enables the cursor to tab to the next unprotected field on the screen. Most smart terminals support serial and parallel printers. Some of the more sophisticated terminals, such as the Tektronix Inc. color display system, provide extensive color display configuration capabilities.

Terminal Emulation Using PCs

PCs are used as standalone peer devices acting as nodes in a communications network, and as emulation devices for smart terminals. The PC has become valuable in environments where both the personal computer and a host-connected terminal are required. Use of the PC in this environment provides a more efficient method for accessing and processing information than a dumb or smart terminal, because it can assume some of the processing activity that otherwise would be performed by the host, including screen formatting and data manipulation. Figure 10.3 illustrates examples of terminal devices.

Computer systems such as those from Digital and Hewlett-Packard (HP) can often communicate directly with PCs using software control. However, because the PC performs data communications and data processing activities somewhat differently than a dumb or smart terminal, it often will require an add-on board for terminal emulation. This is especially true for IBM host computer connectivity because of the way an IBM computer normally communicates with its terminals.

Many add-on boards are manufactured for PC-to-host connectivity. One example is the IRMA board from Digital Communications Associates Inc., which is designed for IBM terminal emulation. Many other communications equipment manufacturers also produce boards to enable the PC to act as either an asynchronous or synchronous terminal in a large computer environment.

A PC emulating a terminal provides many more functions than a dumb or smart terminal, including the capability to communicate with the host as a native device, download files to its own disk storage devices and operate as a standalone computer to process data. Use of the PC as a terminal enables data processing functions to be distributed to remote sites where they can provide the most benefit. Such functions were once available only on the host computer. Today, in many cases, the host computer site is downsized because the processing power required to support large networks can be shifted to LANs and remote PCs.

COMMUNICATIONS PROCESSORS

One solution to the aforementioned problem is to use a communications processor (CP) specifically designed to function as an assistant to the central processing unit (CPU). This CP is used exclusively for data communications and can be interfaced in either of two environments. When the CP is placed local to the host computer system, it is referred to as a front-end processor (FEP). The term "local" implies that the device is hard-wired to the CPU. Figure 10.4 shows a communications processor and some of its components.

When the communications processor is placed at a remote location and communicates with either an FEP or host computer using modems or digital service units, it is called a remote communications processor or remote CP. A remote location is any location that communicates via telephone lines or other communications links back to the host site. The remote CP handles all terminal or communications activities at the remote site and between the remote site and the host system.

Before the introduction of the FEP, the host computer performed all communications functions using memory-resident software. These communications programs had to compete with applications and system programs for processing time and hardware resources. The use of front-end processors may reduce host computer use up to 50 percent by offloading system communications programs that normally would reside in the CPU. This increases host efficiency by having less overhead to conflict with other program operations. An FEP also increases the amount of CPU memory available for other programs to use.

Many companies manufacture communications processors. Among them are IBM, which produces the 3725 and 3745 type processors, and the NCR Corp., which manufactures the Comten communications processors (which support the IBM environment). Other manufacturers such as Digital and HP provide communications processors designed primarily to support their own hardware.

Because channels attached to the host computer are considered to be high speed (1.5 to 5 Mbps) they are usually limited in number. This suggests that it is not practical to attach directly low-speed user terminal devices (9,600 bps) to the host because low-speed devices will substantially slow the host. The front-end processor performs concentrator functions for the terminal network to take advantage of the high-speed I/O channel links. LANs also may be directly attached in the newer communications processors, such as the IBM 3745.

The communications processor, when channel-attached directly to the computer as an FEP, relieves the host of network and terminal communications duties. This gives the host more time for application-processing activities. This does not imply that the CPU does not perform communications activities; in fact, the CPU has extensive responsibilities for managing applications and system programs involved in communications.

The front-end or communications processor software is normally downloaded from the host computer at initial program load (IPL) time, when the operator boots the computer. The communications processor also may contain an optional disk storage device for self-loading the required communications software when power is turned on. When the communications processor is used as a front-end processor, it is channel-attached directly to the computer and also connected to the physical links. These links include twisted pairs, coaxial cable, dial-up or dedicated telephone links, and digital service (DS) links. The communications processor acts as the interface between the outside world and the computer.

The FEP is responsible for link control, which includes connections to the physical communications lines and management of the logical and physical terminal identification

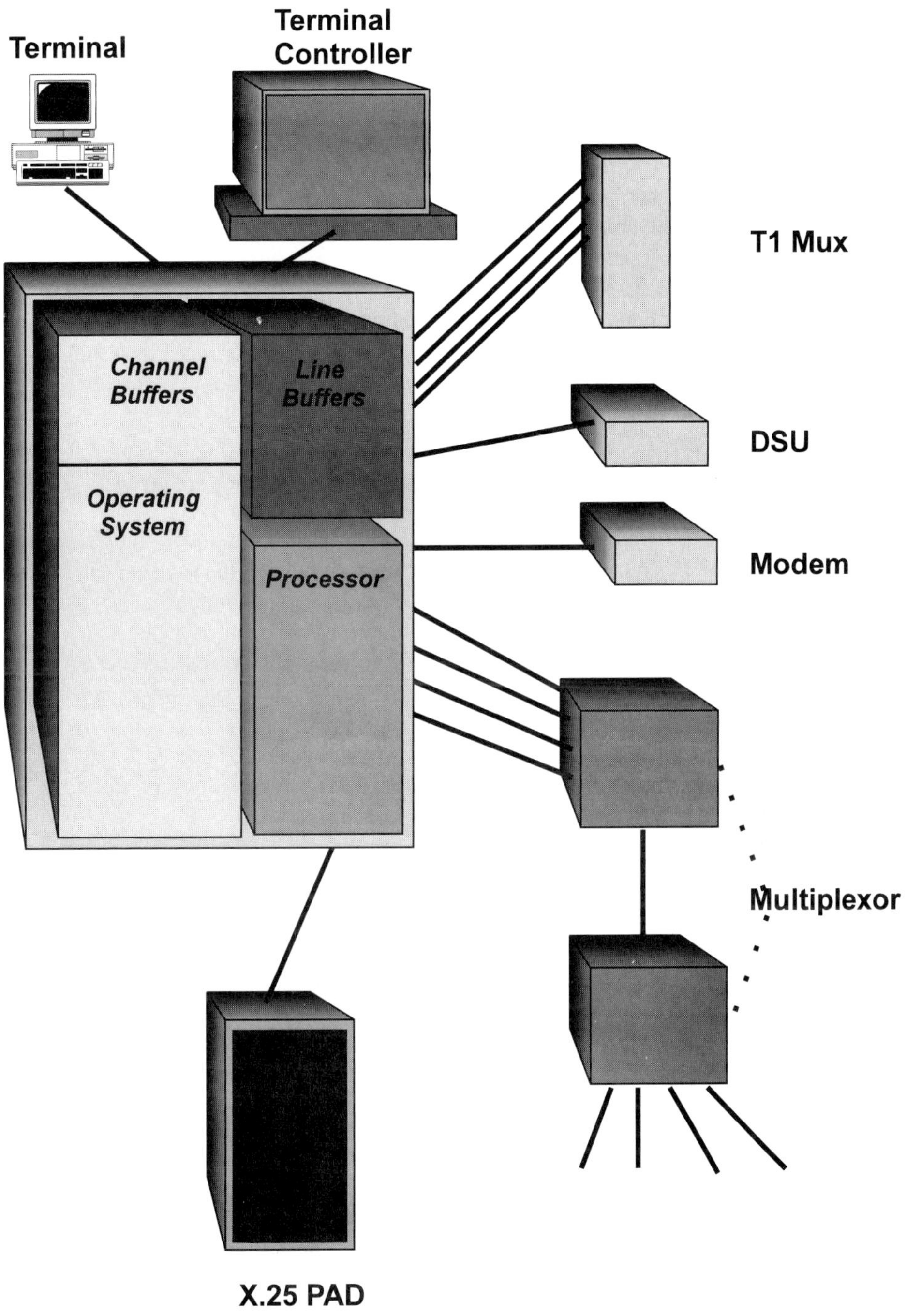

Figure 10.4 Communications Devices Attached to a Communications Processor

(ID) activities. Each terminal has a software-defined ID assigned to it when it is active as well as a permanent hardware address. The hardware and software use these IDs to locate specific terminals in the network.

The FEP also is responsible for establishing the logical links that allow terminals to communicate with the host. Logical linkage is the process of connecting an end-user terminal to a processing program by creating a path from the terminal to the program. This includes all of the physical lines, the system software and the application process. The establishment of a logical link is also referred to as session establishment.

In environments where different types of lines and terminals with different bit speeds and modes (duplex, half-duplex) are attached to the FEP, the communications processor must be able to recognize the bit speed and type of service required by each terminal in the network. Another important responsibility of the CP is polling and selection. This is the process of looking for and recognizing service requests from the terminals on the networks.

The CP can detect certain types of bit and frame errors and either will correct them or request retransmission from the sending device. Some CPs are able to perform data compression and expansion on the data transmitted by reducing the number of bits in a frame on the sending side and expanding the frame back to its original form on the receiving side. The CP also may support multiple binary code sets, such as EBCDIC and ASCII, and provide code sets conversion suitable for host processing. Most FEPs can handle message switching between terminals and broadcasting of messages to all terminals. Often this can be accomplished without the involvement of the host computer.

The FEP also can contain several additional pieces of software to control the interface to other networks. Operating system software to manage communications is also required for the front-end processor. This software is considered to be communications control software and consists of network control software, partitioning supervisors, performance monitoring software, and emulation programs.

Communications processors are often used for LAN control, as well as the more traditional WAN activities.

MULTIPLEXERS

As the need for networking in an organization is evaluated, it becomes evident that a communications processor is not always cost justified because of the network's small size or the limited number of devices or lines it supports. Such a situation creates an opportunity for devices that are inexpensive, support a limited number of devices, and provide functionality similar to that of the FEP. In the past, a concentrator was used to enable multiple devices to use a single line. The concentrator has all but disappeared and has been replaced by the multiplexer.

Multiplexing refers to the combining or interleaving of several data streams (data from multiple terminals) to form one composite data signal. This can also be accomplished by allocating time segments to each device wishing to use the link. The major advantage of incorporating a multiplexer into a network is cost. It is generally less expensive to use a multiplexer than to use direct links or a communications processor line for each terminal.

Multiplexing generally requires either pairs of multiplexers, or a multiplexer and a communications processor capable of communicating with each other. A paired arrangement is needed because the composite data signal must be demultiplexed at the receiving end. Multiplexing devices may be found as standalone devices, combined with modems or as part of a vendor's specific terminal controllers or communications processors.

Multiplexers use the total bandwidth available on the communications link. They operate by dividing the bandwidth into multiple channels or multiple time slots, which allows a group of terminals to communicate with the host computer or communications processor over a single line. Multiplexing techniques can provide a substantial savings in communications link charges by reducing recurring monthly line costs, which can be quite expensive.

Multiplexers interface terminals to communications processors via coaxial cable or to host computer sites via modems or DSUs. They can perform a variety of functions, depending on which type of multiplexer is acquired and how much it costs. In many instances, a multiplexer can eliminate the need for other devices such as concentrators and modems.

The basic function of the multiplexer is to interface many low-speed devices onto a single line that provides a high-speed pathway to the host or communications processor. Very high-speed multiplexers are also used extensively by telephone carriers to handle their large traffic requirements. Many types of multiplexers are offered by a variety of vendors such as Network Equipment Technologies Inc., IBM, Stratacomm Inc., and AT&T. These devices use a variety of multiplexing methods that are generally compatible.

Operational Characteristics of Multiplexers

Multiplexers come in a variety of configurations and can support up to 32 devices depending on the manufacturer and type of multiplexer required. Multiplexers are either intelligent or unintelligent. However, there is little difference between the two. In many cases the multiplexer can be custom configured from basic four-line support to comprehensive support for up to 32 devices. The configuration and operation vary from manufacturer to manufacturer, but they all provide compatible interfaces. The following options are available.

Terminals or computers are cabled to the multiplexer using a connection referred to as a port. The multiplexer is usually microprocessor controlled and programmable to allow technicians to access the configuration parameters. The object of multiplexing is to hold input from the attached devices in a memory buffer for a very short period of time (microseconds) and then, either using time- or frequency-division methods, combine them onto the outbound link. The receiving multiplexer separates the incoming data into its original parts and stores the parts in a port buffer for transmission to the attached device.

Most multiplexers are designed to support low- or medium-speed links (2,400 bps to 19,200 bps). However, with the advent of ISDN and the availability of higher-speed links as well as reduced line costs, many multiplexers support speeds up to DS-3 (45 Mbps). The higher-speed multiplexers may also support fractional T1 links.

Many multiplexers provide individual port configurations that allow installers and technical support personnel to specify explicit character structures (code sets) and bit speeds. Several multiplexers can automatically detect and enable echo suppression on dial-up lines, as well as provide modem holding of the connection upon loss of carrier.

Each port on the multiplexer can be configured for a specific maximum transmit frame rate (the number of frames per second that can be transmitted), which depends on the type of line and the terminal or computer. The multiplexer's memory can be configured for the input queue size and output buffer size for each device. This may be necessary if a retransmission is required because of frequent errors on the line.

Many multiplexers allow for the attachment of a control terminal to provide the support staff with an interface to the multiplexer. The multiplexer contains software to provide statistical capture of operational and performance information via the control terminal and status panel. This information may include:

- The number of free buffers (memory or devices not in use) available
- The number of errors experienced by each port
- Trunk block error rate (the number of errors experienced by line reject commands)
- Trunk frame error rate (the number of frames requiring retransmission)
- Traffic density (the overall bps rate specified in percent of capacity busy)
- Frame per second rates

Multiplexers generally include link diagnostics and test capabilities to check remote and local trunk loopback, which is an end-to-end validation. In addition, tests for remote and local port loopback are used to ensure that the connections to the terminal or computer are intact. Control signal monitoring also is provided along with start-up diagnostics.

Multiplexers offer many additional features, including auto-speed capabilities that constantly monitor the lines for quality, then increase or decrease the bit rate depending on the values detected. The most widely used protocols are: Transmission Control Protocol/Internet Protocol (TCP/IP); High-Level Data Link Control (HDLC), which is an OSI protocol; and Synchronous Data Link Control (SDLC), which is IBMs SNA protocol. Almost all multiplexer manufacturers support them.

Many multiplexer manufacturers also continue to support IBMs Binary Synchronous Communications (BSC) protocol for the attachment of BSC terminals. BSC was developed by IBM as one of their first protocols for communications between terminals and computers, as well as from computer to computer.

Most multiplexers can be configured to support asynchronous (one character at a time) and synchronous (clocked bit strings) devices in the same environment. An additional feature, referred to as piggybacking, enables multiplexer ports to be connected to another multiplexer. This is required when a series of stepdowns to reduce the bit rate are needed because of the high speed of the primary link.

Frequency-Division Multiplexers

Frequency-division multiplexing (FDM) was one of the first multiplexing techniques to be used in the data communications industry. Its technology was taken from the telephone industry. The frequency-division multiplexer is an analog device and is used primarily in broadband LANs where different channels are needed. The LAN requires the use of a frequency-agile modem in a decentralized environment to access specific channels. The frequency-division multiplexer will divide the total available I/O bandwidths into the same number of channels on the link, depending on the number of ports and devices supported. The total input bps rate of all terminals or devices connected to the multiplexer must not exceed the output line bps rate.

If a device attached to the frequency-division multiplexer is taken offline (turned off), no reallocation of the available frequency is made in order for the other attached devices to take advantage of the additional bandwidth. This means that the multiplexer does not have dynamic reallocation capabilities to use all available bandwidth. FDMs have built-in digital-to-analog converters, so the need for modems at both ends of the link is eliminated.

Each channel designated within the total bandwidth is separated from other channels by guard bands. These are subchannels designed to prevent information channels from interfering with one another. This is a waste of bandwidth but is required with analog transmission to prevent crosstalk and interference. The amount of frequency assigned to a given channel depends on the speed of that channel; higher speeds require greater bandwidth. Frequency shift keying is the most common modulation technique used in frequency-division multiplexers. Frequency-division multiplexing is normally used for digitized voice applications and not data. Figure 10.5 illustrates a frequency-division multiplexer with all channels active and shows what happens to the bandwidth when one of the channels becomes inactive.

Time-Division Multiplexers

Time-division multiplexing (TDM) is more efficient than frequency-division multiplexing because it uses the high-speed link more effectively. Time-division multiplexers are digital devices that combine several digital input signals from a computer or terminal into one composite digital output signal. Several time-division multiplexers permit attachment of DDS digital links, but a DSU/CSU is required to interface the multiplexer to the link.

Time-division multiplexers operate by allocating time slots to each terminal attached to the port. Usually the total bit rate for all devices cannot exceed the output line bps rate; however, a higher level of throughput efficiency can be achieved by using data compression.

Compression/recompression techniques, if employed, must be present in each of the multiplexers. To compress the data, the multiplexer uses a binary algorithm to reduce the total number of bits to be transmitted by some percentage factor. The receiving side uses a reversing algorithm and restores the received bit string. In most cases, as with frequency-division multiplexing, the time slot for a device out of service or not currently sending is not reallocated.

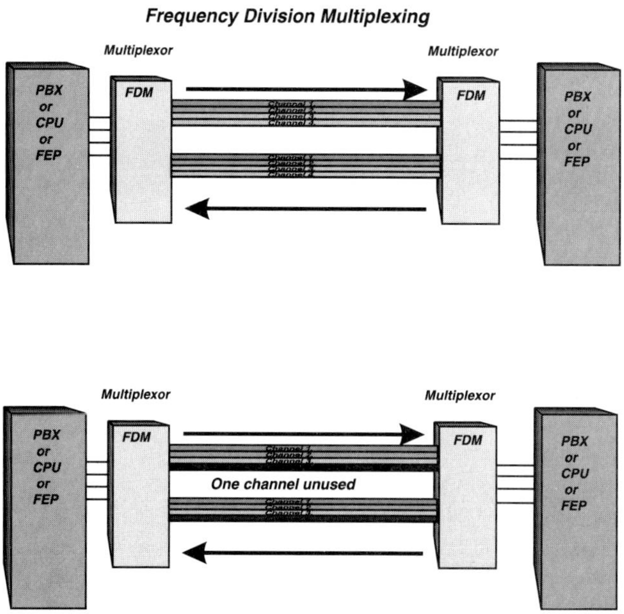

Figure 10.5 Frequency-Division Multiplexor

Some time-division multiplexers allow for the establishment of a fixed order for transmitting and receiving data. The multiplexer can accept as little as one bit or one byte of data from each terminal or computer and assemble the data into a blocked frame within the buffer for transmission. This is referred to as bit or character interleaving. Some newer TDM devices can process packets in a packet-switching environment. Figure 10.6 illustrates a time-division multiplexer servicing four terminals and what the bandwidth will look like when one of the devices is taken offline.

Similar to frequency-division multiplexers, some time-division multiplexers can support both asynchronous and synchronous devices at the same time, but will usually operate in a synchronous mode. As stated earlier, some TDM devices require modems to convert the digital signal into an analog form. However, some time-division multiplexers incorporate modems into their design. Many communications processors and front-end processors include multiplexing circuits, thereby eliminating the need for a multiplexer at the host computer side.

Statistical Time-Division Multiplexers

Statistical time-division multiplexing (STDM) is another multiplexing technique. The statistical time-division multiplexer provides the same capabilities as the time division multiplexers, but it contains additional intelligence to provide time-slot duration allocation dynamically. If a device becomes inactive, the multiplexer will reallocate its time slot to another attached device; this allows the remaining active devices to operate at a higher

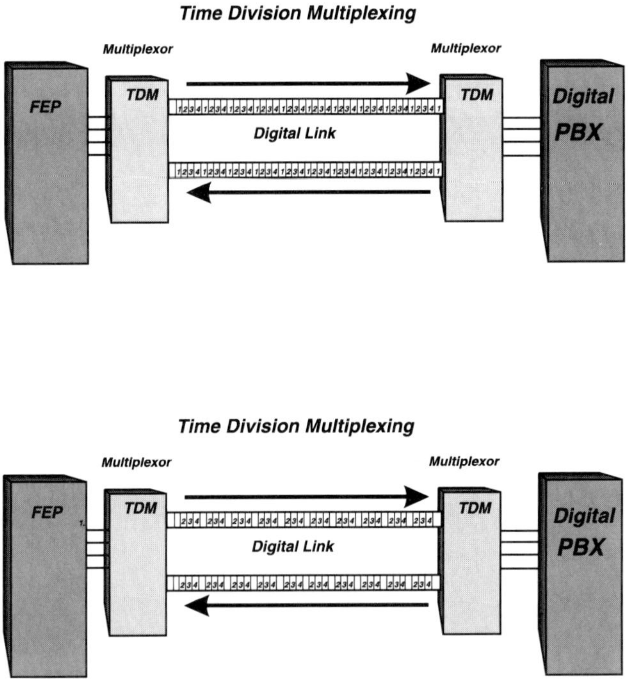

Figure 10.6 Time-Division Multiplexor

speed. A major drawback to using statistical multiplexers is their high cost. Although the cost has been steadily declining, STDM multiplexers are still expensive.

The statistical time-division multiplexer like many time-division multiplexers, has microprocessor-based intelligence and manages all buffering, blocking, and data framing for I/O operations. It also may provide an operator interface for monitoring line performance and configuring lines and ports.

Often, when two or more high-speed devices are linked together, frame flow control management is required. Two types of flow control are used in statistical multiplexers, internal and external. Internal flow control enables the statistical time-division multiplexer to stop the port-attached terminal or computer from sending any more data. This is used primarily when the multiplexer's buffers are full and it cannot accept any more data from a port. Flow control is accomplished by sending an x-off hexadecimal character to the terminal or by lowering the clear-to-send (CTS) signal between the data communications equipment (DCE) and the data terminal equipment (DTE). This is pin 5 on the EIA-232 interface cable. When the statistical time-division multiplexer is ready to receive data from the terminal once more, it will send an x-on hexadecimal character or raise the CTS signal. Figure 10.7 illustrates a statistical time-division multiplexer servicing four terminals and what the bandwidth will look like when one of the devices is taken off-line.

In external flow control, the terminal or computer sends the statistical time-division multiplexer a signal to indicate that it cannot receive any more data. It accomplishes this

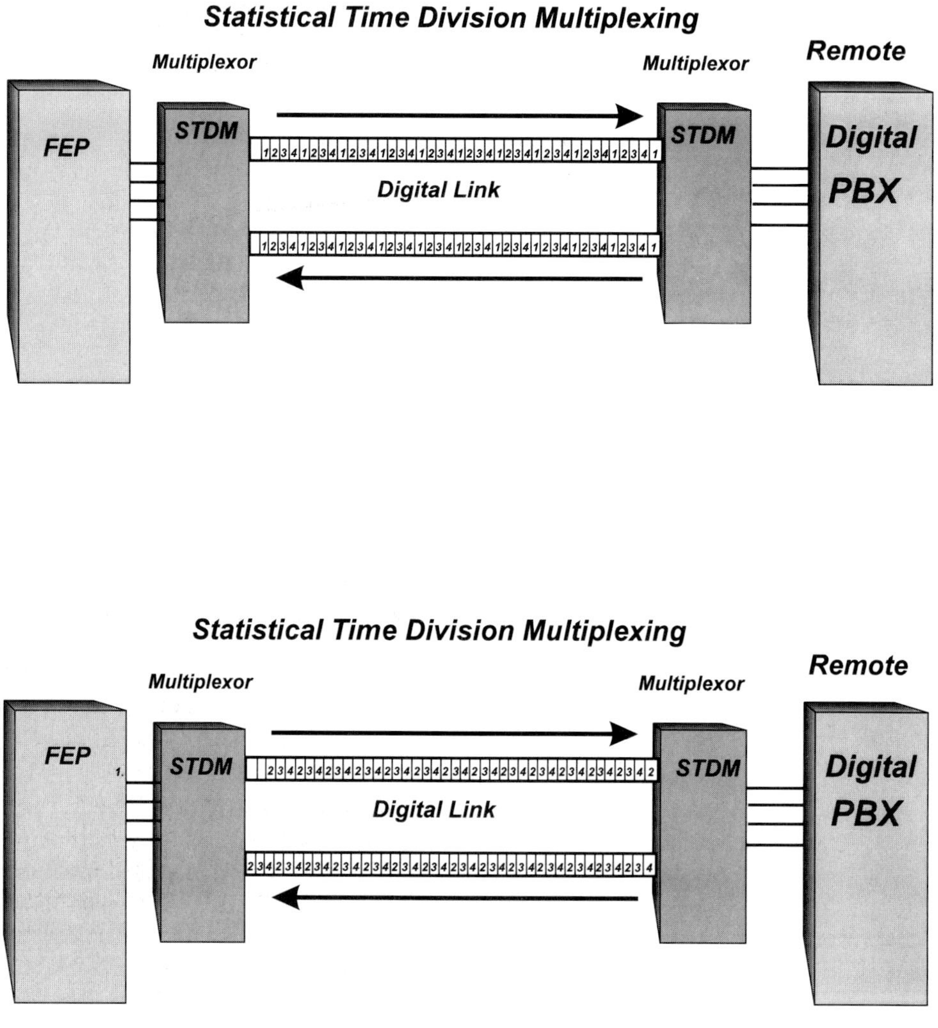

Figure 10.7 Statistical Time-Division Multiplexor

by either using the x-off capability or by lowering the data terminal ready (DTR) signal, which tells the multiplexer to stop sending data. When the terminal is ready to receive again, it performs the reverse operation. Some statistical time-division multiplexers can pass some of the EIA-232 control signals to remote multiplexers or modems. These signals are used to control what are referred to as tail circuits, which allow a local CPU (host computer) to control a remote multiplexer.

Wave-Division Multiplexers

Wave-division multiplexing (WDM) is a relatively new methodology supported by the Institute of Electrical and Electronic Engineers (IEEE) as a standard for multiplexing on fiber-optic cable (see Figure 10.8). WDM permits the use of different wave lengths of light representing different data channels on the same fiber-optic cable. This technique is similar to frequency-division multiplexing. WDM may be viewed as laser beams of different colors crossing each other. Each beam can be distinctly identified even though combined with the other light waves.

WDM technology will require new transport techniques to manage data flow. The Synchronous Optical Network (SONET) standard is supported by the CCITT to handle the very high speeds (in the billions of bits per second range) associated with multifrequency fiber optics. The basic building block of SONET is the Synchronous Transport Signal Level 1 (STS-1). STS-1 is a protocol that employs a synchronous frame alignment allowing DS-1 bit rates to be extracted from the STS-1 frames without the frame being disassembled and reassembled. Most of the STS-1 frame consists of the SONET payload envelope, a large data field that allows STS-1 to carry a clear-channel DS-3 signal or a number of lower bit rate links.

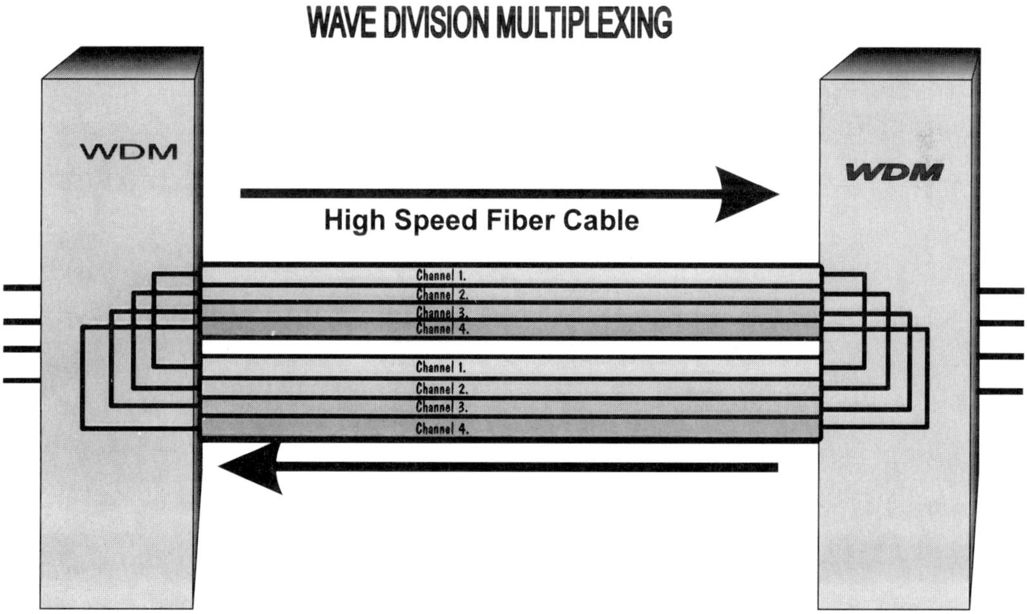

Figure 10.8 Wave Division Multiplexing

Inverse Multiplexing

Inverse multiplexers also may be referred to as bandwidth-on-demand multiplexors in that they have the capability to monitor the data flow from a computer and utilize multiple outgoing circuits when demand is high. This is accomplished by establishing a bit per second and time threshold value that when exceeded, triggers the use of additional circuits for more bandwidth until demand is reduced below threshold levels. From bandwidth-on-demand comes inverse multiplexing. This allows high-speed output to be spread over multiple low-speed lines while maintaining the integrity of high-speed transmission. Figure 10.9 illustrates an example of an inverse multiplexor.

Inverse multiplexing can be accomplished by using a T1 line (made up of multiple 56 Kbps channels) or multiple 64 Kbps links. This type of multiplexor will take two or more 56 Kbps lines and combine them into what appears to the application as one high-speed circuit. Inverse multiplexing typically uses multiple 56 Kbps over leased or switched lines within North America and 64 Kbps lines internationally. ISDN Basic Rate Interface, ISDN Primary Rate Interface, ISDN HO Switched 384, or full T1 circuits also can be used. The benefit is that with inverse multiplexing, the user can get the bandwidth equivalent of a T1 but at a much lower price. Inverse multiplexing is particularly beneficial for internetworking LANs, videoconferencing and traffic overflow.

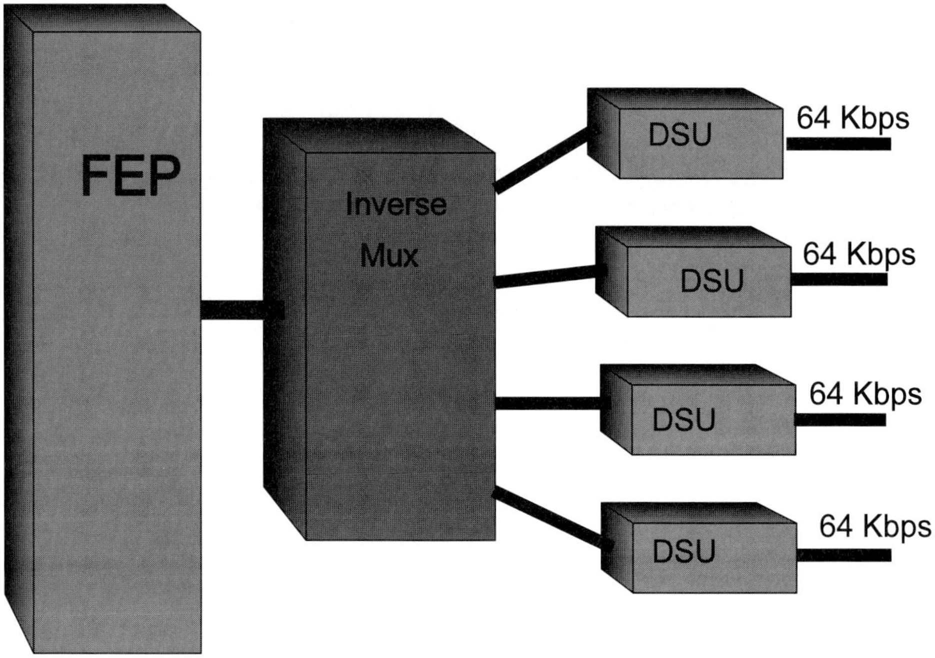

Figure 10.9 Inverse Multiplexor

Multiple data streams are input into the inverse mux. These data streams can come from different ports on the same device such as a router, or they may come from different sources, a PBX, a terminal server, and a video codec. The inverse multiplexor uses time-division multiplexing to split the data stream into multiple 56 Kbps channels. The separate 56 Kbps channels take diverse paths through the public switched network.

Inverse multiplexing improves on a feature some routers offer called load balancing or load sharing. With load sharing, the router tries to balance the traffic across multiple outgoing ports; however, load sharing adds about 30 percent overhead to the transmission. Also, each vendor's implementation is unique, so one vendor's implementation will not work with another.

Inverse multiplexing accomplishes the same function but uses a standard called bonding. This specification was written by the bandwidth-on-demand Interoperability Group, a consortium of 40 manufacturers and has been passed to the ANSI TR41.4 group.

Bonding supports four modes. Mode 0 is a special mode for dual 56 Kbps calls, which offers special economies when both sides are using 56 Kbps lines. Mode 1 works with any number of lines. Mode 2 has an in-band subchannel overhead of 1.6 percent. Finally, Mode 3 uses an out-of-band subchannel. The ISDN H.221 specification also can be used for inverse multiplexing, but it is primarily used for video, not LAN, data.

Multiplexer Framing

In a straightforward statistical multiplexer, the data to be multiplexed from a particular terminal or device is collected and temporarily stored in a terminal buffer. The multiplexer creates a transport frame from each of the terminal buffers. Each frame is n bytes in length and contains several pieces of information such as the location of the sending and receiving devices. The actual framing of the data may employ a standard protocol such as HDLC, described in detail in Chapter 16.

Each frame may vary in length (number of bytes of data) depending on the number of characters accumulated from a particular terminal since the last transmission. If no data is present from a device, no frame is sent. If a single character is present, the frame is transmitted because holding a frame with only a few data characters can cause transmission delay. Some statistical multiplexers allow channel priorities to be established. Typical signals that are passed include request to send (RTS), carrier detect (CD), data terminal ready (DTR), data set ready (DSR), and ring indicator (RI).

Error Control

Most multiplexers provide end-to-end error control and use a technique referred to as cyclical redundancy check (CRC) detection (CRC-16, CRC-32, CRC-64). This error detection field is resident in each information frame and contains the results of a complicated formula for error detection. At the transmitting multiplexer, the CRC bit value is calculated using a series of algebraic calculations based on the total number of data bits in the packet to be transported. The CRC is then transported along with the data inside the transport frame behind the data. At the receiving end, the same formula is used to recalculate the CRC on the received data and the results are then compared to the original CRC sent. If they differ, the receiving multiplexer requests that the frame be retransmitted and discards the data in error.

PACKET ASSEMBLY AND DISASSEMBLY DEVICES

Public data networks are an alternative to dedicated links. They provide error detection, routing and switching functions, relieving the user of those responsibilities. If a PDN is used, it is the responsibility of the user to format the data into packets or frames suitable for transfer over the PDN. If the PDN is used only for carrier activity, computer software in the processors most likely will accomplish the framing activity. Many organizations are not large enough to have a computer dedicated for PDN terminal activity, so they use outside sources. Several manufacturers provide packet assembler/disassembler (PAD) devices to perform the framing and conversion functions. The most common is the X.25 PAD, which provides an interface with a Networks X.25 public data network (see Figure 10.10).

X.25 is a CCITT standard for interfacing DTE and DCE devices. DTE equipment comprises terminals and computers, and DCE equipment consists of PADs, modems and multiplexers operating in a public data network environment. A PAD may be purchased

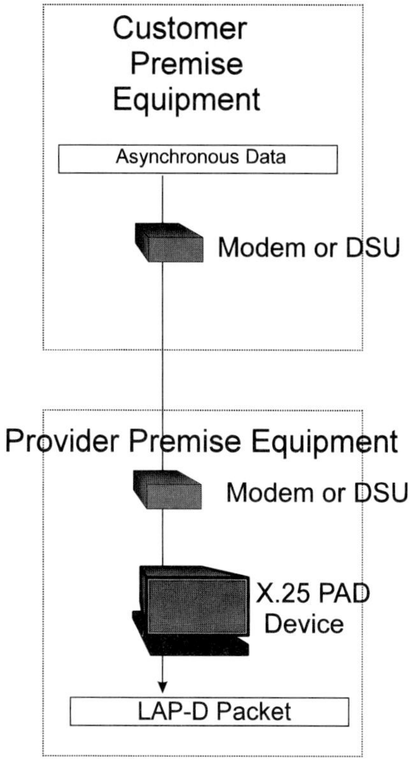

Figure 10.10 Illustration of PAD Communications

by the end user, or provided by the public data network or hardware vendor. Some telephone companies in Europe and Asia only allow the use of equipment provided by themselves. The output from a PAD looks like an OSI-defined HDLC frame and contains a minimum of 128 data bytes, in conformance with the 1980 HDLC standard.

As with most types of communications equipment, there are many options for PADs. For example, many PADs support multiple asynchronous devices and terminals. The PAD communicates via synchronous transmission methods at speeds up to 720 Kbps. All PADs are designed for specific network standards and adhere to one or more of the CCITT standards, including X.3, X.25, X.28 and X.29, as well as the X.121 (universal addressing) extensions.

Many PADs contain intelligence for activities such as password control and security; several PADs also provide statistical and performance reporting. Often a company will use the computer services of the PDN in place of resident computers because of the extensive support provided by the PDN. In this case the company will have only a PAD and asynchronous terminals on-site.

NETWORK CONTROLLERS

As networks grow and become distributed to remote locations, it is increasingly important for data communications technical support to have a clear view of network performance. This ensures that quick action can be instituted if the network begins to experience problems. Along with visibility requirements, network managers also must have the ability to modify configuration parameters that govern the operation of the network.

Network performance determination, as well as problem detection, are often difficult tasks to accomplish because of the number of devices making up the network, their locations, and the wide variety of media in use. Several vendors have developed a type of communications processor for network monitoring and line control referred to as a network controller. Network controllers are designed to provide local and remote monitoring of the communications network and to allow quick detection of errors and rapid problem resolution. Before the introduction of network control centers, substantial degradation of the network had to occur before users or technical support staff had any indication that a problem existed.

Figure 10.11 illustrates a multinational network headquartered in London that requires a network controller. It provides the routing and control for processing and communications nodes at key locations within Europe.

Often the network controller is located in a remote network control center and is designed to act as a hub for all network links. Depending on the size of the network, the controller may support multiple control terminals and enable the setting of definable alarm threshold values. This means that the network controller will notify a control terminal when a programmed value, such as the acceptable retransmit on error rate, is exceeded. Program values can be established by the technical support department for the number of packet retransmissions in a given period of time, parity errors, number of lines active, number of lines idle, or other important operations that should be monitored.

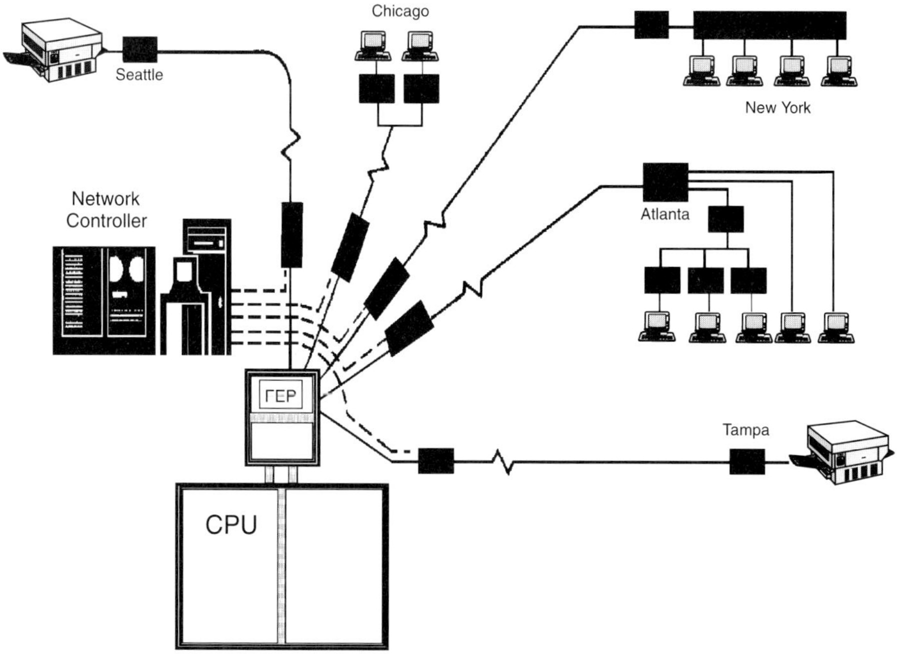

Figure 10.11 Network Control Center

Comprehensive circuit quality monitoring is another function of the network control monitor. Many controllers can monitor links and dynamically shift link resources when errors are encountered. This is accomplished by constantly monitoring and testing all point-to-point and multipoint lines in the network. The controller also can provide a status report on all terminals in the network on demand.

Often, a network controller functions as a value-added device, providing gateway functions for different code sets and protocols. This allows the network to function in a heterogeneous manner without substantial degradation. Many companies (including AT&T, Codex Corp., and Racal Vadic Corp.) offer network controllers that can be configured for specific networks and terminal types.

REMOTE AND NETWORK JOB ENTRY TERMINALS

Remote and network job entry (RNJE) terminals are no longer considered terminals but rather complete processing systems. RNJE has evolved from a pre-PC period when it represented the only method for remote data capture and printing to a current environment, in which it is used for remote ancillary functions such as high-speed printing and terminal access. The role of the RNJE terminal has changed from a remote batch I/O device to one of remote processing capable of supporting the processing needs of the remote sites and then transmitting or receiving the processed information to or from the host computer.

Generally speaking, the RNJE processor provides limited capabilities for independent processing because of its restricted logic and memory. It is typically used as a pathway to a remote host computer for multiple terminals or printers. The RNJE terminal environment consists of several components and is somewhat expensive in comparison to the asynchronous or synchronous terminal, ranging from $20,000 to $75,000.

RNJE terminals are designed primarily for high-speed synchronous data transfer (9,600 bps or greater). They are usually connected via synchronous dedicated links and full-duplex modems. Components of RNJE terminals include the processor (which is usually a minicomputer such as the IBM 5200-8200, a control console that establishes the linkages with the host and supervises the data transfer operations, and a number of attached I/O devices such as readers, data entry terminals, disk drives for local storage, tape drives and printers.

MULTIUSER AND THIN CLIENT/FAT SERVER WORKSTATIONS

The multiuser workstation is an advancement on the data entry workstation concept that has been available for quite some time. It offers a concurrent-window, multiple-access environment so that users can view and access multiple environments such as PC-DOS, UNIX, and ASCII terminal emulation.

This type of system consists of a minicomputer or microcomputer with a reduced instruction set computer (RISC) processor, extensive memory, and I/O capabilities. It is normally designed to use an operating system that has multitasking capabilities, such as UNIX. The system supports a number of semi-intelligent terminals and the processor provides communications links for each of the terminals on request in order to interface with other systems. Most of these systems provide secondary storage disks and tapes as well as printing capabilities for each of the users. Network control is provided for links to outside networks. Manufacturers of multiuser workstations include Sun Microsystems, Hewlett-Packard, Motorola Inc., and IBM.

Multiuser workstations are used extensively in engineering environments. They provide high-processing throughput and are ideal for the CPU-intensive activities required for engineering, design, high-definition graphics, and computer aided design (CAD).

Many manufacturers refer to multiuser workstation environments as client-server or terminal-server environments. They also can be considered LANs; for example, the Sun Microsystems terminal-server environment uses the Ethernet LAN access method for communications between each of the terminals.

GATEWAYS, BRIDGES, AND ROUTERS

Gateways

A gateway allows for the connection of dissimilar networks by providing the following services: provides for codeset and protocol conversion from one network format to another, i.e., an IBM SNA network to a DEC DECNET network; provides for segmentation and reassembly of frames traveling from one network to another; provides secured access between networks which requires password validation; assists in providing dynamic routing through an intermediate network; and provides expedited data services and resequencing services depending upon the intelligence of the gateway. Figure 10.12 illustrates the use of a gateway between two networks.

Gateway Functions. Gateways are designed to provide functions such as protocol conversion, electrical connections, end-to-end translations, network routing, and error detection/correction for packets being transferred between two networks. These activities are accomplished using either gateway-specific protocols or with protocols such as the transmission control protocol (TCP) and the Internet protocol (IP).

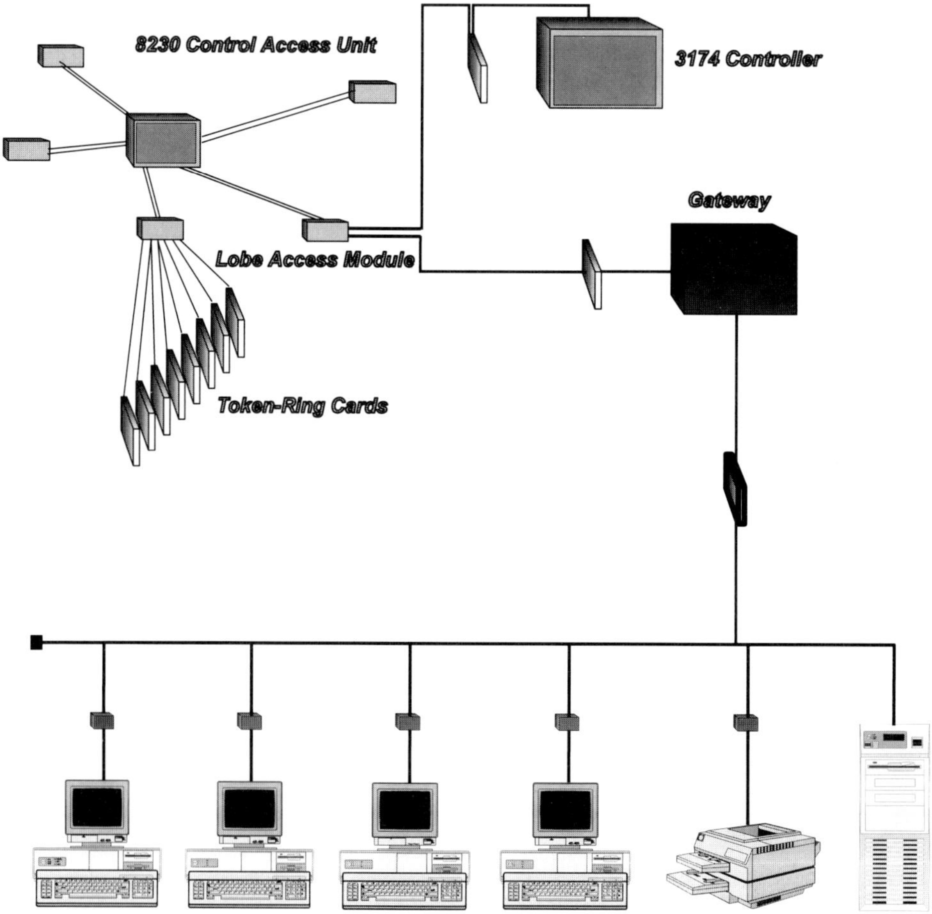

Figure 10.12 Example of a Gateway Connecting Two Networks

Gateways allow data to be shared with other network nodes by using a single access port or specialized communications node to access the other networks. It is important to note that limited protocol support, complexities due to different speeds, and different levels of security implemented in the two connecting networks can all be disadvantages when using gateways. Gateways perform the following specific functions:

Internetwork Communications. Communications between nodes in different networks are accomplished in the gateway using either the open systems interface (OSI) transport/network layer protocols or through one of the available non-OSI internetwork protocols such as those provided by Xerox networking systems protocols. The gateway has the intelligence to deliver the data across networks using the appropriate communications protocol. These protocols provide the foundation for communications between networks.

Routing/Flow Control. A prime function of the gateway is one of routing and flow control. Flow control is needed between networks having different speeds in order to balance and smooth communications. Flow control techniques will be used by the gateway to regulate the packet flow into a network of a differing speed. Routing is important when more than one gateway is involved. The routing schemes may employ a fixed route, or may be dynamic based upon levels of service available (least cost, expedited data, etc.) over several different routes.

Packet Sizing. A gateway is responsible for packet resizing when two networks are designed to handle different packet sizes. When a network designed to support a small data packet size communicates with a network with larger packet sizes, segmentation (breaking packets into smaller physical units), or packet reassembly (creating larger packets) is required. To accomplished this, each gateway must have intelligence and enough buffer space to accommodate the required segmentation or reassembly function.

Addressability. Gateways generally include the ability to maintain and provide addressing information about each of the nodes in the networks with which it interconnects. Each device on the network has a unique address in order to provide a location for delivery of the packets from other nodes and networks. When the gateway contains an address for each of the nodes, this is referred to as a flat addressing scheme. When multiple networks are involved, hierarchical addressing can be employed using one of the nodes in each network for address storage. This technique allows the gateway to only designate the network address. Node destinations are handled by the addressing scheme in each networks addressing database.

Security/Reliability/Priority. Node-to-node and network-to-network security may be a requirement by each of the communicating networks. Connection of secure networks to each other requires the same type and level of security implementation. Networks having incompatible security methods will not be able to communicate with each other, which in networks have compatible security validation software, violation detection can be implemented by gateways and transmission stopped.

Reliability can be built into networks to ensure that packets do not loop indefinitely among gateways. This can be accomplished via time and date stamps or time to live counters attached to packets that decrement as the packet travels through each node. Error information can be sent to gateways to provide packet retransmission or negative acknowledgments. Priority mechanisms such as grades of service may be implemented at a gateway for data of an urgent nature. This also can be employed for monitoring secure networks as well as for flow and congestion control by sending control packets between nodes.

Bridges

Bridges are typically used for the connection of similar networks such as an Ethernet LAN to another Ethernet LAN. They may, depending upon intelligence, provide many of the services that gateways provide such as security, segmentation and reassembly, and expedited data services. Figure 10.13 shows a bridge connecting two similar networks.

Bridge Functions. Bridges are hardware devices employed to interconnect similar networks. Many different types of bridges exist for this function. Typically a

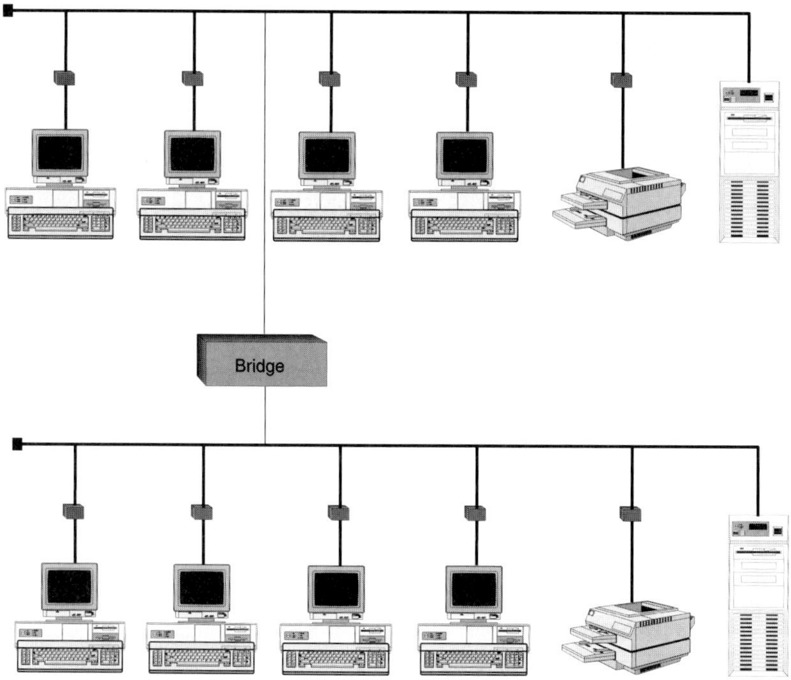

Figure 10.13 Example of a Bridge Connecting Two Networks

bridge would be used to interconnect two Ethernet or Token-ring networks having the same formats and protocols. Many bridges have incorporated several of the functions of gateways. They also provide security, error detection/correction, and address validation.

Like the gateway, a bridge may provide routing and flow control. Routing becomes important when more than one bridge or network is involved. The routing schemes may be over a fixed route, or routing selection may be dynamic based upon levels of service available (least cost, expedited data, etc.) over several different routes. Several LAN routing techniques have been developed and include: Source routing in which each workstation contains all required network addresses and the bridge does no addressing. Transparent routing in which the bridge provides all network addressing and spanning tree routing for Ethernet type networks.

Flow control manages congestion and is used between networks having different speeds in order to balance and smooth the traffic. Networks not employing flow control techniques will usually experience a higher degree of delay than networks employing flow control. A typical bridge is microprocessor based and requires very limited configuration. Many bridges are intelligent with some having the ability to learn node address by examining network traffic and building internal tables. Systematically, it will exam the address table and purge unused addresses. Many bridges support multiple protocols that includes: TCP, IP, HDLC, and SDLC.

Routers

Similar in design to a bridges and gateways but having more intelligence, the router is an interconnection device designed to forward packets and frames to other networks based on the protocol used and destination. The router operates at the network to network level and does not involve itself with node-to-node communications. Routers can be protocol-dependent or protocol-independent and allow for the inter-connection of multiple networks. Today, routers have generally replaced bridges and gateways as the connectivity choice for most networking solutions. Figure 10.14 shows a router connecting four different networks.

CASE STUDY (CONTINUED)

Selecting the required communications equipment was difficult for the bank, because it wanted to make sure that the acquired hardware would be upgradable and have a life of at least three years. The bank's planning team established a set of guidelines for purchasing or leasing hardware that specified that all branch components would be standard from the same vendor. They also wanted local support from the vendor or distributor, and all hardware had to be field upgradable.

For each branch, the planning committee selected medialess PCs for the teller workstations. This will provide a degree of intelligence and will assist in improving transaction response time. Each of the workstations will be connected to a branch LAN controlled by an administration PC with disk storage. This, as well as other administration workstations

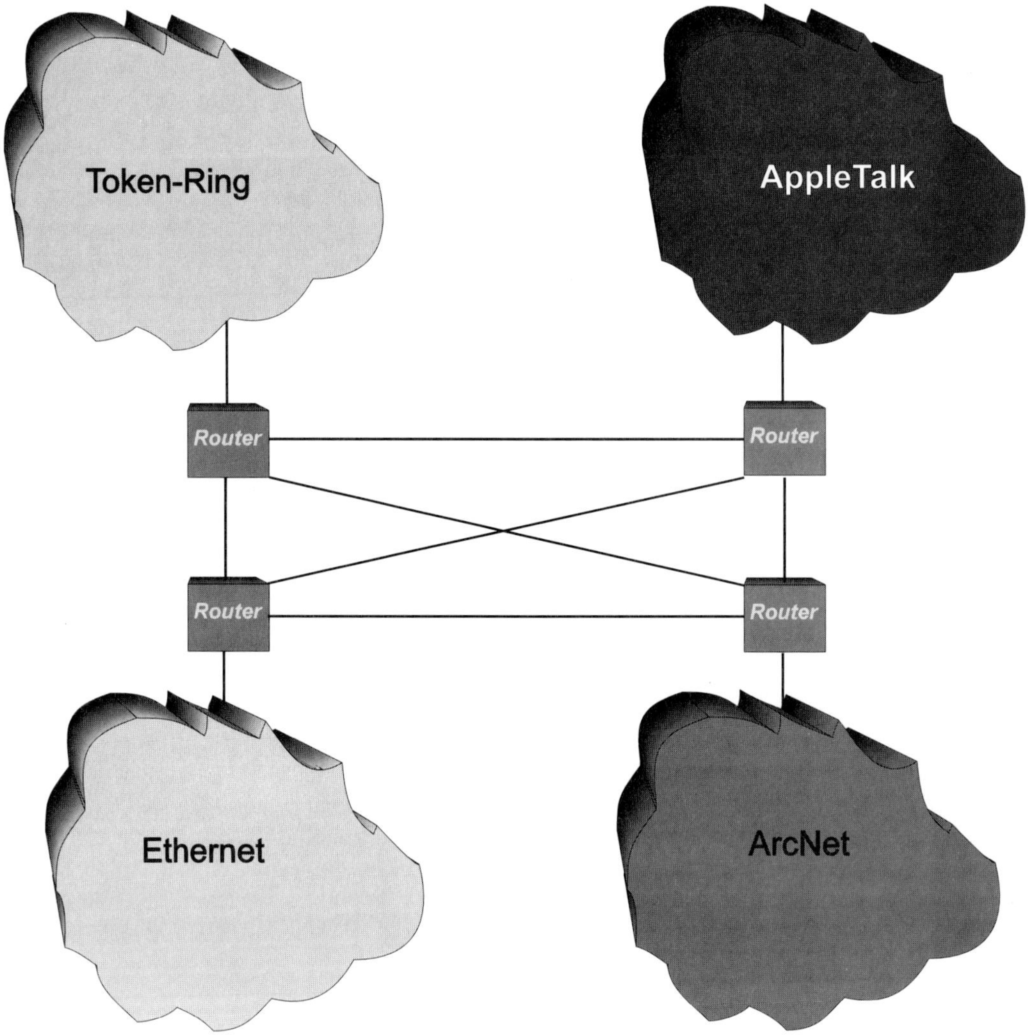

Figure 10.14 Example of Four Networks Connected by Router

used for other branch activities, will be connected through a dedicated link back to a service center. If multiple ATM machines are installed at the branch, they will share a dedicated line using a time-division multiplexer.

The service center does not use a multiplexer because of the small size of the network and the expense associated with multiplexers. However, a front-end processor is used to handle all of the branch links as well as the high-speed link to the other service centers. The front-end processor will be configured to handle 50 dedicated links and up to

10 directly attached terminal controllers. This will provide support for the programming staff, which will use smart terminals for all online activities.

REVIEW QUESTIONS

1. What activities of the host computer are relieved by the communications processor?
2. Name three major functions of the FEP.
3. True or false: FDM stands for frequency-directing modulation.
4. True or false: Multiplexers provide four different techniques for the transfer of data.
5. True or false: STDM is the most current multiplexing scheme.
6. Wave-division multiplexing is used with what medium?
7. True or false: Network controllers provide a focal point for monitoring the communications network.
8. True or false: The terminal controller requires software to function properly.
9. True or false: Smart and dumb are the two classifications for terminal types.
10. True or false: Remote job entry systems are normally used for batch data transfer activities.

CHAPTER 11

Communications Using Modems

INTRODUCTION

The word "modem" is a contraction of modulator-demodulator. Modems attach to computers, terminals, terminal controllers, multiplexers, or communications processors. They convert (modulate) binary digital signals into analog signals (electrical signals representing ones and zeros). The converted signals then can be sent across a telephone network using standard analog telephone lines. A paired modem on the other end of the link reconverts the analog signals into digital signals for input into a device at the receiving side.

Early in the evolution of the modem, users discovered that not all modems are created equal. If an attempt was made to connect modems built to the same standards from different manufacturers, they often would not work together. Today, because CCITT standards are very specific, modems from different manufacturers that conform to a specific CCITT standard will interface properly.

All public telephone operating companies follow the same rules regarding the connection of modems to their networks. These rules were devised by the old North American Bell system and are now covered by international CCITT V series standards for signal strength, modulation technique, interfaces, bit rate, and error correction. These standards govern many types of modem connections to the various worldwide telephone networks.

Unlike computers that transfer bytes in a parallel manner (i.e., eight bits sent over eight wires or channels at the same time), most modems operate in a serial manner (one bit after the other over a single channel). The majority of modems available for use in the world today are of the serial type because of timing problems associated with parallel transmission. A special class of parallel modems has been defined in the old Bell standard; however, due to propagation delay that can vary the bit times from one channel to another, they are not generally available and are considered experimental. Modems are manufactured by many companies, including Racal-Vadac, Codex, IBM, Hayes Microcomputer Products Inc., and U.S. Robotics Inc.

MODEM HISTORY AND DEVELOPMENT

Originally, AT&T, the worldwide telephone companies and the regional Bell operating companies referred to modems as datasets or data termination equipment (DTE) devices. Customers who wanted to use the public telephone network for data communications were required to use equipment provided by those telephone companies.

The telephone carriers required customers to use a dataset supplied by a telephone or telecommunications company, as well as a direct access arrangement (DAA) device. This device was an external registered protective circuitry device installed between the dataset and the line. Its function was to protect the telephone companies' lines and central office equipment from high-voltage surges, out-of-bandwidth frequencies, and high signal levels. In other words, it kept the dataset from putting out voltages high enough to cause the lines to catch fire.

While control of the attachment of datasets to the telephone network was and still is necessary, it was also construed as a monopolistic practice because the telephone companies would not allow datasets manufactured by anyone other than themselves to be used. Due to activities started by various modem manufacturers that ultimately led to a supreme court ruling in 1968, restrictions on the use of non-Bell attachments were lifted. DAA devices are no longer required because their function has been incorporated into the various modems now available. Internationally, many telephone companies outside North America still require customers to lease or purchase modems from them rather than from an independent source.

MODEM STANDARDS

Bell Standards

AT&T set the de facto standards for modems, and many of those standards are still valid in North America. Most Bell standards have a compatible CCITT V standard that is used worldwide, while some Bell standards have become obsolete. Currently, most modems installed in North America and Europe conform to the V series standards.

CCITT V Standards

The CCITT is the major international organization responsible for modem standards. Many of those standards have replaced the Bell standards. In many cases, CCITT and Bell standards are compatible or identical. Because modem technology has advanced rapidly since the Bell standards were developed, many CCITT standards do not have a Bell counterpart.

The CCITT is also responsible for many other standards, including modem interface, error correction and data compression standards. These are electrical, mechanical, and interface standards for the connection of modems (DCE) to terminal equipment (DTE); the CCITT V.34 and V.21bis standards are good examples. The V.24 interface standard (similar to EIA-232) and the V.28 standard are also commonly used. Another class of standards has also emerged from the CCITT for modem error correction and data compression, and currently consists of V.42 and V.42bis. The suffix "bis" or "ter" following a standard means that the standard has an authorized variation.

Table 11.1 lists the standards for most CCITT V series modems. Where applicable, corresponding Bell standards have been included.

The following V recommendations are some of the more important standards that apply to modems and data transmission.

Table 11.1 ITU-T Standards

CCITT	Data Rate Norm Fall	Operational Modes	Modulation Method	Sync/Async	Bell Version
V.21	300	Duplex	FSK	Async	–103
V.21Ter	300	Duplex	FSK	Async	–212
V.22	1,200	Duplex	PSK	Sync/Async	–212
V.22Bis	2,400 1,200	Duplex	QAM/DPSK	Sync/Async	None
V.23	1,200 600	Half	FSK	Async	*202
V.26	2,400	Half	DPSK	Sync	201
V.26Bis	2,400 1,200	Half	DPSK	Sync	
V.26Ter	2,400	Duplex	TCM or QAM	Sync	
V.27	4,800	Half-Duplex	DPSK	Sync	–208
V.27Bis	4,800 2,400	Half-Duplex	DPSK	Sync	
V.27Ter	4,800 2,400	Half-Duplex	QAM	Sync	
V.29	9,600 4,800	Half	QAM	Sync	–209
V.32	9,600 4,800	Half-Duplex	TCM/QAM	Sync	None
V.32Bis	14,400	Half-Duplex	MNP Prot.	Sync/Async	None
V.33	14,400–12,200	Half-Duplex	TCM/QAM	Sync	None
V.34	33,300–14,400	Half-Duplex	TCM/QAM	Async/Sync	**
	Variations and enhancements raise the maximum speed to 56 kbits per second				
V.36	72K	Dublex			
V.37	72K	Duplex			

*Some compatibility

–Similar not compatible

Part 1—General Interfaces

V.1	Equivalence between binary notation symbols and the significant conditions of a two-condition code.
V.2	Power levels for data transmission over telephone lines.
V.3	International Alphabet No. 5.
V.4	General structure of signals of International Alphabet No. 5 code for data transmission over public telephone networks.
V.5	Standardization of data signaling rates for synchronous data transmission in the general switched telephone network.
V.6	Standardization of data signaling rates for synchronous data transmission on leased telephone-type circuits.
V.7	Definitions of terms concerning data communication over the telephone network.

Part 2—Interfaces for Voice-band Modems

V.10	Electrical characteristics for unbalanced double-current interchange circuits for general use with integrated circuit equipment in the field of data communications.
V.11	Electrical characteristics for balanced double-current interchange circuits for general use with integrated circuit equipment in the field of data communications.
V.13	Simulated carrier control.
V.14	Transmission of start-stop characters over synchronous bearer channels
V.15	Use of acoustic coupling for data transmission.
V.16	Medical analogue data transmission modems.
V.17	New 14,400 bps fax speed.
V.19	Modems for parallel data transmission using telephone signaling frequencies.
V.20	Parallel data transmission modems standardized for universal use in the general switched telephone network.
V.21	300 bps duplex modem standardized for use in the general switched telephone network.
V.22	1,200 bits per second duplex modem standardized for use in the general switched telephone network and on point-to-point two-wire leased line telephone-type circuits.
V.22bis	2,400 bits per second duplex modem using the frequency division technique standardized for use in the general switched telephone network and on point-to-point two-wire leased line telephone-type circuits.

V.23	600/1,200 baud modem standardized for use in the general switched telephone network.
V.24	List of definitions for interchange circuits between data terminal equipment (DTE) and data circuit-terminating equipment (DCE).
V.25	Automatic answering equipment and/or parallel automatic calling equipment on the general switched telephone network including procedures for disabling of echo control devices for manually and automatically established calls.
V.25bis	Automatic calling and/or answering equipment on the general switched telephone network (GSTN) using the 100-series interchange circuits.
V.26	2,400 bps modem standardized for use of four-wire leased telephone-type circuits.
V.26bis	2,400/1,200 bits per second modem standardized for use in the general switched telephone network.
V.26ter	2,400 bits per second duplex modem using the echo cancellation technique standardized for use on the general switched telephone network and on point-to-point two-wire leased telephone-type circuits.
V.27	4,800 bps modem with manual equalizer standardized for use on leased telephone-type circuits.
V.27bis	4,800/2,400 bps modem with automatic equalizer standardized for use on leased telephone-type circuits.
V.27ter	4,800/2,400 bps modem standardized for use in the general switched telephone network.
V.28	Electrical characteristics for unbalanced double-current interchange circuits.
V.29	9,600 bps modem standardized for use on point-to-point four-wire leased telephone-type circuits.
V.31	Electrical characteristics for single-current interchange circuits controlled by contact closure.
V.31bis	Electrical characteristics for single-current interchange circuits using optocouplers.
V.32	A family of two-wire duplex modems operating at data signaling rates of up to 9,600 bits per second for use on the general switched telephone network and on leased telephone-type circuits.
V32bis	Extension to V.32 with speed up to 14,400 bps.
V.33	14,400 bps modem standardized for use on point-to-point four-wire leased telephone-type circuits.
V.34	28,800–33,300 bps modem for use over the dial-up public switched telephone network.

Part 3—Wideband Modems

V.35 Data transmission at 48 Kbps using 60–108 Khz group band circuits.

V.36 Modems for synchronous data transmission using 60–108 Khz group band circuits.

V.37 Synchronous data transmission at a data signaling rate higher than 72 Kbps using 60–108 Khz group band circuits.

Part 4—Error Control

V.40 Error indication with electromechanical equipment.

V.41 Code-independent error control system.

V.42 Error-correcting procedures for DCE using asynchronous-to-synchronous conversion.

V.42bis Data compression procedures for DCEs using error correcting procedures.

Part 5—Transmission Quality and Maintenance

V.50 Standard limits for transmission quality of data transmission.

V.51 Organization of the maintenance of international telephone-type circuits used for data transmission.

V.52 Characteristics of distortion and error-rate measuring apparatus for data transmission.

V.53 Limits for the maintenance of telephone-type circuits used for data transmission.

V.54 Loop test devices for modems.

V.55 Specification for an impulsive noise measuring instrument for telephone-type circuits.

V.56 Comparative tests of modems for use over telephone-type circuits.

V.57 Comprehensive data test set for high data signaling rates.

Part 6—Internetworking with Other Networks

V.100 Interconnection between public data networks (PDNs) and the public switched telephone networks (PSTN).

V.110 Support for data terminal equipment (DTEs) with V-series type interfaces by ISDN.

V.120 Support by an ISDN of data terminal equipment with V-series type in terfaces with provision for statistical multiplexing.

V.230 General data communications interface layer 1 specification.

MODEM INTERNALS

Modems are designed to transmit as well as to receive data therefore, they must contain circuits to handle both activities. Two primary circuits are called the transmitter/modulator circuit, which manages the output activities, and the receiver/demodulator circuit, which controls the line input (see Figure 11.1).

The transmitter/modulator circuit is responsible for converting the digital bits originating from the terminal into electrical signals compatible with the analog transmission facility. This circuit consists of several components. The data encoder is used in conjunction with transmitter timing to determine the modulation changes required (i.e., dibits, tribits, etc.), depending on the modulation technique employed by the modem. The transmitter control is used to control the on/off cycle of the carrier and the request-to-send/clear-to-send delay times needed to keep the terminal in synchronization.

The transmit timing source is a component used to output a transmit clock (TC) signal for synchronous modems only. It provides the bit clocking signal for the terminal if the terminal does not contain a clocking source of its own. All modems must contain a modulator, which changes the carrier waveform using one of the CCITT or Bell modulation standards to represent data as determined by the data encoder.

Modems also contain a band-limiting filter that shapes the frequency spectrum and ensures that all signals are placed on the correct frequencies within the analog spectrum generated by the transmitter/modulator. A line amplifier also may be found in some modems, which allows the output signal to be adjusted for more or less strength. Most

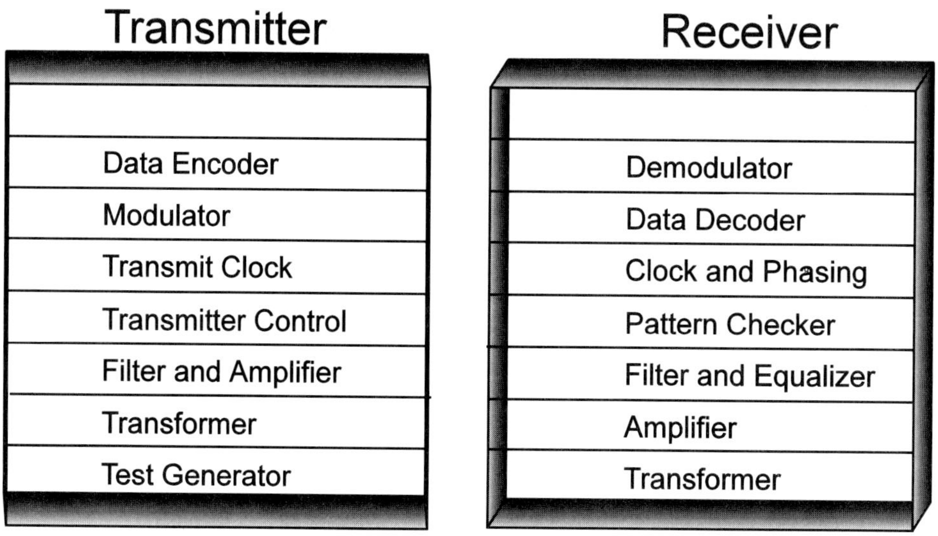

Figure 11.1 Modem Internals

modems contain a transformer that provides a connection between the modem and the line; it is used for both the transmit component and the receive component.

On some modems, a test generator may be incorporated to generate a pseudo-random data pattern for diagnosis when in the test mode. This pattern is used to validate the modem's integrity.

The receiver/demodulator circuit accepts analog signals from the communications link, and converts them back to their original digital format. After conversion, the formatted digital information is transferred to the terminal or computer via an interface cable for processing. The receiver comprises several components. A data decoder is used in conjunction with the demodulator to format the received data into serial digital form. The modem's receive circuit may also contain clock and phasing circuits that generate a receive clock signal from the demodulated signal for output on the V.24/EIA-232 interface to the terminal. The modem receive circuit also contains a demodulator that extracts the part of the spectrum signal containing information as it is received from the telephone link.

Most modems also contain an equalizer that compensates for channel distortion allowing higher data rates and better performance, because the data signal is cleaner and requires less error handling. Another component is the AGC amplifier, which provides automatic gain control of the incoming signal. The AGC amplifier allows the modem to compensate for amplitude variations in the telephone line; this ensures that the signal strength stays constant. AGC has been used in television for quite some time to limit the strength of the signal so as to better manage and convert it to sound and images. A band-limiting filter may be included to shape the frequency spectrum of the analog signal received from the telephone line. It eliminates any extraneous frequencies that are outside the bandwidth of the analog data stream.

Finally, if the modem incorporates a test generator, it will also have a pattern checker. This is used to validate the received pseudo-random data pattern from the pattern generator for errors when the modem is placed in the test mode.

TRANSMISSION METHODS

Modems use one of two transmission methods for transporting data across a communications link: asynchronous and synchronous. Modems are primarily designed to support one or the other method; however, some modems can be configured to support either type of transmission. Each method is used with specific network architectures and transmission protocols.

Asynchronous Modems

Asynchronous modems were in the past considered low-speed devices (50 bps up to 2,400 bps) and were normally less expensive than synchronous modems. This is no longer true because of changes over the past few years as this type of modem can now operate at speeds up to 115,000 bps. Because of their low costs, they generally offer fewer features than synchronous modems. But this is rapidly changing because of the popularity of the PC and the market it represents for this type of modem.

The asynchronous modem is used for start/stop, character-oriented transmission where 10 or 11 bits are sent at a time. Some of the more expensive modems can transmit and receive asynchronous data at rates up to 115,200 bps by using error detection and correction logic and bit compression. Asynchronous modems are supported by standards from the ANSI and the ITU.

With asynchronous transmission, no clocking or timing signals are required by the transmitting or receiving modem because only a single character is transported at a time. Bit counts are used in place of timing circuits to delineate one character from another. Every character sent consists of a start bit (0 bit), seven or eight data bits, a parity bit, and a stop bit (1). When the line is idle and no data is being transmitted, the modems will exchange one bits. Buffering and blocking activities are a function of the DTE. Therefore, the modem usually receives a steady stream of bits for transport and rarely has to wait for the terminal or computer.

Synchronous Modems

Synchronous modems are considered high-speed devices with bit rates in excess of 64 Kbps. While it appears that little difference exists in the bit rates for asynchronous and synchronous modems, the true measure of a modems performance is in its effective bit rate which is the number of data bits transmitted. The synchronous modem has an effective bit rate that can be as much as 35 percent higher than the asynchronous modem. Synchronous transmission implies the use of clocks to time the data bits being sent and received. Clocking is derived from the incoming bit rate and from a modem or terminal clocking source. Clocking is required to ensure synchronization between the DTE (computer) and the DCE (modem). The clocking signal is a digital square wave exchanged between the modem and the terminal at the same rate as the transmission bit rate of the modem.

Two clocking sources may be used: the internal clock signal and the external clock signal. The internal clock signal is generated in the modem and provides the terminal with transmit signal timing information (via pins 15 and 17 in the EIA-232 interface cable). The external clock is a timing signal sent from the terminal that provides signal element timing information to the modem (transmit clock) via pin 24 of the EIA-232 interface.

A third type of timing option is referred to as slave or loop timing. Slave timing is derived by taking clocking information from the timing recovery circuit in the demodulator and passing it to the modulator. It also passes the clocking information to the DTE's transmitting circuits and to all other devices requiring timing. This type of timing is vulnerable to line impairments such as phase hits and phase jitter, which were described in Chapter 8. One case in which it might be employed is when a single clocking source for the entire network is required to ensure that the bps rate is exactly the same in both directions of the link, such as with statistical multiplexers.

Synchronous modem transmission is used primarily for high-speed data transmission. Data rates are normally greater than 19,200 bps and can exceed 56 Kbps. Telephone carrier clocking signals derived from the telephone carrier's master clock may be used to synchronize the receiver with the distant transmitter and to provide the clocking signals to the terminal. Additional error checking and correction abilities may be provided with modems using synchronous transmission.

Synchronous modems are usually more expensive than asynchronous modems. In many cases, asynchronous modems operating at speeds higher that 1,200 bps are really operating in a synchronous mode (i.e., an asynchronous and synchronous converter is incorporated into the modem).

OPERATIONAL MODES AND OPTIONAL FEATURES

A number of different types of modems are available for a particular specification, in other words, not all 14,400 bps modems are identical. The following options are either software-configured in microprocessor-based modems, or strap-configurable. (Straps or jumpers within the modem have to be manually changed.)

The leased-line modem may be configured by a technical user or the data communications support group to use a continuous carrier. With this option, the modem will transmit a carrier 100 percent of the time regardless of the state of the terminal request-to-send. A modem configured with continuous carrier may also be positioned with switched request-to-send. In this case, the RTS-CTS handshake is provided to the DTE for timing purposes.

The switched noncontinuous carrier option turns the carrier on or off in accordance with the request-to-send signal. This option must be used for remote locations in a multipoint environment. Some half-duplex modems provide a low-speed reverse channel for the transfer of control information. A reverse channel provides a very low-speed circuit, which is a benefit in that the line does not have to be turned around using valuable transmission time. The Bell Series 202 modems have such a reverse channel, allowing up to 50 bps to be sent or a reverse channel tone for circuit assurance.

Dial-up modems do not usually have carrier options. Full-duplex modems provide continuous carriers in both directions, and half-duplex modems provide switched carriers.

Some modem manufacturers provide additional options for customizing the modem for a specific application. The request-to-send control option can be set either to be controlled by the terminal or to function as constant carrier output. The clear-to-send delay times can be varied depending on the use of the modem as a master or slave device. A two-wire or four-wire option permits the user to configure the modem for either two-wire or four-wire operations. If the data set ready option is provided, it can be set to indicate that the modem is off-hook in the dial-up mode.

Another option allows the user to adjust the carrier detect level. Analog pre-equalization permits a user to configure the modem based on the use of conditioned or unconditioned lines. This option would normally be operating when conditioning is present on the line. Private line disconnect/line current disconnect is an option that allows the modem to disconnect (turn off) when it detects no line current for a specified period of time. Another option enables the user to increase or decrease the transmit output level.

The one-second holdover option permits the receiving modem to maintain timing during periods of momentary dropouts not exceeding one second. This option is used with modems having a continuous carrier from a distant transmitter and cannot be used by a multipoint master station.

An additional option for dial-up modems is the send space disconnect. This option permits a modem at the other end of the connection to disconnect from the line while in an unattended state by sending approximately three seconds of space signals. Also, receive space disconnect permits the dial-up modem to disconnect from the line in response to receiving approximately two seconds of space signals.

Optional line-testing capabilities are provided in some modems depending on their cost and the manufacturer. These tests may include loopback tests, frequency tests, and power level tests. Several manufacturers offer modems with switches or intelligence that enable them to handle different data speeds. Another option, known as auto-speed sensing, enables the modem to sense the incoming speed of the line and change its own receiving speed. Several modem manufacturers provide modems that can automatically dial a telephone number and automatically answer incoming calls.

While computer system security is primarily a software problem, line access security is also a major concern. Restricted access to the modem is often difficult to implement, but it is the first line of defense against unauthorized system access. One optional feature available for modems from several manufacturers is intelligence programmed to validate the user on the incoming line. One method stores the telephone numbers of all authorized users and calls them back after validation using the stored numbers. Modems with this feature are referred to as call-back modems.

Some leased-line modems contain logic for dial-up back-up capabilities. This feature allows the modem to automatically dial a number stored within the modem's memory and is used for reconnection of a failed leased circuit.

TYPES OF MODEMS

Types of asynchronous or synchronous modems include external or standalone, integrated, line drivers, and modem eliminators. These can be divided into two major classes, dial-up (switched-line) and dedicated (leased line). DSUs and CSUs are the digital counterparts to analog modems.

Switched-line or dial-up modems are generally considered to be low-speed modems supporting normal speeds of up to 33,300 bps. They may operate using either the asynchronous or synchronous transmission method. The dial-up modem generally provides for the attachment of a telephone handset for call placement and a switch for voice or data. Many types of switched-line modems are available and comply with either Bell or CCITT standards.

Dedicated or leased-line modems operate over leased lines that are in place full time. They operate at speeds of up to 64 Kbps in a synchronous mode. This class of modem provides an extensive array of options, including loopback testing and in many cases data compression. Like the switched modem, this type may have a handset attached for voice communications. Many dedicated modems require C- and D-level conditioning, described in a previous chapter, to operate correctly.

The following sections describe some of the many different types of switched-line and dedicated modems (see Figure 11.2).

Acoustical Couplers

The acoustical coupler is rarely used today, having been replaced by the integrated modem, but some equipment still requires them. These devices allow you to position a standard telephone handset transmitter and receiver into a box with an ear and mouth piece after you have dialed up the computer and have received a carrier tone. They are normally used with some of the older portable computers or terminals such as the Texas Instruments Silent 700. They often are used in a low-speed environment and are limited to 300 bps. Acoustical couplers are designed to use voice-grade dial-up lines only and are normally used asynchronously.

Standalone Modems

Standalone modems, or external modems (as opposed to an internal modem such as those that fit into a PC), are self-contained devices with a standard telephone and the line at-

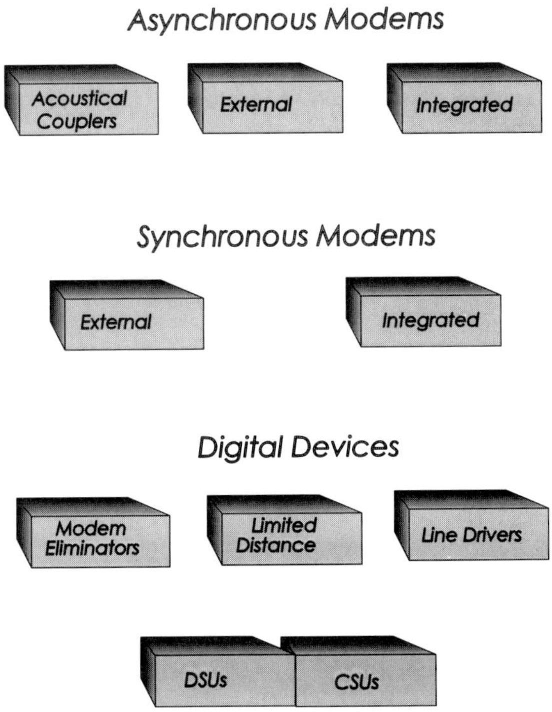

Figure 11.2 Modem Types

tached to it. The computer or terminal device is attached on the other side of the line using an interface cable such as the EIA-232. This type of modem is used with all types of computers.

The standalone modem supports bit speeds of 300 to 28,800 bps (asynchronous) and 1,200 to 64 Kbps (synchronous). No distance limitation exists for this type of modem. They normally support half- or full-duplex transmission and may provide internal or external clocking for synchronous transmission. Standalone modems support both switched (voice grade) and dedicated lines.

Integrated Modems

A dial-up or leased-line integrated modem is built into the terminal or computer. It may be either part of the communications controller or an expansion board that plugs into a PC card slot. The Hayes model 14,400 Smart Modem is an example of an integrated modem used in microcomputers. The integrated modem can be asynchronous or synchronous and can support speeds from 300 bps to 64 Kbps. Integrated modems can operate using either switched or dedicated lines and may have auto-answer/auto-dial capabilities. They often can be provided in customized configurations for synchronized networkwide activities.

Limited-Distance Modems

Short-haul or limited-distance modems (LDMs) are another class of standalone modems that are advisable or necessary when distances between devices requiring interfacing exceed the allowable cable lengths supported by EIA-232-D and other interface cable standards. While this is considered a modem, in fact, it is a digital device. Normally, an LDM is used in large computer environments where distances exceed 500 feet but are less than several miles, depending on the type of wire being used. Generally, private lines that pass through a central office are not suitable for an LDM device because they do not support the wider bandwidth that short-haul modems require. This dictates that most LDM devices be connected using private metallic twisted-pair wiring. Characteristics of the LDM include 1,200 to 64 Kbps rates, a maximum of 50 miles distance (speed dependent), half- or full-duplex operation, internal and external clocking, point-to-point or multidrop, and loopback testing.

Line Drivers

Line drivers are not really modems in the classic sense, but rather are considered to be digital interface converters. They do, however, perform most functions of a standard modem. One area in which they differ is the modulation techniques they use to drive the data across the line. Line drivers convert the digital bits received from the interface cable into a low-voltage, low-impedance signal that is better suited to the twisted-pair media being used.

The characteristics of the binary signal remain unchanged, making the line driver insensitive to the code set or protocol. The principal difference is that they do not normally use the telephone network, requiring dedicated twisted-pair metallic lines. Characteristics of the line driver are a 20-mile distance limitation, half- or full-duplex capability, internal and external clocking, point-to-point or multidrop operation, and dedicated metallic lines only (no loading coils). Line drivers are typically used in factories or office buildings because the wiring is usually easy to install and distances are limited.

Fiber-Optic Modems

Fiber-optic modems are also referred to as fiber-optic drivers. They provide the same capabilities as standard line drivers plus the added benefits associated with fiber-optic technology. Fiber-optic modems usually provide full-duplex high-speed links. These links are plug-compatible V.24, EIA-232-D or RS-422 links that support speeds of up to 50 Mbps and of distances up to 10 km. Another capability of fiber-optic drivers is the provision of asynchronous or synchronous transmission (if synchronous, they will provide clocking signals). In addition, the error rate is less than 10.9 or better at speeds up to 20 Mbps.

The fiber-optic driver is in a constant state of change due to the rapid advancement of fiber-optic technology. At the present time, many fiber-optic drivers operate as direct replacements for devices used with conventional transmission, offering little additional benefit. With the advent of new high-speed fiber protocols and curbside fiber, the fiber-optic driver is quickly maturing and more benefits are being realized.

Modem Eliminators

Modem eliminators are not considered modems at all but interface blocks. A modem eliminator allows hard-wire attachment of computers or terminals to terminal controllers or communications processors over greater distances than allowed in a direct connection using coaxial cable.

The interface blocks can be considered EIA-232-D interface connectors configured for crossover wiring. The transmit and receive leads (pins 2 and 3) and the clear-to-send and request-to-send leads are crossed from one side to the other. For synchronous transmission, additional crossovers are required for pins 15, 17, and 24.

Modem Sharing or Pooling Devices

These devices are also referred to as line splitters. They can be viewed as bridging units that allow several inputs to share a single output. Their primary purpose is to reduce the need for multiple modems at the remote location (see Figure 11.3).

Split-Stream Modems

Split-stream modems, also referred to as multiplexing modems, perform many of the same functions as a multiplexer on a much simpler level. They can connect to multiple DTEs, and multiplex and demultiplex the signals going to and coming from each DTE (see Figure 11.4).

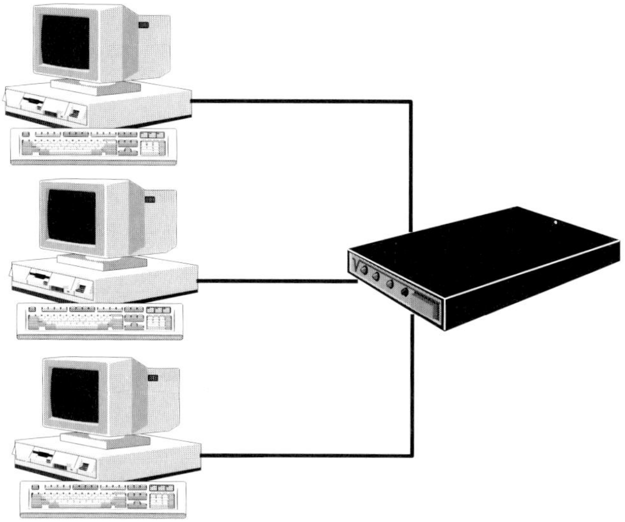

Figure 11.3 Modem Sharing

Extended Circuit/Tail Circuit Modems

Tail-circuit modems are used to capture and extend one channel of a multiplexing modem. They may be used in cases such as multiplexing four channels from one city to another. As an example, two modems having four channels, each with multiplexing capabilities, can be located in different cities. Via the use of tail-circuit modems at each location, any one of the four channels can be extended to other locations. The end points can be in the local area or in a different city (see Figure 11.5).

An extended modem must be able to pass through some of the EIA-232/V.24 standard control signals, including request-to-send, carrier detect, data-terminal-ready, data-set-ready, and the various timing circuits in a synchronous environment. These circuits are used to control tail-circuit configurations, allowing the local terminal full control of the remote modem.

Also available are tail-circuit adapters, which are used to perform the null modem function of reversing the data leads and crossing the control signals as required to attach two modems together. Each modem must be able to transfer or cross the send and receive clocks as well as buffer data for reclocking when necessary.

DIGITAL SERVICE UNITS AND CHANNEL SERVICE UNITS

DSUs and CSUs provide a digital connection for the computer or front-end processor communications port to the telephone company's digital telephone network. The DSU and CSU are combined into one DCE-type device but originally consisted of two independent devices provided by the carriers.

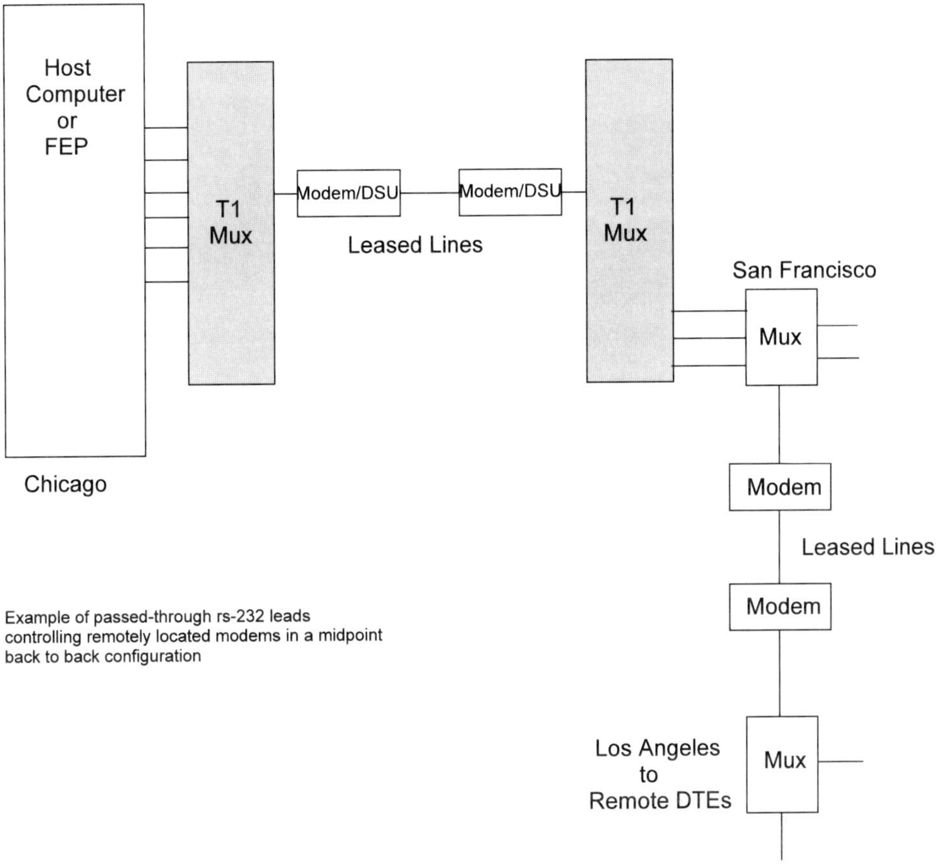

Figure 11.4 Split-Stream Modem

At the beginning of North American AT&T digital service in 1962, the local Bell telephone companies provided user sites with a device referred to as network channel terminating equipment (NCTE). The NCTE provided impedance matching, loop back testing, and diagnostics. The NCTE connected to a DSU supplied by AT&T that provided the digital line interface to the terminal or computer. The NCTE has been replaced by the DSU/CSU device.

The DSU connects to the DTE using one of several interfaces, the most common of which are the EIA-232-D and the CCITT V.35 interfaces. The CSU attaches to a DDS link over four wires using a bipolar digital signal. DSUs and CSUs are designed to support bit speeds of up to 1.544 Mbps and interface to T1 links. CSU/DSU come in two-wire and four-wire versions (see Figure 11.6).

CSU/DSU used for Switched 56 service are more expensive than dedicated DSU/CSU because they can dial. Some units have a full dialpad and the local, long dis-

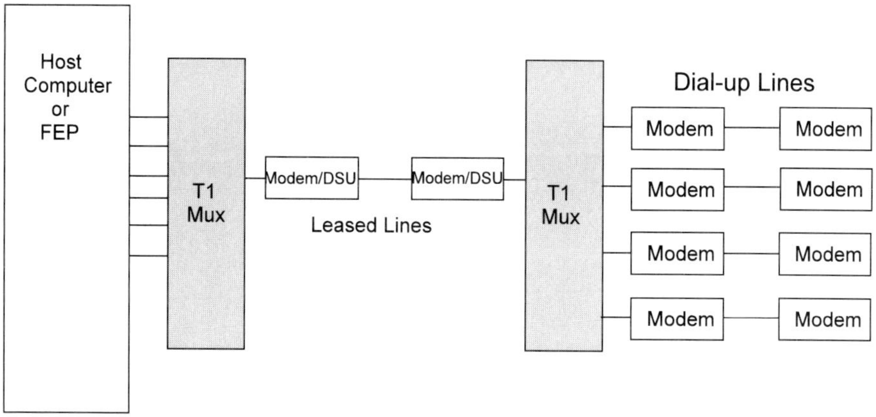

Figure 11.5 Modems Connected Via Extended-Tail Circuit

tance or international number is entered. Other units store numbers in memory and support a speed dial capability. Another dialing alternative is to attach the CSU to a computer and do all your dialing from the computer.

CSU/DSU generally cost between $500 and $2,500. Low-end units may not have a keypad. Instead, they might have just a button for scrolling back and forth until you get to the function you want. Less expensive units may also have limited programmability and storage capacity.

High end units have such features as: a full keypad; an LCD that shows diagnostic and call progress information (status of calls); and plain English commands (as opposed to number codes that have to be deciphered).

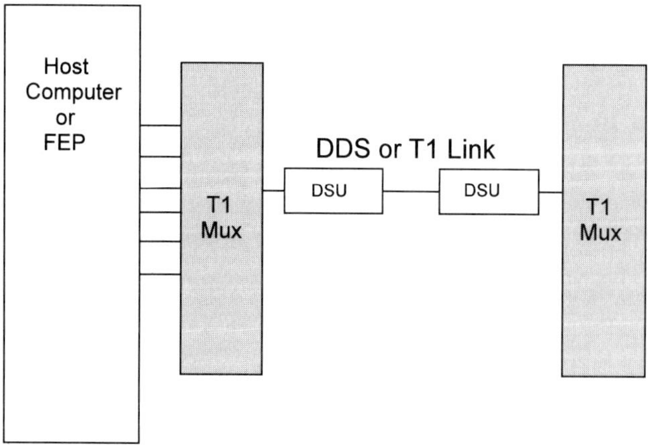

Figure 11.6 DSU's used to Connect Remote Points

DIAGNOSTICS AND TESTING

When a network experiences a failure, the first step in resolving the problem is to isolate and identify it. A series of diagnostic tests are usually provided as built-in functions in most modems to identify interface, line, and transmission problems. These tests can be invoked by the user to ensure the quality of the modem's operations. If the modem is operational, additional testing services may be provided by the various Bell operating companies or AT&T to determine the operational status of the telephone company equipment and hook-ups.

Some modems have a self-test capability that tests all clock, transmitter, and receive circuits in the modem. Additional testing circuits may perform noise and frequency checks. Some of the more intelligent modems have tests to determine if noise levels (signal-to-noise ratios) on the line are within telephone company specifications. They also may check that the modems are using the frequency properly.

Loopback tests are normally provided as part of the modem's diagnostics. The local analog loopback test function tests the local terminal or computer, connecting cable, and local modem; it does not test the interface to the telephone lines (see Figure 11.7). The remote analog loopback test allows the DCE to test the local digital cable, its own digital and analog circuits, the telephone circuits, the telephone interface to the remote modem and the analog circuit of the remote modem. The local digital loopback test allows the DCE to test itself, the interface cable, and the digital circuits of the local modem. The remote digital loopback provides the same functions as the remote analog loopback but also tests the remote modem's digital circuitry (see Figure 11.8).

ERROR CORRECTION

A select class of modems provide error correction. These modems are microprocessor-based and have sophisticated logic, buffers, and memory management. Error detection and correction are usually accomplished using an error-correcting protocol provided by the modem manufacturer, such as the Codex 2270/2280 error-correcting data modem or the Microcom Inc. error-correcting modem using the Microcom Networking Protocol (MNP). MNP is a proprietary protocol that Microcom licenses to other modem manufacturers. A protocol defines the rules for handshaking between two computers, two networks, or from end-to-end. MNP supports computer-to-computer handshaking.

The MNP logic included in the modem supplements the protocol provided by each computer in the point-to-point link. If you were to look at data communications as a hierarchical function, MNP would be placed at the physical layer, which is transparent to other protocols at higher layers. MNP provides up to 11 classes of error correction that offer a 70 percent to more than 200 percent increase in data transmission efficiency. A key feature of MNP protocol is that it is transparent to communications hardware and software activity at the host and remote locations. It provides increased throughput using real-time adaptive data compression techniques and incurs no overhead.

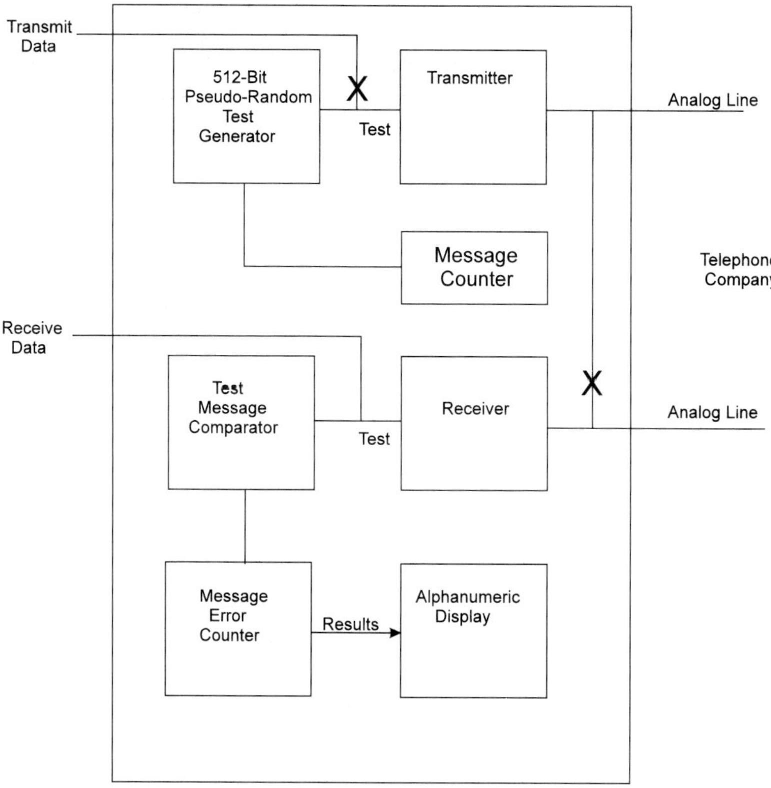

Figure 11.7 Analog Loopback Self Test

Another method of providing error correction is to include a vendor-supplied error controller at each end of the link between the DCE and DTE devices. This type of device uses MNP or CRC for error-correction.

Modems that provide error correction are microprocessor-based and contain sophisticated logic, buffers, and memory management for performing those functions. This is usually accomplished using an error correcting protocol provided by the modem manufacturer. An example of this would be the Codex 2270/2280 error correcting data modem or the Microcom error correcting modems using Microcoms networking protocol (MNP).

Data Compression

All current high-speed dial-up modems support data compression protocols. The sending modem compresses the data on-the-fly (as it transmits) and the receiving modem decompress the data (as it receives it) to its original form.

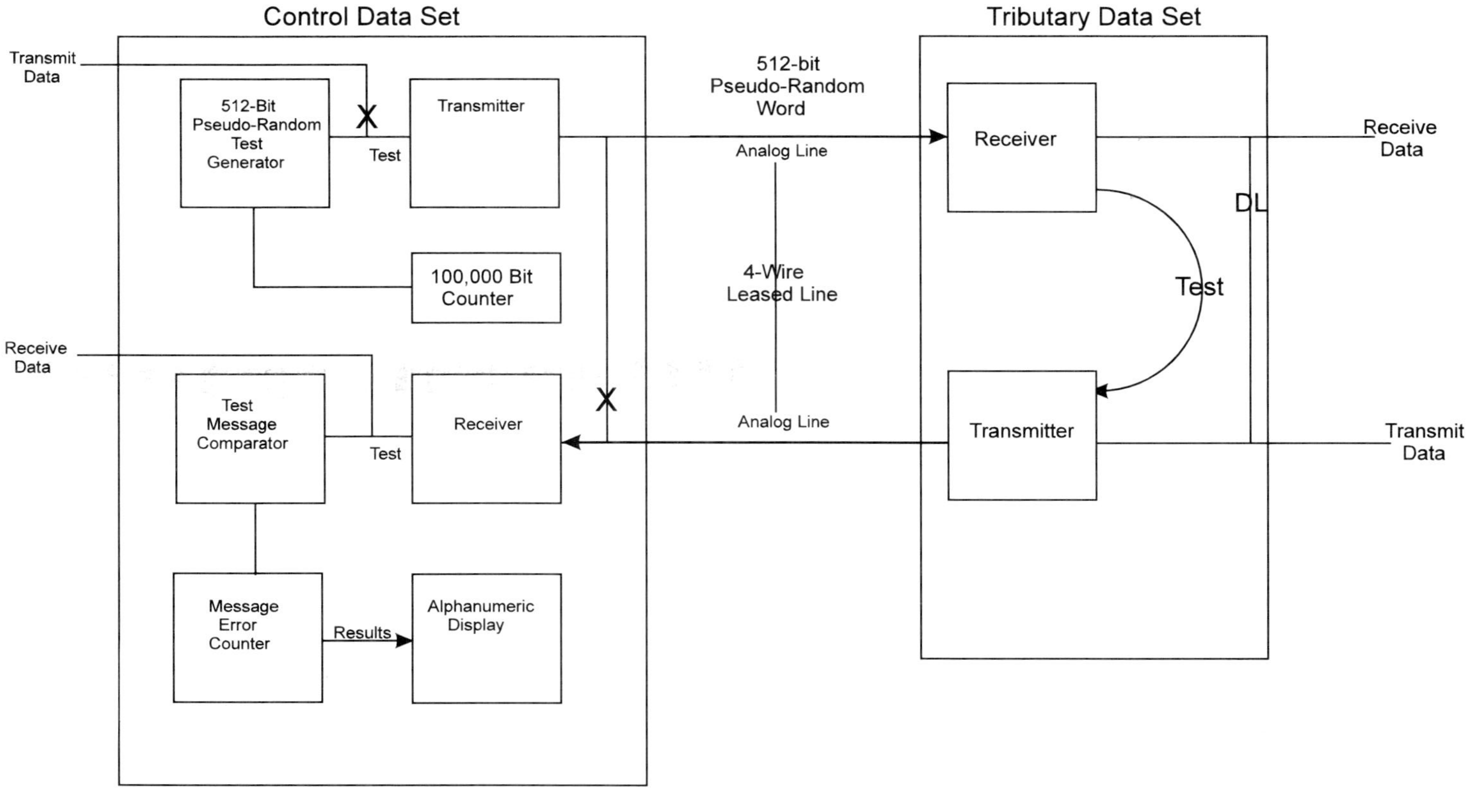

Legend

DL Data Link
MUX Multiplexor

Figure 11.8 Remote Digital Loopback Test

There are two standards for data compression protocols, MNP-5 and CCITT V.42bis. Some modems also use proprietary data compression protocols.

A modem cannot support data compression without using an error control protocol, although it is possible to have a modem that only supports an error control protocol but not any data compression protocol.

A MNP-5 modem requires MNP-4 error control protocol, and a V.42bis modem requires V.42 error control protocol.

Note that although V.42 includes MNP-4, V.42bis does not include MNP-5. However, virtually all high-speed modems that support CCITT V.42bis also incorporate MNP-5.

The maximum compression ratio that a MNP-5 modem can achieve is 2:1, a 9,600 bps MNP-5 modem can transfer data up to 19,200 bps. The maximum compression ratio for a V.42bis modem is 4:1.

Microcom Network Protocol. The Microcom Networking Protocol (MNP) is a de facto standard protocol that provides error correction and data compression in dial-up modems. The protocol's design allows for a broad range of services to be implemented, while maintaining compatibility among modems with different levels of MNP capabilities. For example, a modem capable of MNP Class 5 and Class 7 data compression can talk to a modem that lacks MNP data compression.

MNP is an error correction protocol accepted by international standards authorities (CCITT V.42 standard). MNP offers a reliable and widely accepted method of correcting errors in transmissions over dial-up communications lines. MNP incorporates three different data compression methods, including the CCITT recommendation, V.42bis.

Since its original definition, MNP has evolved through nine classes of enhancements. Of those nine classes, the first four provide error control and are in the public domain. Classes 5 through 7 may be licensed from Microcom. Currently MNP error control (Classes 2, 3 and 4) has been adopted, along with the LAPM protocol, as mandatory elements of the CCITT V.42 recommendation for modem error control.

Class 1. This is the first level of MNP performance. MNP Class 1 uses an asynchronous byte-oriented half-duplex method of exchanging data. MNP Class 1 implementations make minimal demands of processor speeds and memory storage. MNP Class 1 makes it possible for devices with few hardware resources to communicate error-free. Class 1 implementations are no longer included in modems.

Class 2. MNP Class 2 uses asynchronous byte-oriented full-duplex data exchange (i.e., data goes in both directions at once). All microprocessor-based modems are capable of supporting MNP Class 2 performance.

Class 3. This class uses synchronous bit-oriented full-duplex data exchange, eliminating the overhead of start and stop bits used in byte-oriented asynchronous communications. The user still sends data asynchronously to the modem while communications between modems is synchronous.

Class 4. Class 4 introduces Adaptive Packet Assembly and Data Phase Optimization, which both enhance performance. Adaptive Packet Assembly alters the size of the packets in which data is sent between modems according to the quality of the physical link. Higher the line quality allows for larger the packets. Larger packets, while more efficient (the ratio of user data to control data is higher), are also more susceptible to errors. Data Phase Optimization means that repetitive control information is removed from the data stream to make packets more efficient. Both techniques, when combined with Class 3, yield a protocol efficiency of about 120 percent. (A V.22bis 2,400 bps modem will realize approximately a 2,900 bps throughput.)

Class 5. This class implements MNP basic data compression to realize a net throughput efficiency of 200 percent on average. (A 2,400 bps modem will realize 4,800 bps.) Class 5 uses a real-time adaptive algorithm to compress data. The real-time aspects of the algorithm allow the data compression to operate on interactive terminal data as well as file transfer data. The adaptive nature of the algorithm means data compression is always optimized for the user's data. The compression algorithm continuously analyzes the user data and adjusts the compression parameters to maximize data throughput.

Class 6. This class implements Universal Link Negotiation and Statistical Duplexing. The first feature allows a single modem to operate at a full range of speeds between 300 and 9,600 bps, depending on the maximum speed of the modem on the other end of the link. Modems begin operation at a common slower speed and negotiate the use of an alternative high-speed modulation technique.

Class 7. This class implements a more efficient data compression method than the one used in Class 5. The difference between the two classes is that Class 5 realizes an average 200 percent speed improvement over a non-MNP modem, versus an average 300 percent improvement for Class 7. Class 7 data compression uses Huffman encoding with a predictive algorithm to represent user data in the shortest possible Huffman codes.

In addition to Class 5 and Class 7 data compression, MNP also supports V.42bis data compression. Based on the Lempel Ziv data compression model, V.42bis supports an average 400 percent efficiency improvement.

Class 8. Not defined.

Class 9. This class reduces the amount of time required for the modem to perform two frequently-occurring administrative activities: to acknowledge that a message was received and to retransmit information following an error. Message acknowledgment is streamlined by "piggy-backing" the acknowledgment in its own dedicated packet. Retransmission is streamlined by indicating in the error or Negative Acknowledgment Packet (NAK), the order sequence number of each of the failed messages. Rather then sending all the messages over again (even the good ones) from the point of the error, as is usually done with error correcting protocols, only the failed messages are resent.

Class 10. MNP Class 10 consists of Adverse Channel Enhancements that optimize performance in environments with poor or varying line quality, such as cellular telephones, international telephone calls and rural telephone service. These enhancements fall into four categories:

1. Multiple aggressive attempts at link setup
2. Adapting packet size to accommodate varying levels of interference
3. Negotiating transmission speed shifts to achieve the maximum acceptable line speed
4. Dynamically shifting to the modem speed most suitable to transmission line conditions.

An additional method of providing error correction is by including an error controller at each end of the link between the DCE and the DTE devices. This type of device will use MNP or CRC for error correction activity.

INTERFACES

When interfacing modems to the telephone network, it is important to note that most central offices require an operating signal strength level about 12 dBm. Normal line loss between the customer's site and the telephone central office is usually 3 to 6 dBm. The signal level of the modem must be high enough to maintain a good carrier with the central office. Many types of data jacks that plug into the modem are certified for use with the public switched telephone network, as well as for dedicated line use. The following sections describe those most commonly used in North America. Many countries have established their own standards for data jacks.

Permissive Connections

This data jack permits a fixed level of transmission over switched public network facilities. It is designed to limit the line signal strength from the subscriber's site to the central office to no more than 9 dBm. The permissive connection does not guarantee the signal level at the central office, nor does it guarantee line quality. Two-wire circuits use a six-pin modular jack referred to as RJ-11. This is the standard jack found installed in residences within North America.

Programmable Connections

Programmable connections installed at customers' sites provide a means of adjusting the signal level that is sent from the modem to the central office over switched facilities. This type of connection allows for a maximum signal level to be transmitted, and modems requiring this type of connector must be specified by the equipment vendor. The "programmable arrangement" data jack uses a resister to adjust the signal level. The fixed loss loop arrangement output signal level is attenuated to a signal strength of 4 dBm. The local telephone company balances the output of the data jack to the signal level received at the locally attached central office. These connections are known as RJ-41 and RJ-45 data jacks.

Private-Line Interfaces

Modems using private-line facilities require a network termination interface (NTI) device. Each country providing leased-line capabilities specifies the terminating devices permitted for use with its telephone network.

CASE STUDY (CONTINUED)

The Bank of Scotts Valley uses both dedicated-line modems and dial-up modems for its branch operations. The connection of each branch's teller network back to its service center necessitates the use of 19.2 Kbps dedicated-line modems employing both C level and D level conditioning. The ATM network also uses a dedicated-line modem for its communications. But because it requires very little data transfer, line speeds are limited to 2,400 bps. If the primary link fails, the ATM will instruct a standby dial-up modem to establish a new connection using auto-dial capabilities.

The bank has decided not to use digital links between the service centers and their branches because of the low initial traffic requirements and the expense for digital transmission lines. The bank's data communications planning group also intend to install modem-pooling equipment at each of the service centers because traffic is light enough to use this type of modem. As the communications volume grows, the bank will install additional modems to manage the traffic.

As part of the overall communications system, the bank wants to provide several dial-up links that will include 9,600, 14,400, and 28,800 bps asynchronous lines. At the present time, these dial-up lines are available to the programming staff and for backup connectivity to the branches should a primary link fail.

Communications between the front-end processors located at each service center is handled by T1 digital service units.

REVIEW QUESTIONS

1. What are the basic functions of a modem?
2. True or false: The two classes of modems found in use today are asynchronous and synchronous.
3. Why does synchronous data transmission require a clock?
4. Describe auto-dial and auto-answer.
5. What CCITT-type modem would be used in high-speed data communications over full-duplex lines?

6. True or false: Standard modems do not have a distance limitation.
7. True or false: Some modems are limited to half duplex only.
8. True or false: Modem eliminators are not really modems at all.
9. True or false: Security is a function of the computer, not the modem.
10. True or false: Auto-answer will automatically make a call for the user.
11. What are the four loopback tests available for testing lines and modems?
12. Name two diagnostics available with most modems.

CHAPTER 12

Hardware and Logical Interfaces

INTRODUCTION

An interface connects intelligent devices together and provides a means by which they can communicate with each other. Before the development of standards that govern the interconnection of electronic or electrical devices, it was often impossible to connect certain types of equipment together. This was primarily because manufacturers defined their own interfaces and connections, which were often incompatible with other manufacturers' equipment.

When data communications was in its infancy, the typical interface (the cable) for terminals and dataset connections often used a proprietary wiring method. Often, a 20-milliampere current loop wire or other similar low-level wiring technique was used as the interface. Bit rates were limited, and little or no flow control was provided. This permitted equipment interfaces to be quite simple but also created the potential for transmission errors because the modem could lose its carrier and did not have the ability to inform the terminal or computer to stop sending data.

The rapid growth of data communications and the need for reliable, higher-speed devices dictated that a new type of interface be designed. Separate circuits were needed for control and timing activities to create a higher degree of efficiency. These circuits were developed to transport information between the terminal and modem, taking into account timing, status conditions, and carrier availability. Data communications networks consist of many hardware components that require physical and electrical interfaces at different points in the network. Several different types of interfaces were developed during the 1960s to satisfy those requirements. These interfaces have since evolved into standards used throughout the world. The main standards in use today originate from the EIA, ANSI, and the CCITT.

This chapter discusses several standard electrical and physical interfaces, including the EIA-232-D and the V.24 low-speed interfaces and the RS-449, X.21, and V.35 high-speed interfaces.

INTERFACE RESPONSIBILITIES

A node defines an entry and exit point for network communications. Each node in a network consists of one or more intelligent devices, such as computers and one or more modems or DSUs used for the actual transfer activity. For communications to occur between any two nodes on a network, certain activities must occur between the modems or DSUs at each end of the link, as well as between the computer and modem or DSU at each of the nodes. A DTE-to-DCE interface provides a physical and electrical connection for the computer to the modem or DSU and communications line. (DTE refers to a computer or other intelligent device that creates or processes data for transmission across a network; DCE is used to describe a modem or other type of transmission equipment that actually performs the data communications.) Figure 12.1 illustrates an example of a DTE-to-DCE interface.

The interface must accomplish four primary functions for communications to take place successfully:

- Data bits must be transferred between the computer and the modem.
- Control signals must be transferred between the DCE and DTE to ensure that both devices are ready.

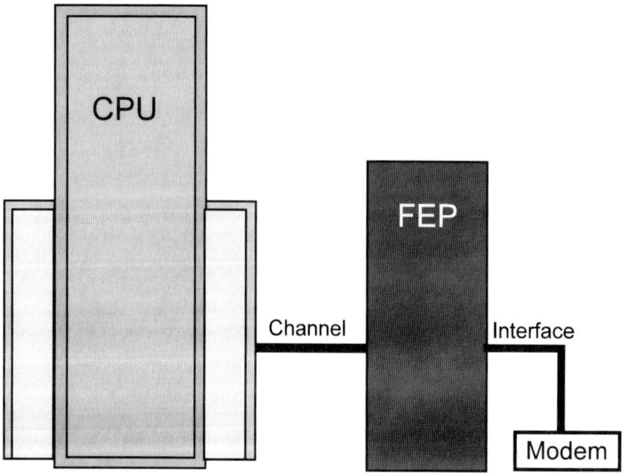

Figure 12.1 DTE-to-DCE Interface

- The transfer of timing signals between the DCE and DTE must occur to ensure synchronization between both devices.
- Electrical grounds must be present.

TYPES OF INTERFACES

The use of a certain type of interface depends on the specific environment that it will be required to support, such as low- or high-speed communications, serial or parallel communications, analog or digital communications, or combinations thereof.

EIA-232-D Interface

The EIA-232-D is perhaps the most common interface in use between modems and computers around the world today. It has been a standard connectivity method for many years. Developed by Bell Laboratories, the EIA-232-D interface is used for most low- and medium-speed communications links. Outside of North America, it is referred to as the CCITT V.24 interface. The interface may be used in a variety of ways; therefore, how the EIA-232-D is used will determine which signals and connector pins are incorporated.

Typical uses of the EIA-232-D interface include:

- Modem-to-computer connections
- Storage device, tape drive and disk drive connections to the computer
- Connections between two microcomputers
- Connection of printers and plotters

The EIA-232-D interface supports half- or full-duplex transmission over dial-up or leased telephone links either asynchronously or synchronously.

The EIA-232-D interface uses a 25-pin connector referred to as the DB25 connector, which supports up to 20 concurrent signals. Figure 12.2 illustrates a DB-25 connector used for the EIA-232-D interface. It may also use a 15- or nine-pin connector because many interfaces require less than 20 circuit signals. Not all EIA-232-D interfaces are wired the same; wiring specifically depends on the devices, the transmission mode, and the computer or modem configuration. The CCITT V.24 and EIA-232-D specifications are identical except for pin usage descriptions. The most commonly used signals and pins are illustrated in Figure 12.3.

EIA-232-D is certified to support a maximum length of 50 feet. EIA-232-D signals are capable of going farther than 50 feet; however, this limit is established by the CCITT and EIA to reduce error rates and ensure integrity. The maximum bit rate supported is 20 kilobits per second, but this depends greatly on the distance the data must travel. Typical voltage levels are –3 volts to +25 volts, with a nominal operating voltage of 12 volts. Table 12.1 lists the comparable V.24 circuit functions and pin designations.

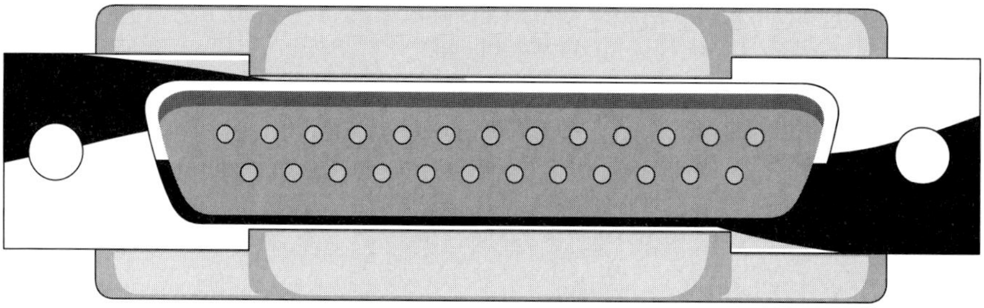

Figure 12.2 EIA-232-D Interface

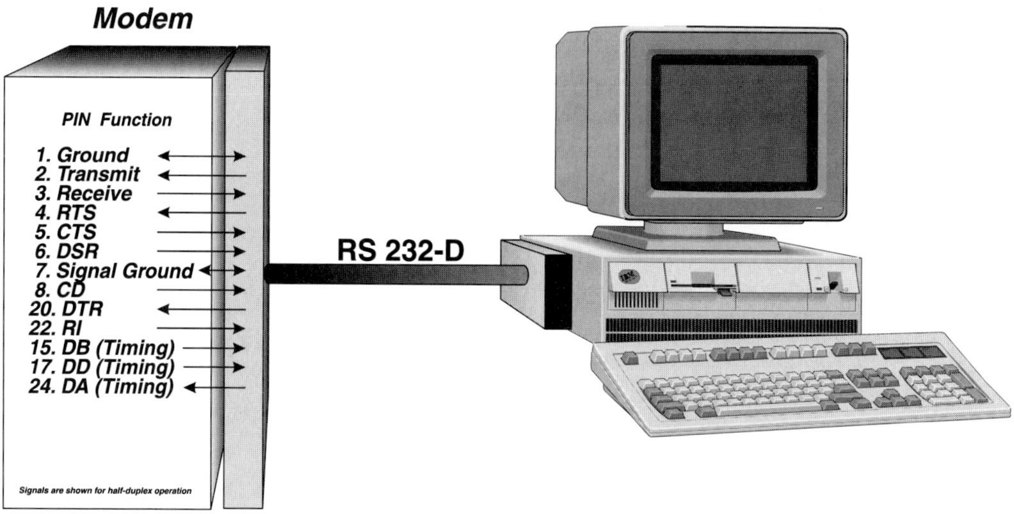

Legend

RTS Request to Send
CTS Clear to Send
DSR Dataset Ready
CD Carrier Detect
DTR Data Terminal Ready
RI Ring Indicator
DB Transmission Signal Timing
DD Received Signal Timing
DA Transmit Signal Element Timing

Figure 12.3 Modem, Computer Handshaking Sequence

Table 12.1 EIA232-D–V.24 Circuit Functions

EIA232-D—V.24 Interface		
Circuit Function	***Pin***	***CCITT–V.24 Designation***
Protective Ground	A	101/AA
Signal to Ground	B	102/AB
Request to Send	C	105/CA
Clear to Send	D	106/CB
Data Set Ready	E	107/CC
Carrier Detect	F	109/CF
Test Mode	K	142/
RX Data A (Received Data)	R*	104/BB
RX Data B (Received Data)	T*	104/BB
RX Clock A (Received Timing)	V*	115/DD
RX Clock B (Received Timing)	X*	115/DD
TX Clock A (Transmit Clock)	Y*	114/DB
TX Clock B (Transmit Clock)	AA/a*	114/DB
TX Data A (Transmit Data)	P*	103/BA
TX Data B (Transmit Data)	S*	103/BA
External Clock A	U*	113/DA
External Clock B	W*	113/DA
Remote Loopback (Test)	BB/b	140/CG
LL Loopback	J	141/
RT Loopback	EE/f	126
Test Pattern	L	125/CE

*Balanced Pairs

EIA-232-D Signal Sequence Phases

Link Activation Phase. This phase creates the connection between the DCE and the DTE, generally at the time the carrier is being established. The following activities normally take place in a standard modem/terminal interface sequence:

• Data terminal ready	(terminal)	indicates to the modem that it is ready
• Data set ready	(modem)	informs the terminal that it is ready
• Request to send	(terminal)	issues an RTS to the modem
• Clear to send	(modem)	responds with a clear-to-send to the terminal
• Carrier detect	(modem)	informs the terminal it has a carrier

The following additional signal is used for dial-up lines:

• Ring indicator	(modem)	indicates to the terminal that a call is incoming

Data Transfer Phase. This phase involves the actual data transfer and acknowledgment between the modem and the computer. This phase includes the following signals:

• Request-to-send	(terminal)	must be on for it to send
• Clear-to-send	(modem)	must be on to receive from the terminal
• Transmit data	(terminal)	indicates that data is being transferred
• Receive data	(modem)	indicates data is being transferred
• Dataset ready	(modem)	informs the terminal that it is alive and well

Link Deactivation Phase. Link deactivation involves the orderly termination of the data link between the modems and terminals. The following signals are required:

• Data terminal ready	Issued by the terminal
• Dataset ready	Issued by the modem
• Received data	Issued by the modem
• Carrier detect	Issued by the modem
• Transmit data	Issued by the terminal

RS-449 Interface

The EIA interface RS-449 was developed as an eventual replacement for EIA-232-D. However, it appears to be slow in catching on, despite the U.S. government's requirement that it be available on all newly purchased equipment. It is important to note that adapting a new interface standard requires a re-engineering of the hardware components to support the new interface. This is often both expensive and time-consuming. Therefore, it may be necessary to support the old interface as well.

The RS-449 interface is a mechanical interface supported by two different electrical standards. This differs from the EIA-232-D, which is both a mechanical and an electrical standard. The electrical standards are referred to as RS-422-A and RS-423-A. These standards are also referred to as the CCITT V.10 and V.11 interfaces, or the X.26 and X.27 interfaces. While the RS-449 standard was designed to replace the EIA-232-D standard, it incorporates many of the EIA-232-D circuits in order to maintain compatibility.

The RS-422-A electrical standard applies to higher-speed native RS-449 usage and uses different circuits and different voltage levels than the RS-423-A electrical standard. The RS-423 a electrical standard is used for interfacing EIA-232-D devices to RS-449 interface device signals and provides compatible electrical signal strength and circuits.

The RS-449 interface uses a 37-pin connector or a secondary nine-pin connector when only basic functions are required. This interface provides much better noise and crosstalk isolation and, when used with the RS-422 electrical standard, provides balanced transmission between each end. The RS-449 interface is functionally equivalent to the CCITT V.10/X.26 and V.11/X.27 standards. RS-449 supports a data bit rate in excess of 2 Mbps.

A maximum of 31 circuits are available, and the interface cable length can be extended to 200 feet. When RS-449 is operating with the RS-422 electrical standard, it provides additional interface control and loopback control functions over the EIA-232-D interface. Figure 12.4 shows the pin assignments for the RS-449 interface. Table 12.2 provides a comparison between the EIA-232-D, V.24, and RS-449 interfaces.

CCITT X.21 Interface

X.21 is a CCITT standard released in 1976 that defines the interfaces for public circuit-switched data networks. This interface provides functions between the modem and terminal using a 15-pin connector. When X.21 was released, the CCITT intended to make it a universal synchronous interface. With the advent of very high-speed communications devices, X.21 is now considered an interim interface until a newer standard replaces it.

The design of X.21 was based on minimizing the number of physical circuits while adding to the interface's functions. The interface uses the data circuits for control when they are temporarily idle and relies on the intelligence that must be built into the modem and the computer.

X.21 supports balanced and unbalanced configurations and consists of circuits for signal ground, common return, transmit, receive, bit timing, and byte timing. Control functions use transmit and receive circuits when no data is present.

CCITT V.35 Interface

The CCITT V.35 interface is used to interface DTE devices to DSU/CSU (digital modems) on wideband circuits. Where the V.24 interface is used for speeds below 20 Kbps, the CCITT V.35 interface establishes the standard for speeds above 20 Kbps. This standard specifies an interface cable that uses a balanced twisted-pair arrangement with

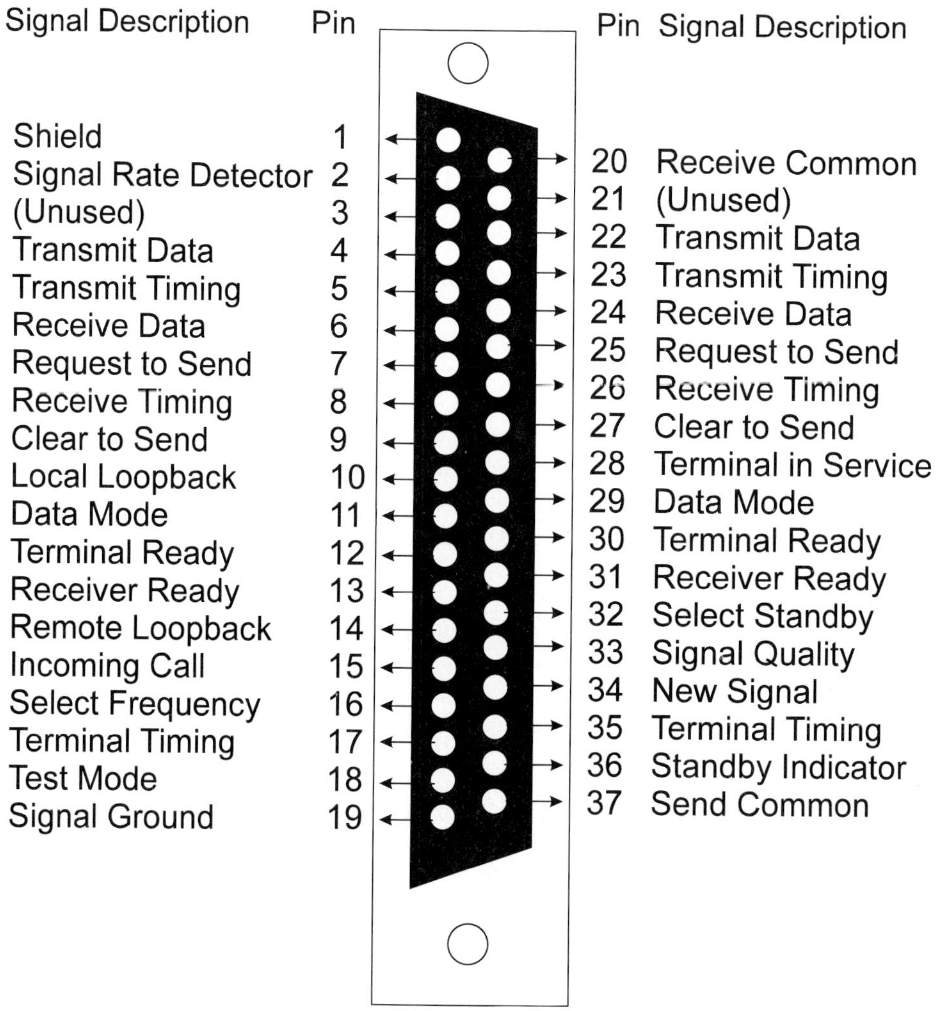

Figure 12.4 RS-449 Pin Assignments

characteristic impedance of around 100 ohms. Differential voltage across the transmitter pair is 0.55 volts.

The V.35 interface uses a 34-pin connector designated as ISO 2593 and consists of balanced pairs for reduced noise, crosstalk control, data transmit and receive, and timing circuits. Its operational range is up to 48 Kbps, and it is used primarily with wideband analog transmission facilities and synchronous terminal communications (see Figure 12.5).

Table 12.2 RS-449, EIA-232-D, and CCITT V.34 Interface Equivalency

RS-449		EIA-232-D		CCITT V.24	
SG	Signal Ground	AB	Signal Ground	102	Signal Ground
SC	Send Common			102A	DTE Common
RC	Receive Common			102B	DCE Common
IS	Terminal in Service				
IC	Incoming Call	CE	Ring Indicator	125	Calling Indicator
TR	Terminal Ready	CD	Data Terminal Ready	108/2	Data Terminal Read
DM	Data Mode	CC	Dataset Ready	107	Dataset Ready
SD	Send Data	BA	Transmit Data	103	Transmitted Data
RD	Receive Data	BB	Receive Data	104	Received Data
TT	Terminal Timing	DA	Transmit Timing (DTE)	113	Transmit Signal Element Timing (DTE)
ST	Send Timing	DB	Transmit Timing (DCE)	114	Transmit Signal Element Timing (DCE)
RT	Receive Timing	DD	Receive Timing	115	Receive Signal Element Timing (DCE)
RS	Request to Send	CA	Request to Send	105	Request to Send
CS	Clear to Send	CB	Clear to Send	108	Ready to Send
RR	Receive Ready	CF	Receive Line Signal Detector	109	Data Channel Receive
SQ	Signal Quality			110	Signal Detector
NS	New Signal	CG	Signal Quality Detector	128	Select Transmit Frequency
SF	Selected Frequency	CH	Rate Selector (DTE)	111	Rate Selector (DTE)
SI	Signaling Rate Indicator	CI	Rate Selector (DCE)	112	Rate Selector (DCE)
SSD	Secondary Send Data	SBA	Secondary Transmit Data	118	Reverse Channel Transmit
SRD	Secondary Receive Data	SBB	Secondary Receive Data	119	Reverse Channel Receive
SRS	Secondary Request to Send	SGA	Secondary Request to Send	120	Reverse Transmit Line Signal
SCS	Secondary Clear to Send	SCB	Secondary Clear to Send	121	Reverse Channel Read
SRR	Secondary Receive Ready	SCF	Secondary Line Signal Detector	122	Reverse Receive Line Signal
LL	Local Loopback			141	Local Loopback
RL	Remote Loopback			140	Remote Loopback
TM	Test Mode			142	Test Indicator
SS	Select Standby			116	Select Standby
SB	Standby Indicator			117	Standby Indicator

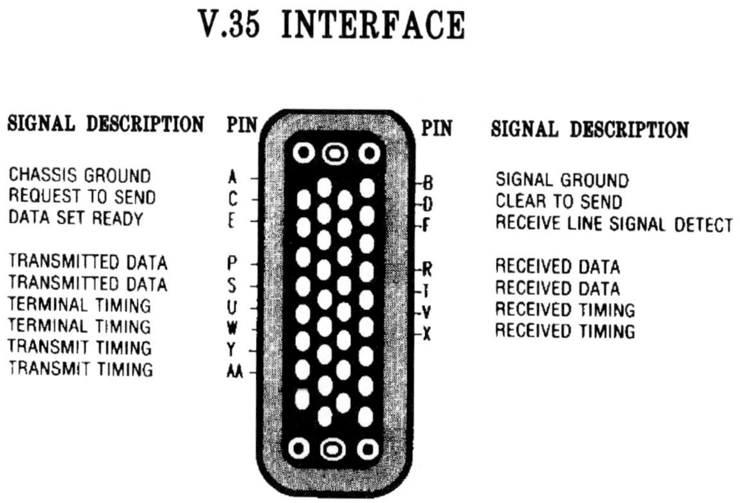

Figure 12.5 V.35 Interface

EIA RS-530 Interface

The EIA RS-530 interface is a new interface that was designed to supersede the RS-449 interface as the RS-449 did not gain wide acceptance with the various manufacturers. It complements the RS-232 interface and has the following capabilities:

- Uses the same DB-25 pin connector.
- Provides greater throughput than the RS-232 interface, up to 2 Mbps.
- Provides bit serial data transfer.
- Is compatible with CCITT V.10 and V.11, X26 and RS-449 standards.
- Is not compatible with the RS-232 interface.

Centronics Parallel Interface

The Centronics 100 interface cable is used primarily for printer connections, but it also may be used for data communications. It consists of a pair of 32-pin connectors carrying 15 pairs of signal wires. The interface cable is made up of braided shielded twisted pairs and can have a maximum length of 25 feet. The Centronics interface is used for the connection of parallel devices and supports bit rates up to 50 Kbps having either positive or negative logic. Its nominal voltage levels are 0 to +5 Volts DC, and it uses Amphenol industrial connectors 57-40360 and 57-30360. Figure 12.6 illustrates the Centronics interface connector. Table 12.3 describes the Centronics interface pin assignments and functions.

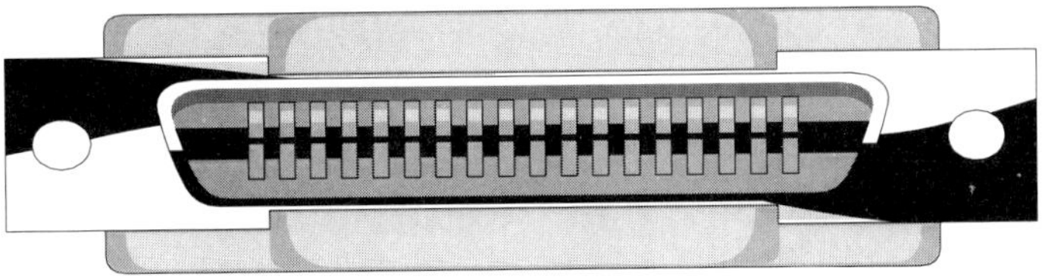

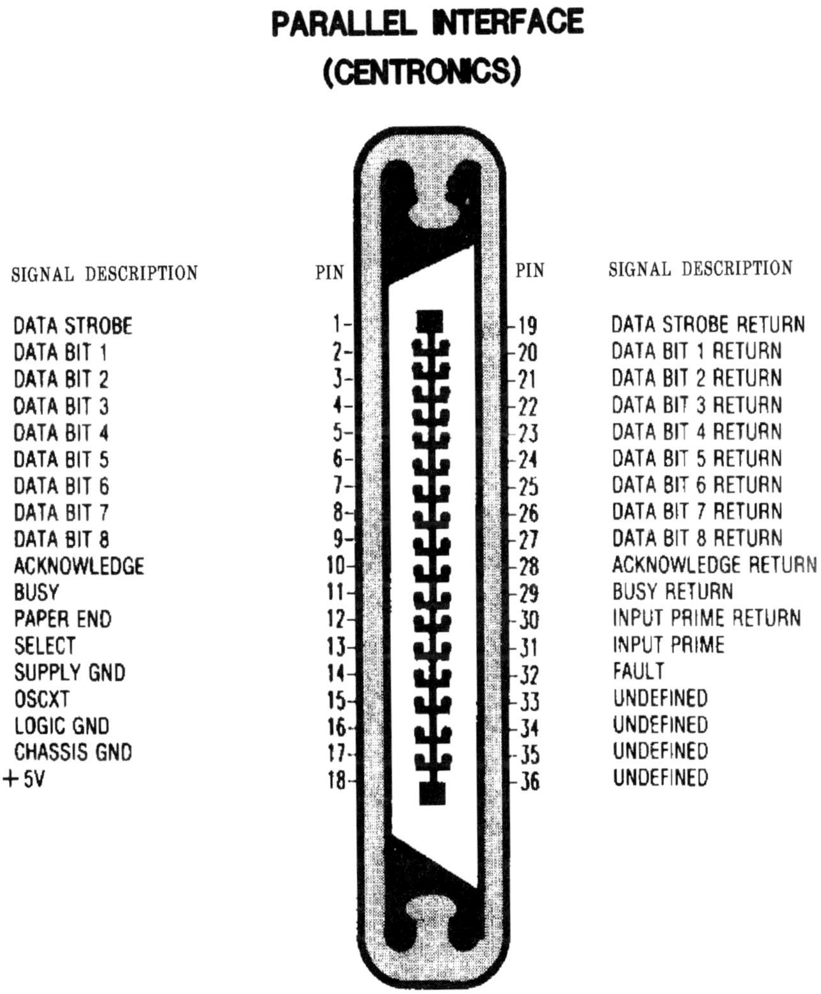

Figure 12.6 Centronics Interface

Table 12.3 Centronics Interface Pin Assignments

Signal Name	Pin No.*	Source	Description
DATA STROBE	1,19	Host	A 1.0 msec pulse (minimum) used to clock data from the processor to the printer.
Data 1	2,20	Host	Data 1–Data 8: Input data levels. A high represents ONE, a low represents ZERO.
Data 2	3,21	Host	
Data 3	4,22	Host	
Data 4	5,23	Host	
Data 5	6,24	Host	
Data 6	7,25	Host	
Data 7	8,26	Host	
Data 8	9,27	Host	
ACKNLG	10,28	3700	Acknowledge pulse indicates input of a character into memory or the end of a functional operation.
BUSY	11,29	3700	Indicates that the printer cannot receive data.
FAULT	32	3700	After warmup, this signal is always ONE.
PE	12	3700	This signal is always ZERO.
SELECT	13	3700	Printer selected; remains at ONE after warmup indicating on-line and no fault condition.
+0V	14,16,35	3700	Signal ground.
Chassis ground	17	3700	Frame ground.
FAULT	32	3700	After warmup, this signal is always ONE.

The following pins are not used: 15, 18, 30, 31, 33, 34, 36.

*Second pin of pair is the twisted signal return (+0V).

Dataproducts Printer Interface

The Dataproducts interface cable is almost identical to the Centronics interface cable except that it consists of a pair of 100-pin connectors carrying 14 pairs of signals. It is made up of braided shielded twisted pairs and can have a maximum length of 25 feet. This cable is also used primarily for the connection of parallel devices. The interface uses the equivalent of a Winchester MRA 50 S D5J connector. It supports bits rates of up to 50 Kbps with either positive or negative logic. The nominal voltage levels are 0 to +5VDC and Amphenol Connectors 57-40360 and 57-30360 are used. Table 12.4 shows the pin assignments and functions of the Dataproducts interface.

IEEE-488 Interface

The IEEE-488 bus standard defines the digital interface for programmable devices and test instrumentation connected to a computer. This interface is an internationally accepted connection method that defines the connection of no more than 15 devices at maximum distances of 60 feet. The maximum bit rate cannot exceed 1 Mbps; actual speed depends

Table 12.4 Dataproducts Interface

Signal Name	Pin No.*	Source	Description
Data 1	B, D	Host	Data 1-Data 8: Input data levels. A high represents ONE, a low represents ZERO.
Data 2	F, J	Host	
Data 3	L, N	Host	
Data 4	R, T	Host	
Data 5	V, X	Host	
Data 6	Z, b	Host	
Data 7	n, k	Host	
Data 8	d, f, or p, s	Host	
DATA STROBE	j, m	Host	A 0.5 msec pulse (minimum) used to clock data from the processor to the printer logic.
DEMAND	E, C	3700	Indicates that the controller is capable of receiving a character.
READY	CC, EE	3700	Indicates that the controller has powered up successfully.
ONLINE	y, AA	3700	After warm up, this signal is ONE, unless the print file is full.
Interface connect	x	3700	Pins x and v provide electrical continuity only.
Verify	v	3700	

The following pins are not used: A, H, K, M, P, S, U, W, Y, BB, DD, FF, HH, a, c, e, h, r, t, u, w, z.

*Second pin of pair is the twisted signal return (+0V).

**Either d,f or p,s may be used for the eighth bit; if VFU is to be used, p, s must be selected. This option is selected through the Interface Setup, Parallel Settings, Interface, and Dataproducts menu dialogues.

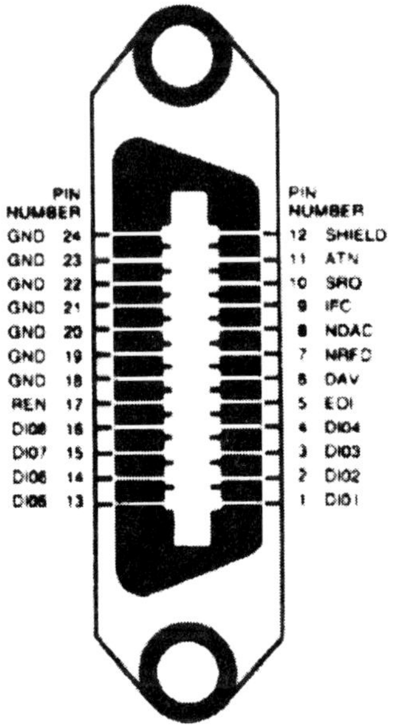

Figure 12.7 IEEE-488 Interface

on the length of the bus. This interface transmits and receives data in a byte-serial, bit-parallel manner between system components. Figure 12.7 illustrates the IEEE-488 Interface.

REVIEW QUESTIONS

1. What CCITT interface is considered to be the industry standard for DCE-to-DTE connections?
2. What is the bit rate for the RS-449 interface?
3. How many connector pins does the EIA-232-D interface have?
4. What interface uses balanced pairs?
5. To what is the V.24 interface equivalent?
6. What are the distance and speed limitations for the EIA-232-D interface?
7. Name the three phases of activities associated with EIA-232-D.
8. Why is the RS-449 interface better suited to high-speed data communications than the EIA-232-D interface?
9. Name the four signal types that are used for the transfer of information between DTE and DCE equipment.
10. What is CCITT V.35?

PART V

STANDARDS, NETWORK ARCHITECTURES, TOPOLOGIES, AND PROTOCOLS

CHAPTER 13

Regulatory Agencies and Standards

THE NEED FOR STANDARDS

A major issue facing the data communications industry and the user community today is worldwide compatibility. Buyers and users of data communications and computer equipment would like equipment from different vendors to be compatible because of their need for global communications. Requirements dictate that all nodes in a network be able to communicate with each other, regardless of who manufactured the products.

In North America and in several other countries, users can select data communications products from a variety of vendors. By choosing products manufactured to the same specification, users can achieve interoperability within and among networks. Unfortunately, this is not the case worldwide because of the different interface standards or software protocols permitted by some telecommunications authorities and licensed manufacturers.

In the early 1970s, users and vendors began to realize that standards were needed for hardware and software interfaces. Until that point, almost every manufacturer built equipment according to its own specifications.

In 1976, the ISO (International Standards Organization) was established to address the worldwide compatibility issue and develop international standards. Over the years, several additional international and national standards organizations were formed to address various areas of concern, such as software and language compatibility, hardware electrical interfaces, media compatibility, communications signal compatibility, and formatting.

Software and Language Compatibility

A major area of importance to the international standards groups is software and programming language compatibility. Software compatibility is required to ensure that control information transported over the network can be understood by each node in the network. This control information consists of addressing, acknowledgment, and network routing. Language compatibility enables users to move software from one type of computer to another. An example is the UNIX operating system developed by AT&T Bell Laboratories. UNIX is rapidly becoming a supported operating system on most computers today. The term "supported operating system" means that a specific computer vendor has adopted UNIX for use on its computers, and ensures interoperability at some level with other manufacturers' UNIX adaptations.

Hardware Electrical Interfaces

Electrical and cable interfaces also necessitated the attention of the standards organizations. Standardizing electrical and cable interfaces ensures that equipment manufactured by different vendors can communicate using the same specifications. These specifications include the designation of maximum voltages for electrical signals, nominal operating voltages, binary encoding standards, and designation of control information.

Media Compatibility

As early as 1964, media compatibility became a major issue evidenced by problems such as the differences in widths of the magnetic computer tapes used by IBM and Honeywell Inc. Tape media compatibility relates to tape width, reel and hub sizes, recording densities, and certification. Another media compatibility concern is disk media interoperability between different types of drives.

Communications Signal Compatibility

Worldwide standards are essential when it comes to communications signal compatibility. Since 1911, standards have existed for ensuring that telephone and communications networks located around the world are able to communicate with each other. The standards for communications signal compatibility include the specification of wire type, voltage levels, carrier frequencies, and bandwidth.

Format Standards

Another area of emerging data communications standards is formatting. Standards are required because not all screens are alike in format, nor are all electronic mail messages the same. In addition, standards are needed so that images and text can be understood by different brands of equipment.

While it appears that major headway is being made to address and rectify the issue of incompatibility, different standards still exist. In many cases, however, the standards can be interfaced by incorporating bridges and gateways. Such adjuncts that mitigate compatibility problems most likely will remain active for quite some time.

INTERNATIONAL STANDARDS ORGANIZATIONS

International standards organizations are voluntary organizations consisting of representatives from the United Nations' member countries. These organizations accept input from members and develop recommendations based on common needs and requirements. While the standards that these organizations develop do not bind the manufacturers to compliance, they do set forth a worldwide recommendation.

International Telecommunications Union (ITU)

The ITU is an international agency of the United Nations established to assist in creating international standards for communications, data processing, and data communications.

Its main area of focus is equipment compatibility and transborder data communications. The ITU has four standards committees: the General Secretariat, the International Frequency Registration Board (IFRB), the (CCIR) Consultative Committee for International Radio (CCIR), and the ITU-T, formerly known as the CCITT.

- The General Secretariat is responsible for the administration of the Union's headquarters.
- The IFRB is responsible for ensuring technical cooperation between member countries on radio frequency assignments. It is also responsible for standards dealing with the use of the frequency spectrum and the positions of satellites.
- The CCIR is responsible for the standards concerning radio communications.
- The Consultative Committee for International Telephone and Telegraph is now referred to as the ITU-T.

The ITU-T is one of the four permanent parts of the International Telecommunications Union, based in Geneva, Switzerland. It is primarily responsible for developing telephone and data communications standards among participating agencies. However, the scope of its work is much greater than just data communications and telephone standards. It also includes telematics, new services, systems, and digital networks such as ISDN.

The activities of the ITU-T divide into three areas: various study groups to set up standard recommendations for telecommunications equipment, systems, networks, and services; plan committees for developing general plans for development of networks and services; and specialized autonomous groups to produce handbooks, strategies, and case studies for supporting developing countries.

Each of the study groups draws up standards for a certain area—for example, Study Group XVIII specializes in digital networks, including ISDN. Members of study groups are experts from government administrations, manufacturing companies, scientific, and other international organizations. All standards developed have to be agreed upon by consensus.

The ITU-T is responsible for the following: the V series which defines analog standards for data communications networks; and the X series that defines digital hardware and electrical interface standards between different manufacturer's equipment.

International Standards Organization (ISO)

The ISO is a voluntary international group chartered in 1946 to achieve worldwide agreement on international standards. Its members are representatives from national standards bodies and interested nonvoting organizations. The main responsibilities of the ISO are:

- Interfacing with national organizations such as ANSI for distribution of proposed standards for worldwide review
- Interfacing with other international organizations such as the ITU-T to decide standards that affect data communications networks
- Assisting the ITU-T in defining the X and V series interface standards
- Defining worldwide standards for procedures, parameters, and formats for data transfer

The ISO comprises groups whose responsibilities include the development of standards over a wide area of technical issues. A standard generally takes seven steps to reach maturity and requires a time frame of approximately four years. These seven development steps are listed below:

1. Creation of the draft proposal
2. Review by the membership
3. Modification of the draft proposal
4. Submission of the second draft
5. Review by the membership
6. Final draft submission
7. Acceptance as a standard

EUROPEAN STANDARDS ORGANIZATIONS

European Computer Machinery Association (ECMA)

In 1961, the ECMA was formed to promote the development of international standards. It has 15 voting members and seven nonvoting members who manufacture data processing and communications equipment in Europe. The ECMA oversees a set of technical committees that make recommendations to the ITU-T regarding data communications equipment.

European Conference of Posts and Telecommunications Administrations (CEPT)

Founded in 1969, this organization's charter is to promote cooperation between post, telephone, and telecommunications providers within Europe. Membership is open to the administrations of European countries that belong to the Universal Postal Union or the ITU.

NORTH AMERICAN STANDARDS ORGANIZATIONS

North American standards organizations include manufacturers and users of data processing and data communications equipment and services. A major goal of these organizations is to develop national standards that eventually will be adopted as worldwide standards. Many of those standards have been adopted by the ITU and its member organizations.

American National Standards Institute (ANSI)

ANSI is a volunteer organization comprising equipment manufacturers and users of data processing equipment and services. ANSI publishes national standards but does not develop them. Its responsibility is to submit to the ITU-T committees proposed standards and suggested changes from its members.

Electronics Industries Association (EIA)

The EIA is a trade association of American equipment manufacturers responsible for hardware and electrical interface standards used around the world for modem and communications equipment. See Appendix A for a list of EIA standards.

Institute of Electrical and Electronics Engineers (IEEE)

The IEEE is the world's largest professional society, with a membership that exceeds 200,000 people. The IEEE is an international organization that was founded in the U.S. The IEEE works closely with ANSI to develop many types of standards in the areas of communications and information processing. Among the types of standards the IEEE is responsible for are LAN standards. See Appendix A for a list of IEEE standards.

National Institute of Standards and Technology (NIST)

The NBS is the former United States federal agency responsible for many areas of standards development. It has been replaced by the National Institute of Standards and Technology (NIST). Many states have adopted NBS/NIST standards as their own in the absence of local testing facilities.

Government Open Systems Interface Profile (GOSIP Standards)

GOSIP is a set of standard recommendations designed to promote compatibility within government environments. It is patterned after the ISO standards and was created to insure that different agencies could communicate electronically with one another.

U.S. Federal Communications Commission (FCC)

The FCC is responsible for rules, regulations, and standards pertaining to telephone equipment and radio frequency transmission as well as for establishing radiation emission standards. The FCC works closely with the international CCIR and the IRFB, as well as ANSI, to ensure that standards are compatible worldwide.

Corporation for Open Systems (COS)

In 1984, when COS was formed, it had a membership of more than 20 manufacturers of computers and semiconductors. COS was created to make some order out of the multitude of existing standards. It accomplishes this primarily by providing a certification service for companies that want to have their products certified as ISO compliant.

Figure 13.1 illustrates the major standards organizations and how they relate to each other.

INTERNET STANDARDS

The Internet is not the responsibility of any single agency however, standards do exist and are constantly evolving. Several voluntary organizations as well as the U.S. government play active roles in establishing and maintaining Internet standards.

While the DDN is in reality a separate network (MILNET) from the Internet, the Department of Defense's Defense Systems Information Agency (DSIA) maintains an important coordination role. It is responsible for maintaining and determining official DDN protocols, architectures, policy guidelines, management strategies, and operations procedures.

Several other groups also share responsibility for establishing standards, maintaining, and enhancing the Internet. They include:

DDN Network Information Center (NIC)

The Department of Defense Network Information Center (DoD NIC) is located at Government Systems Inc. (Chantilly, Va.), and is funded by the Defense Information Systems Agency's Defense Network System Organization (DISA DNSO) to provide general user services to DISN users via telephone and U.S. postal mail.

The DoD NIC works closely with the DISN Host Administrators, Node Site Coordinators, Domain Administrators, Network Administrators, and other service administra-

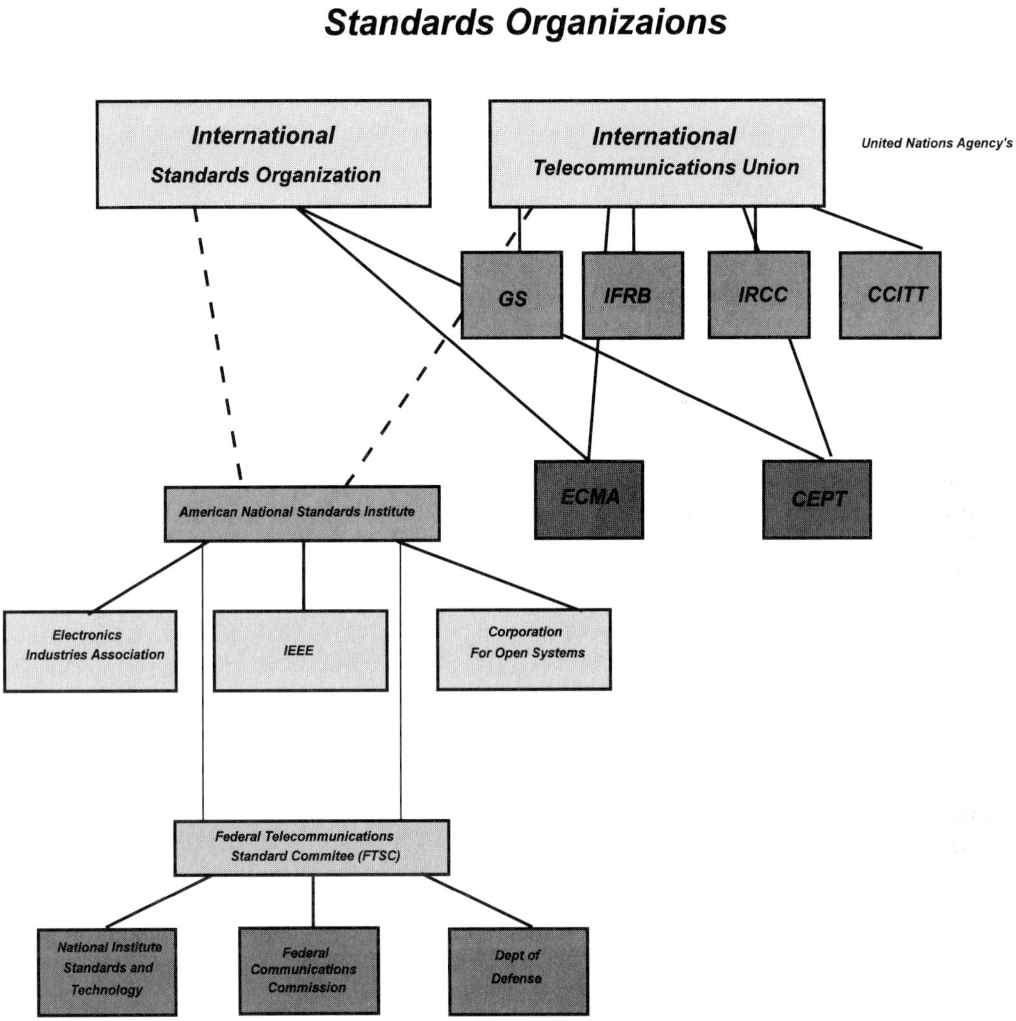

Figure 13.1 Standards Organizations

tive authorities to assist new and existing users in obtaining pertinent network information.

The DDN NIC collects and distributes information about protocols. They also maintain a database of documents that can be obtained by electronic file transfer. One of the NICs major areas of responsibility is the assigning of names to networks and hosts and for distributing blocks of addresses to organizations.

The Internet Society

The Internet Society is a non-governmental international organization for global cooperation and coordination for the Internet, as well as its internetworking technologies and applications. The Society's individual and organizational members are bound by a common stake in maintaining the viability and global scaling of the Internet. They comprise the companies, government agencies, and foundations that have created the Internet and its technologies, as well as innovative new entrepreneurial organizations contributing to maintain that dynamic. The Society is governed by its Board of Trustees elected by its membership around the world.

The Internet Architecture Board (IAB). The Internet Architecture Board (IAB) is a technical advisory group of the Internet Society. Its responsibilities include: governing the TCP/IP protocol suite. Each member of the IAB is responsible for specific working groups and are known as the Internet Engineering Task Force (IETF). Responsibilities include:

1. Internet Engineering Steering Group (IESG) Selection. The IAB appoints a new IETF chair and all other IESG candidates from a list provided by the IETF nominating committee.
2. Architectural Oversight. The IAB provides oversight of the architecture for the protocols and procedures used by the Internet.
3. Standards Process Oversight and Appeal. The IAB provides oversight of the process used to create Internet Standards. The IAB serves as an appeal board for complaints of improper execution of the standards process.
4. Request for Comment (RFC) Series and Internet Assigned Number Authority (IANA). The IAB is responsible for editorial management and publication of the Request for Comment (RFC) document series, and for administration of the various Internet assigned numbers.
5. External Liaison. The IAB acts as representative of the interests of the Internet Society in liaison relationships with other organizations concerned with standards and other technical and organizational issues relevant to the worldwide Internet.
6. Advice to ISOC. The IAB acts as a source of advice and guidance to the Board of Trustees and Officers of the Internet Society concerning technical, architectural, procedural, and (where appropriate) policy matters pertaining to the Internet and its enabling technologies.

IETF

The Internet Engineering Task Force (IETF) is the protocol engineering and development arm of the Internet. The IETF is a large open international community of network designers, operators, vendors and researchers concerned with the evolution of the Internet architecture, and the smooth operation of the Internet. It is open to any interested individual.

The actual technical work of the IETF is done in its working groups, which are organized by topic into several areas (e.g., routing, network management, security, etc.). Much of the work is handled via mailing lists, however, the IETF also holds meetings three times per year.

The internal management of the IETF is handled by the area directors. Together with the chair of the IETF, they form the Internet Engineering Steering Group (IESG). The operational management of the Internet standards process is handled by the IESG under the auspices of the Internet Society. The Internet Architecture Board (IAB) is a body of the Internet Society responsible for overall architectural considerations in the Internet. It also serves to adjudicate disputes in the standards process. IETF responsibilities include:

Management of Internet Documents. There are two types of Internet documents: Internet-Drafts and Request for Comments (RFCs). Internet-Drafts have absolutely no formal status and can be changed or deleted at any time. RFCs are the official document series of the IAB, and are archived permanently (i.e., they are never deleted, and once an RFC is published, it will never change); however, it is important to note that not all RFCs are standards. The Database and Directory Services, part of the InterNIC, provided by AT&T, has an RFC index.

Internet Address Assignment. The Internet Assigned Numbers Authority (IANA) is the central coordinator for the assignment of unique parameter values for Internet protocols. The IANA is chartered by the Internet Society (ISOC) and the Federal Network Council (FNC) to act as the clearinghouse to assign and coordinate the use of numerous Internet protocol parameters.

Secretariat. The Corporation for National Research Initiatives (CNRI) runs the IETF Secretariat with funding from the US government.

DE FACTO STANDARD SETTERS

A de facto standard is a procedure or method that at one time was a specific manufacturer's method of building a product or specifying its connectivity. Over time, this method becomes an unwritten standard (de facto) because the product is seen as an advantageous means of achieving connectivity among different vendors' equipment. Due to widespread use of these methods, they became the standards by which one manufacturer's equipment becomes compatible with another.

IBM

IBM has been instrumental in establishing many of the standards in existence today. Examples are the ASCII and EBCDIC code sets. These code sets are in use around the world and have become both ANSI and ISO standards. Another IBM standard is the Binary

Synchronous Communications protocol (BSC). This early protocol is widely used as a handshaking technique to connect with IBM communications equipment. SDLC, which is the foundation for another protocol supported by the ISO and the ITU-T, is another standard that originated with IBM. Finally, IBM developed the Systems Network Architecture (SNA), a set of standards for connectivity among IBM WANs and LANs.

AT&T Bell Laboratories

AT&T Bell Laboratories has been responsible for many different standards. Among them are modem standards, telephone network interface standards, the original RS-232 standard, and the UNIX operating system, which is becoming an open operating system standard and has been implemented by a large number of computer manufacturers.

Digital Equipment

Digital Equipment, a major computer, software, and communications product manufacturer, has played a role in the creation of many standards as well. Its primary contribution was in the development of ASCII terminal standards and the Ethernet LAN standards, which were jointly developed with Intel Corp. and Xerox Corp.

Xerox Corporation

Xerox, in addition to many printing and document standards, was a major player in the development of the Ethernet standard along with Digital and the Intel Corporation. Xerox continues to lead in the development of standard based methodologies in the areas of open document exchange and flat panel display technology.

The Microsoft Corporation

Over the past decade, the Microsoft corporation has demonstrated its leadership in establishing de facto and official standards for a wide area of technologies. They include: The DOS and Windows operating system environment, the graphical user interface (GUI), object linking and embedding (OLE) open document standards, telephony standards, and software interface standards such as application program interfaces and the Windows sockets interface for communications.

CASE STUDY (CONTINUED)

At some point in the future, the Bank of Scotts Valley will require internetworking capabilities to permit customers such as other banks to access certain types of data. The bank's planners believe that internetwork communications will necessitate an open systems approach because they want to be able to communicate with non-IBM banking networks. Despite this need, they have selected IBM as their principal vendor because of IBM's ability to provide the required hardware and a wide variety of banking software. IBM also meets the bank's requirements for 24-hour support, which can be provided locally.

The planners believe that any internetwork connectivity required in the near term can be satisfied using one of the many available bridges, gateways, and protocol converters that attach to IBM equipment. The major requirement is that all communications equipment support international standards, and IBM can provide hardware and software interfaces that conform to those standards. A system that provides compatibility with other networks using international standards is important to the bank because, at some point, it may want to open a branch in another country.

Another reason the bank wants its data communications activities to conform to international standards is that it will have to exchange information frequently with other local banks and federal agencies. When requested to do so, the bank will exchange information with the U.S. Internal Revenue Service, Social Security Administration, and Federal Reserve Bank.

REVIEW QUESTIONS

1. Name the two ITU-T standard interfaces most commonly used today.
2. What is the ITU-T responsible for?
3. Why was the International Telecommunications Union created?
4. What major contribution has the IEEE made to data communications?
5. To what activity do the X.21, X.25, and X.29 standards apply?
6. What are the two main standards organizations in the world today?
7. Name two organizations responsible for de facto standards.
8. Name a standard that ANSI developed.
9. What does the EIA do?
10. Why are standards necessary?

CHAPTER 14

Network Configurations and Network Design

INTRODUCTION

The design of most networks incorporates a variety of different topologies. The word *topology* refers to the physical design and connectivity of a network. It is important to stress that networks should be designed to meet the needs of the network users rather than based on a specific topology. Many network designers use basic topology concepts to form the basis of a design that meets the network user's criteria. In some cases, specific network topologies must be used to support the technology of the hardware. In LAN configurations, the star, bus, and ring (loop) topologies are frequently used because of the way the networks must be wired and the components required.

The basic function of any network is to provide access paths by which an end user at one location can access end users at other locations. A node can be considered a computer, an intelligent terminal, an RJE terminal, or a network switch. A topology can be defined as the physical arrangement of nodes and links to form a network, including the connectivity pattern of the network elements.

In the past, it was common to find large computer centers set up to support batch environments. Today, we see a combination of batch, online, and distributed environments in use by many companies around the world. These networks are referred to as hybrid networks.

CENTRALIZED, DECENTRALIZED, AND DISTRIBUTED NETWORKS

There are three basic types of communications networks: centralized, decentralized, and distributed.

Centralized Networks

The centralized network is a self-contained entity that manages independent operations, such as those found with a computer system and its terminal network. The computer generally has primary/secondary relationships with each of its connected communications devices. This type of network requires dedicated hard-wired devices that operate using high-speed parallel or serial transmission techniques. The hardware engineering of all components within the network provides native interfaces to other components (see Figure 14.1).

Decentralized Networks

Decentralized organization implies some processing distribution function using a host processor to control a number of remotely located processors. The host processor can off-load activities to other processors, but the host processor still retains control. The normal mode of data transfer is via serial links in a synchronous or asynchronous transmission

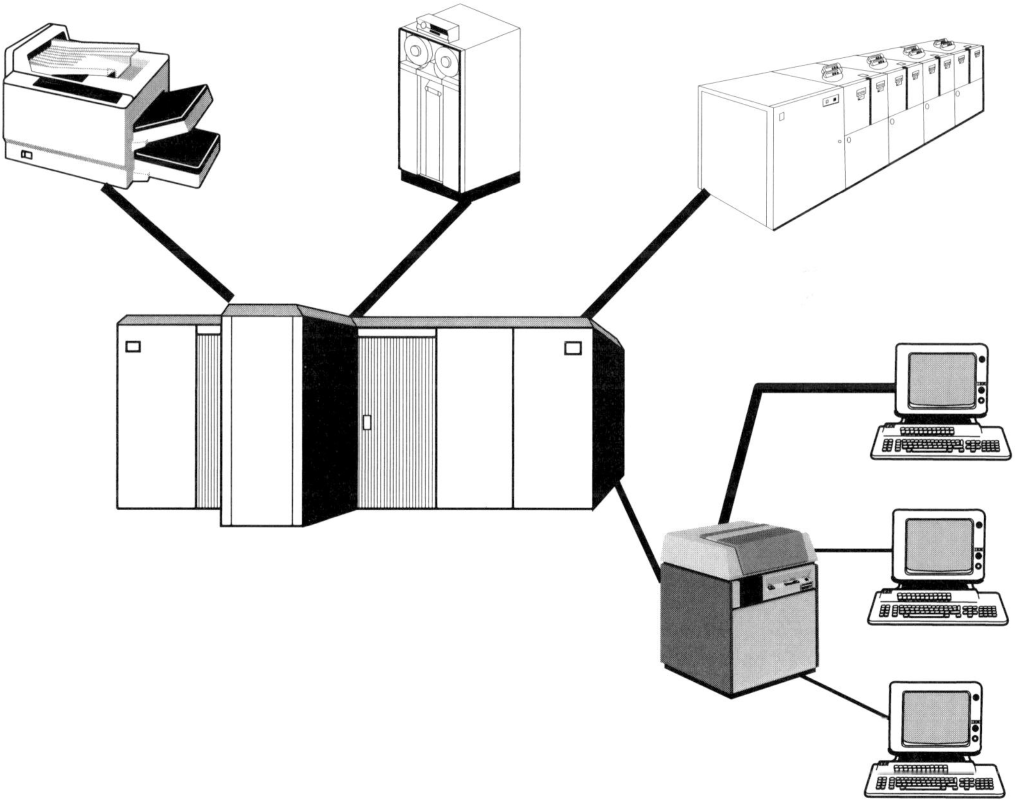

Figure 14.1 Centralized Networks

environment. The host processor and communications processor can interface directly with a public communications network (see Figure 14.2).

Distributed Networks

This type of network is generally considered to be a peer-to-peer network spread over a large area (see Figure 14.3). It usually consists of many different types of equipment, such as computers and communications processors. In distributed networks, all nodes in the network share responsibility for application processing. In many cases, this requires that data format conversion and extensive communications protocol handshaking occur in each node.

POINT-TO-POINT, MULTIPOINT, AND PEER CONFIGURATIONS

Point-to-point, multipoint, and peer-to-peer are terms that describe the methods of connectivity within a network.

Point-to-point configurations involve a primary station and a secondary station operating either in a contention mode or a controlled mode. Contention mode means that both points have equal status and must listen before attempting to transmit. This is illustrated by two computers exchanging data. Controlled mode means that both computers share in controlling the link.

In *multipoint configurations*, a primary station (the host computer) and multiple secondary stations (terminals, workstations, or PCs) function in a selection or polled environment (see Figure 14.4 and Figure 14.5). This requires a communications link configuration in which several terminals share access to the master on a single link. The secondary stations must have enough intelligence to recognize their addresses in a data or information packet sent from the host.

Peer-to-peer connectivity enables computers to establish a peer-level connection. No primary or secondary station is designated when the link is established. This was developed mainly to permit peer-level communications among IBM host computers running under the control of Systems Network Architecture and with non-IBM computers such as those of Digital Equipment and Tandem Computers. Prior to the development of peer-level connectivity, non-IBM computers that required connection to an IBM host were treated as if they were terminal controllers and their activities were restricted to that level.

Some manufacturers have offered peer-to-peer communications for quite some time using a connectivity procedure called asynchronous balanced communications mode. This allows the linked computers to treat each other as peer-level devices.

METHODS OF COMMUNICATING

Two distinct methods for communicating with each network node in point-to-point or multipoint configurations are selection and polling. Selection involves the primary computer sending an "I have something to send to you" packet that instructs the secondary

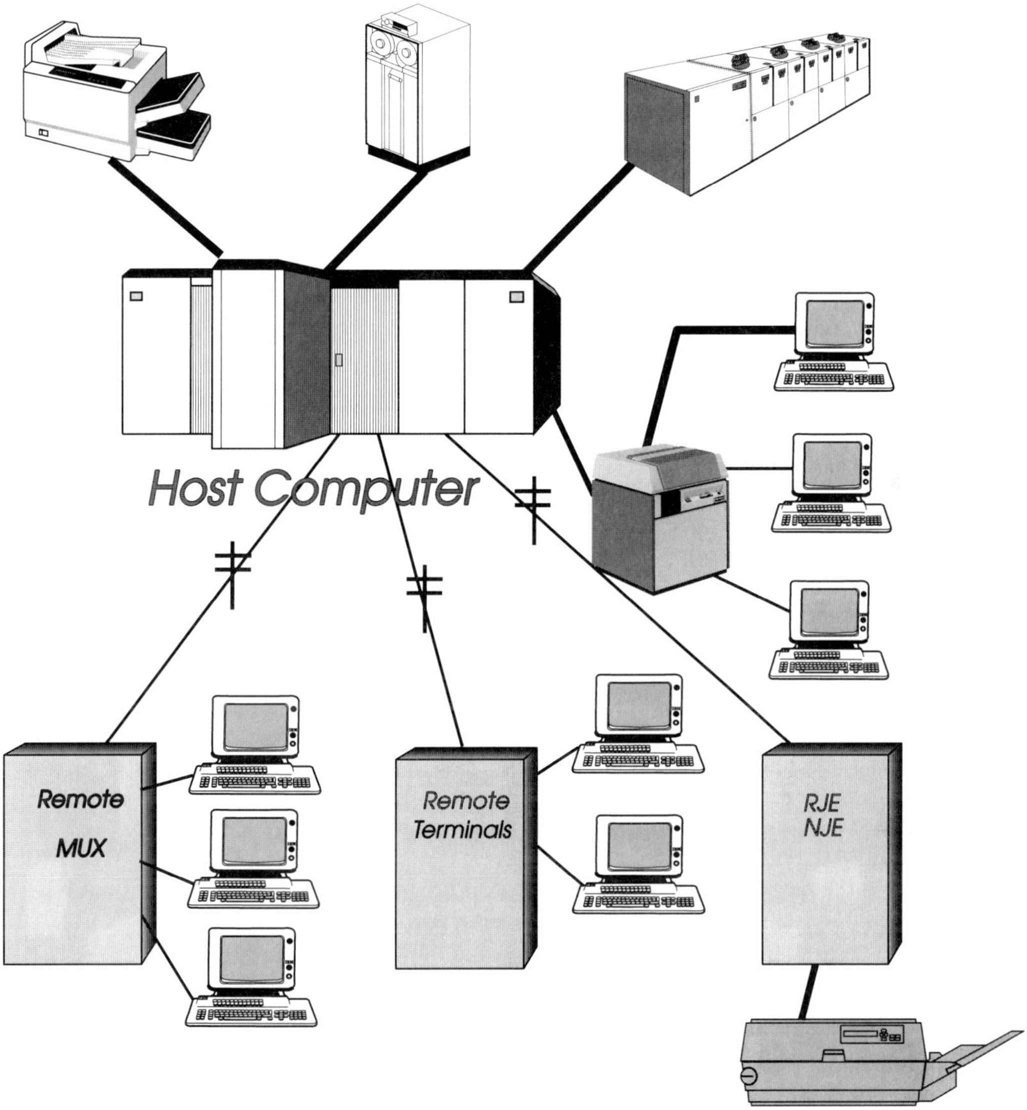

Figure 14.2 Decentralized Networks

computer to get ready to receive. The selection technique requires a controlling point (primary station/secondary station environment) for network control. Another form of selection, broadcast selection, allows the primary computer to address all secondaries at the same time. This selection method is used for activities such as sending user start-up welcome messages and communicating network status to each terminal user.

Polling implies that the primary (master) station will ask each of the secondary stations if it has something to transfer by sending each of them a request. If a receiving ter-

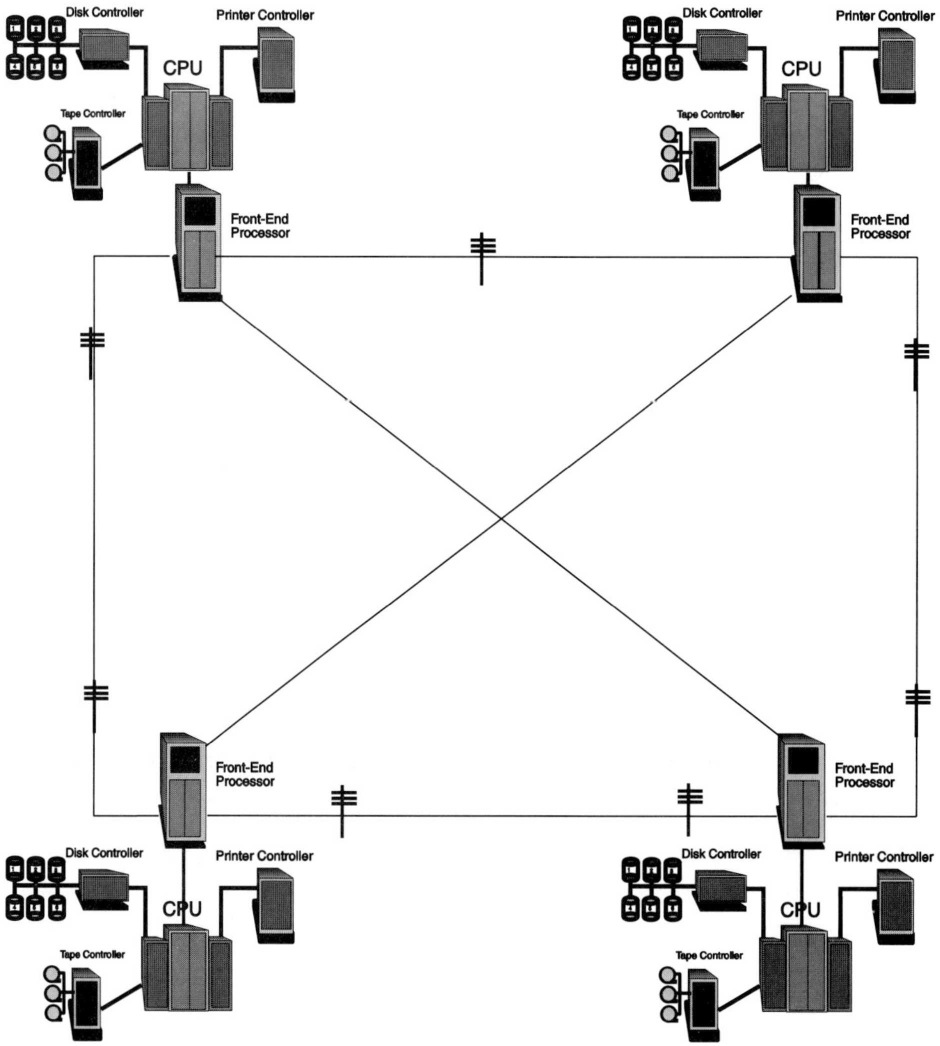

Figure 14.3 Distributed Environment

minal has traffic for the primary computer, it will return a positive response and the computer will establish a dialogue with the terminal.

There are two main methods of polling communications: list polling and hub polling. In list polling, the computer queries each secondary node to see if it has anything to send. It also can prioritize its request to the secondary stations so that some secondary stations can be accessed more frequently. In hub polling, the primary node polls only the first secondary device on the link. When that session is complete, the secondary device

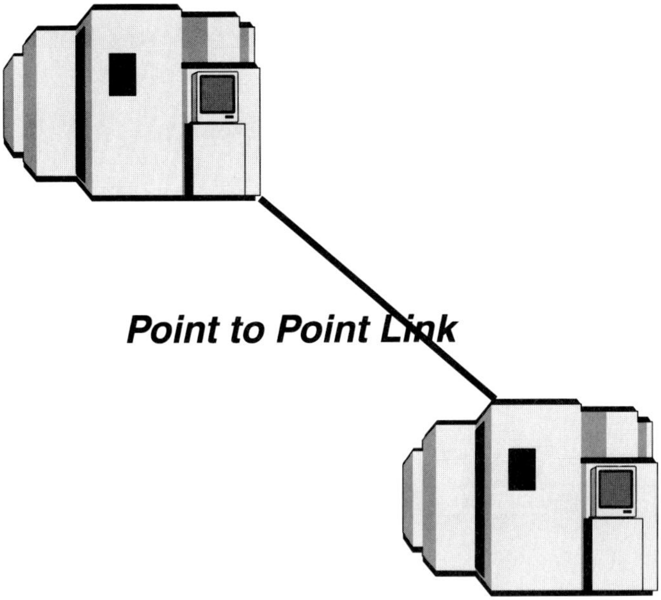

Figure 14.4 Point-to-Point Link

passes on the poll to the next secondary device on the line. Hub polling appears to be more efficient than list polling because of the reduction of message traffic between the primary node and each secondary node in the network.

Selection and polling can be contention based or control based. Contention mode implies that each intelligent device (computer) monitors the link and contends for use of the link. It does this by sending a message to the other devices attached to the link indicating that it wants to use the link. This also may be referred to as a peer-to-peer or balanced arrangement. Peer-to-peer arrangements mean that each computer on the network operates within the network as a peer-level device. Controlled mode means that the network has a primary station, usually a host computer, and a number of secondary terminal stations. The primary node operates by alternately polling and selecting all secondary nodes in the network.

TYPES OF TOPOLOGIES

Star Topology

The star topology is used primarily for connecting communications devices that are considered local (dedicated hard-wired) to the host processor or switching hub. If the star is organized in a centralized manner, it usually will be simple in design and have minimal

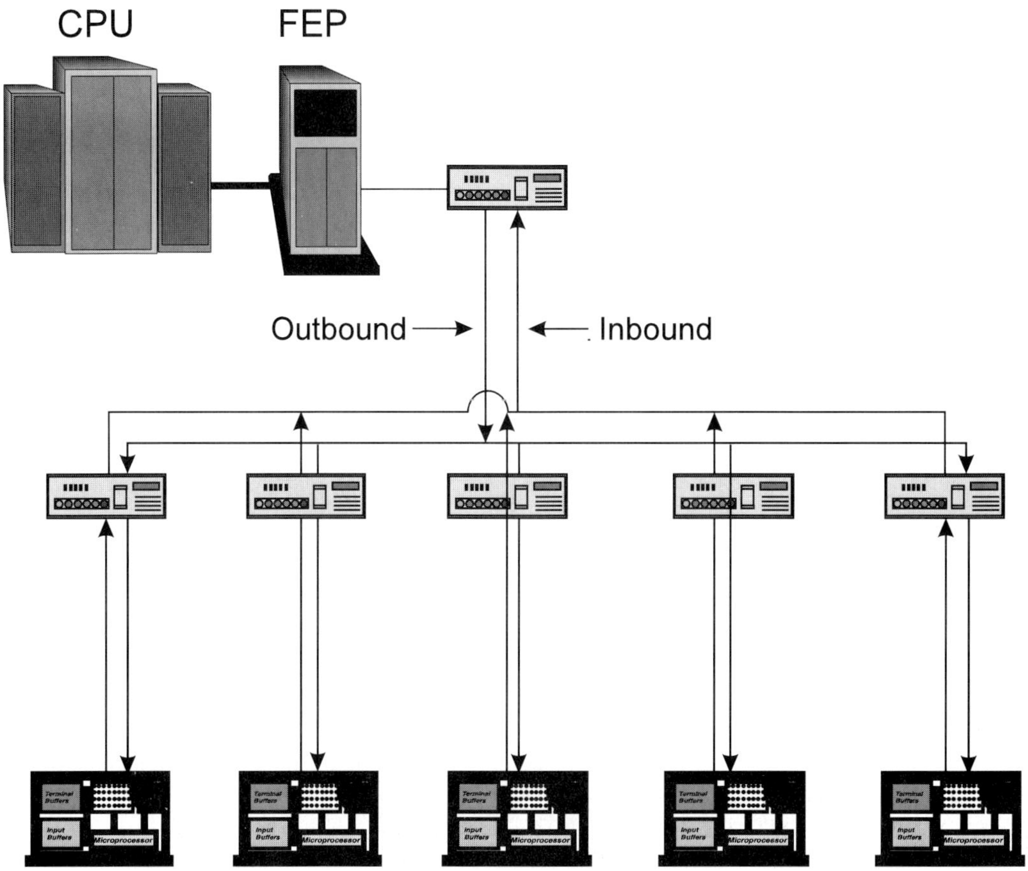

Figure 14.5 Multipoint Link

software requirements (see Figure 14.6). Decentralized stars are more complex in both design and software.

The centralized star topology permits easy routing because the central or hub node maintains the path to all other nodes in the star network. The intelligent central point easily controls access to and from the network nodes. The centralized star configuration also allows for a priority status mechanism to be used by any node when required.

The star usually has dedicated circuits to each of its nodes and will experience few problems. Any contention problems that do occur on the star network are normally resolved by the central node. The central node in the star topology must be very reliable because any failure at this node will bring down the entire network. The connection of each outer node to the central node is accomplished using a point-to-point line configuration. A failure at one of the end nodes would be considered an isolated failure. However, if the

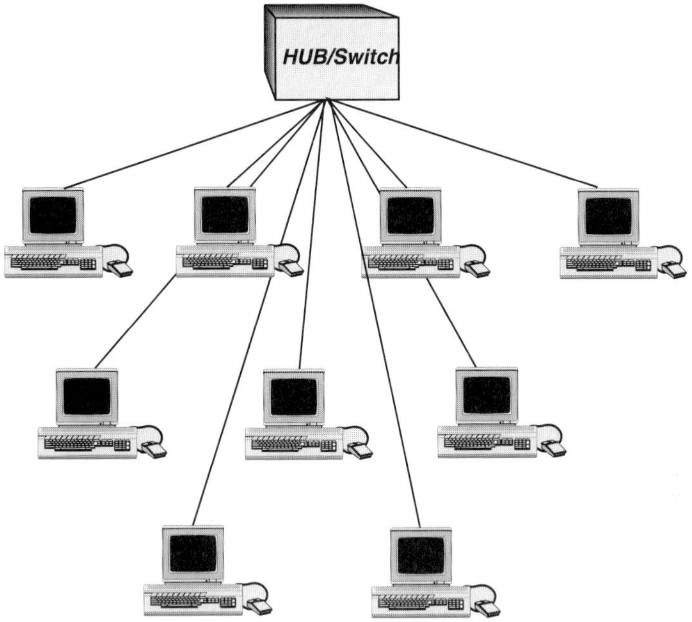

Figure 14.6 Star Configuration

central node were to fail, the entire network would be down, unless a redundant controller and links to each node were established.

The star topology implies the use of polling and selection techniques in a controlled (primary/secondary) communications environment. The star network is normally designed to support limited distances (usually within a mile), and bit transport performance is usually limited to less than 3 Mbps. It is commonly used for the connection of terminals to a host computer. The star topology also is frequently used with LANs in a centralized or decentralized environment. An example of the star topology in a centralized environment is a 10BaseT Ethernet system.

Ring Topology

The ring (loop) is another topology used for creating simple communications networks. It is principally used for connecting nodes in a LAN (generally within a five-mile radius), but it is not limited to the LAN environment (see Figure 14.7).

The ring topology permits an easy routing path because each node receives what is transmitted and decides if the data or information packet is for itself. Rings can be implemented either as a one-way system or as a bi-directional system. Bi-directional systems are the preferred method, if a failure occurs within the ring at some point, the traffic can be routed in the other direction. Many LAN manufacturers use this type of topology for their products.

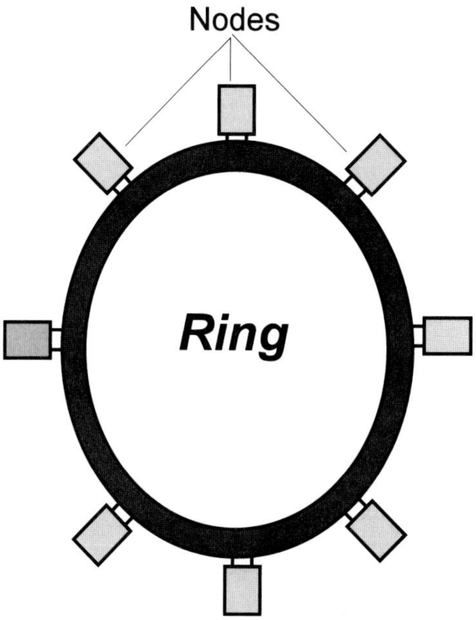

Figure 14.7 Ring Topology

The ring topology uses broadcasting concepts in a controlled/contention environment. Each node operates in a controlled mode until one node wishes to use the ring. The node wishing to transmit then assumes the role of the primary station and broadcasts its data to a specifically addressed station. All stations on the network will receive the data transmitted. However, those stations not addressed will ignore the traffic. While this topology is typically used with LANs, it can be used in other types of networks as well. Additional nodes can be easily connected to the network using current LAN technology.

Bus Topology

The bus topology requires a medium that allows traffic to flow in both directions at the same time. Unlike the star or ring topologies, nodes associated with a bus do no routing at all. This is primarily because the bus is a broadcast medium in which all nodes receive all transmissions (see Figure 14.8).

In many bus systems, all nodes contend with each other for use of the medium, which requires that all devices use a method for avoiding collisions. The lack of centralized control and routing makes the bus topology very reliable because it does not depend on a central node for routing and control. If a node failure occurs, it usually does not disrupt the network, and it is a simple task to add nodes to the network.

As with the ring topology, the bus topology uses the broadcast mode in a controlled environment. Each node operates in a contention mode until one node wishes to use the bus. It then assumes the role of the primary station and broadcasts the data to a specifically addressed station. All stations receive the data, but ignore it if they are not the intended receivers.

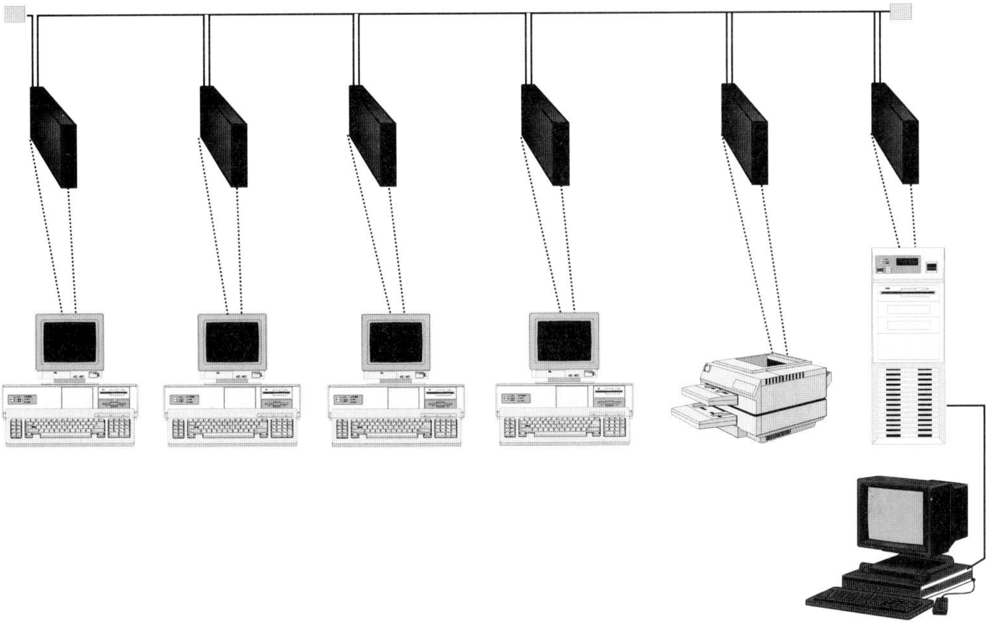

Figure 14.8 Ethernet Bus Topology

Multipoint Bus Topology

The multipoint bus topology consists of a primary or master node and a number of secondary or slave nodes attached to the same media (see Figure 14.9). This topology operates in a highly controlled environment. The multipoint bus master polls or selects the node with which it wishes to communicate.

In a multipoint arrangement, the master invites each of the secondary nodes to send data in a predetermined rotation. No secondary can communicate with the master unless invited to do so. As explained earlier, two main methods of communication in a multipoint polling environment are list polling, where a specific device is polled using its unique address, and hub polling, a form of group pooling. The master also can use a selection technique, whereby the secondary node is informed that the master has something to send to it.

Two subcategories of multipoint communications are multipoint broadcast polling and conference multipoint polling. In multipoint broadcast polling, anything sent by the master is received by all remote stations and anything sent by a remote station is received only by the master. Remote nodes cannot communicate directly with each other. In conference multipoint polling, all stations can send directly to one another.

One major advantage of this topology is that it works well over long distances. Conversely, it can cause long delays because the master must check each secondary station continuously, regardless of the activity.

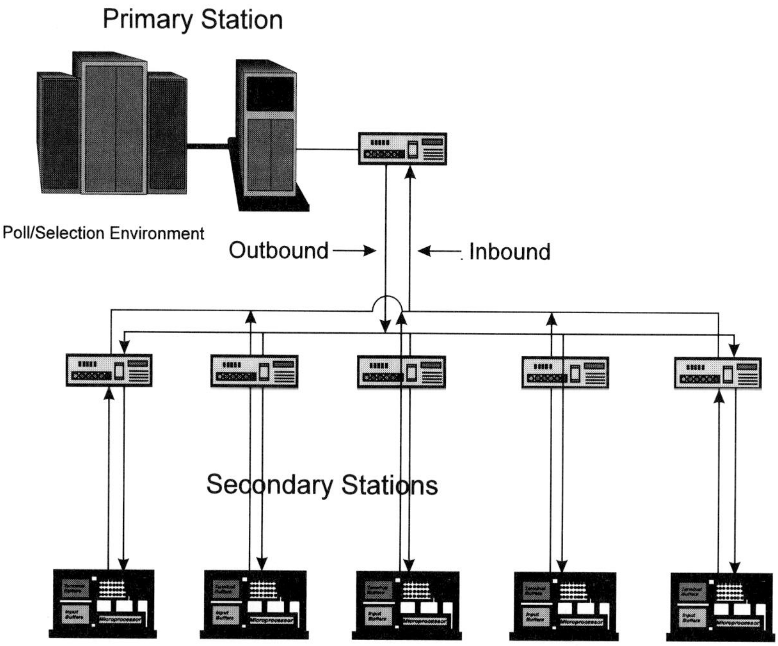

Figure 14.9 Multipoint Bus Topology

Tree Topology

The tree (hierarchical) topology is really just an extended star topology in which several stars are networked together. The tree topology is designed to support different types of configurations at the end of each tree and branch, making it a very flexible networking topology (see Figure 14.10).

The primary environment for the tree topology consists of clusters of end nodes at various remote locations. The tree's branches are connected using point-to-point lines. The tree topology allows centralized as well as decentralized networking, depending on the configuration of the nodes in the network. This topology uses a combination of controlled, contention, polling, selection, and primary/secondary combinations.

Mesh Topology

The mesh topology is excellent for long-distance networking, because it provides extensive backup, rerouting, and pass-through capabilities. Each node can communicate with any other node on the network and also can act as a pass-through node if necessary. This function is needed in the event of a line or node failure elsewhere in the network. The mesh topology is commonly used in large internetworking environments with stars, rings, and buses attached to each node (see Figure 14.11).

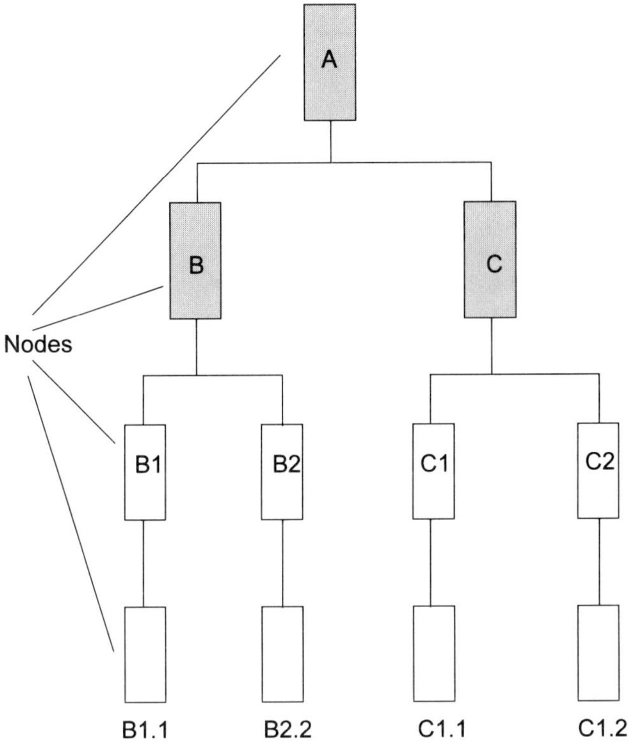

Figure 14.10 Tree Topology

The mesh topology is ideal for distributed networks, because it allows for resource sharing and load balancing by making all nodes available via software control. Resource sharing allows user programs requiring resources on another node to be transparently switched to the other node. Load balancing permits a node to shift program activities to other nodes when its load factor exceeds a certain percentage. Resource-sharing and load-balancing software must be implemented in each node and require the exchange of control and status information.

The mesh topology also is used extensively in packet-switching networks supplied by value-added public data network vendors. With the mesh topology, suppliers can use hardware and software resources so as to achieve the maximum benefit.

Mesh networks can be implemented using a variety of transmission methods that include DS-x links, fractional T1 links, backup dial-up capabilities and voiceband dedicated links. As the costs associated with network communications drop, the mesh topology appears to be one of the more feasible and cost-effective topologies available.

Mesh of Trees Topology

The mesh of trees is a hybrid organization combining both the mesh and tree topologies (see Figure 14.12). The internal network is actually a mesh, and functions as previously

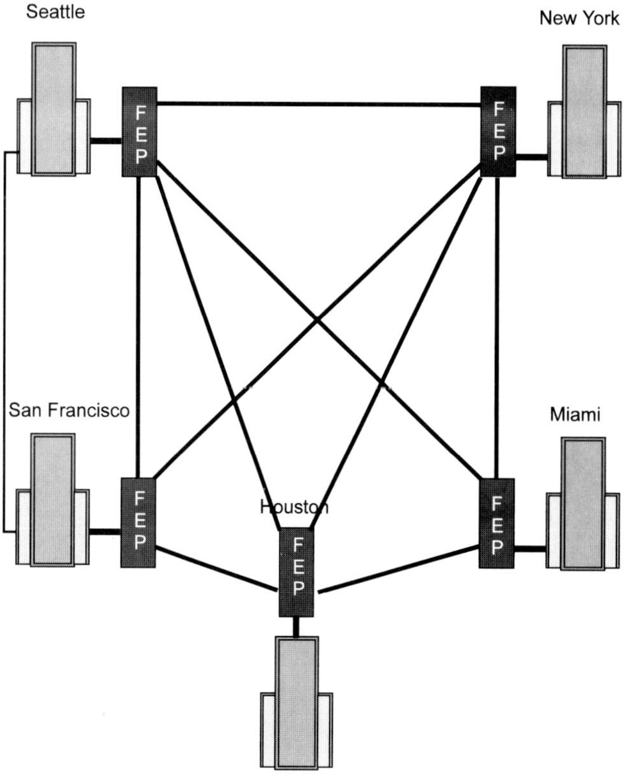

Figure 14.11 Mesh Topology

described. The mesh of trees topology can be considered a distributed network system, allowing any of the end nodes access to the resources of any other node within the mesh via any available resource path. The end user is usually unaware of the actual route taken to make a connection to another node. The public switched dial-up network employs the mesh of trees topology.

NETWORK DESIGN AND SELECTION

Anyone involved with planning the implementation of a communications network must consider the overall design of the network, how it is to be implemented, and the ongoing growth and maintenance of the network. The following sections outline key elements to consider during the design phase.

Computer/Terminal System Resources

The number of planned or required resources plays an important role in the design of the overall network and often determines the types of components that will be used. Another consideration is geographical location: Will the needed equipment be available for use in

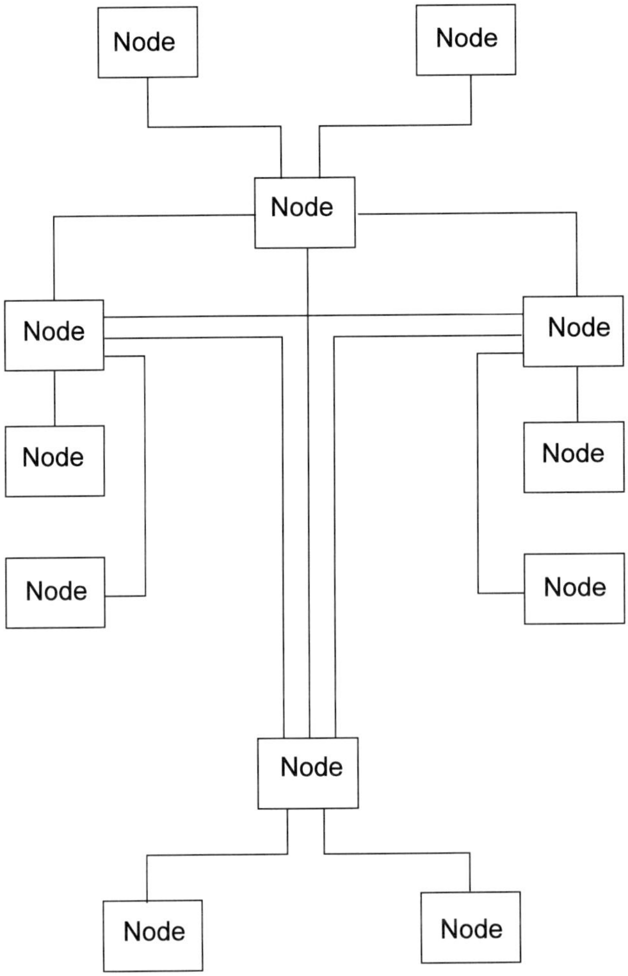

Figure 14.12 Mesh of Trees Topology

all the locations? It is also important to determine support policies before purchasing equipment because the availability of equipment does not always imply support from the vendor. System functions (point-to-point or multipoint operation) need to be explored because support may not be available from the telephone company network. It is important to determine what the data communications equipment requires prior to contracting for link services.

Customer or User Base

When designing a network, consider the number of end users as well as any planned expansion of the network. The location of the end users plays a major role in how the cables will be run, where the equipment is to be located, and how link services will be provided. It also is important to understand the type of user (e.g., customer, employee, etc.) that the network will primarily serve. This is vital information for terminal or computer selection and for network configuration. There is a direct relationship between the type of user, the type of terminal or computer equipment, whether the network is interactive- or transaction-based, and response time.

Centralized and decentralized communications systems provide different levels of support and throughput. If the normal processing requirements are local, a distributed approach may provide the best operational environment. Data and access requirements also play an important role in response time.

Economic Considerations

Two basic premises for the implementation of data communications networks are that unlimited budgets do not exist and that growth is a contingency that must be planned for. It is important to forecast long-range requirements that include accurate, realistic costs. In most companies, growth is a problem because it can vary widely from year to year. To provide some degree of control, try to budget for equipment whose upgrade capabilities are built on a basic frame rather than on replacing equipment as growth becomes inevitable.

Carrier Considerations

To control costs, review public carrier and public network facilities. Do not assume that the telephone or network carrier will have all levels of service. Proper planning involves the consideration of communications link requirements. The availability of switched, dial-up, T1 and DS-x links, and their associated costs, plays a key role in determining whether private, public, or dedicated transmission facilities will be employed. The most expensive aspect of any network activity is the monthly recurring line costs. The expenses associated with maintenance must also be considered.

Staffing for Maintenance and Administration

Staffing is another factor often overlooked in the planning of a communications network, yet it is important from both a budgeting and control standpoint. It may be a mandatory requirement for larger networks in order to monitor and maintain acceptable network performance.

The best-designed systems are as transparent to the end user as possible and are always designed with future needs in mind. It is crucial to isolate the end user from the physical network and not add another layer of complexity to online services. Without proper network planning, growth typically causes major disruptions to data communications activities and also can necessitate additional training programs for the end users. Bear in mind that the typical end user is not concerned with the communications network itself but with its ability to provide adequate access and processing capabilities.

NETWORK INTEGRITY

Network integrity is a key concern to network users and administrators. Integrity ensures that network configurations are well maintained and that only authorized users will have network access. In other words, maintaining the integrity of the network means that changes will not open the way for unauthorized access to network management resources. Network management resources include the communications control software, hardware access, and configuration tools. Simply put, the addition of new nodes, the deletion of existing nodes, and access to applications must be controlled from a central point to ensure that the network will not be corrupted.

Corruption can occur if management is not restricted to a limited number of people and an online management and tracking system is not in place. Online management and tracking allows central control of the network's access and configuration, provides visibility across the network, and enables personnel to perform software maintenance. Online management should include access to information about all attached network hardware, their locations, any authorized applications the nodes have access to, and all authorized users at each node or on the network.

NETWORK SECURITY

In today's internetworking environment, network security is a critical consideration because of the variety of access methods available. Several levels of security must be scrutinized if networks are to be made secure from internal as well as outside forces. To provide a safe and secure network, you must restrict access on the hardware, line, and software levels.

Configuration and Control Software

System maintenance programs must be password protected to ensure that only authorized personnel have access to them. Passwords should be frequently changed as well as properly managed. Often, passwords for employees who leave the company are left on the system for an extended period of time. Procedures must be established and well managed so that security maintenance is performed on a regular basis.

User Access

User access security is of major importance, particularly with respect to dial-up access. Both the system software and application program must incorporate multiple levels of password protection. This can be accomplished by implementing multiple levels of security, including specific application access, read-only access, update access, timed access, and multiple passwords.

Hardware Access

Hardware access security is difficult to implement other than on a key-lock basis or by tying user logon passwords to specific terminals. Another method of protecting hardware is to restrict the terminal or node to a specific set of applications. This requires that the security code be placed in communications control programs to validate logon identification.

Data Security

Data security is primarily a data processing problem, but it can cause concern in a data communications environment. Data transported in a non-encrypted format such as standard binary, ASCII or EBCDIC can be accessed via taps on the communications line. This type of intrusion can be protected against by using one of the many data compression or encryption security techniques available. These techniques operate by scrambling the bit patterns on the sending side and reassembling them on the receiving side. This type of security will work only if intelligence and memory are available at both ends of the link.

NETWORK BACKUP

In organizations that require a high percentage of network availability, communications links, and hardware are possible failure points. Networks do fail, and network backup has always been a thorny issue because of the expenses associated with implementing and maintaining backup systems. Knowledge of which backup system (if any) is appropriate for your network and what your company's budget limitations are is mandatory. Occasional dial-up users probably would have no need for a backup system; however, large, multinational corporations need to include backup capabilities as an integral part of their operations. Several types of backup should be considered, including redundant communications links, standby communications processors, and fault-tolerant processors. Redundancy means that alternate paths are available for the movement of data.

Communications Links

Several methods of backup can be used to guarantee that communications link failures do not bring the network down. The degree of backup required depends on the needs of the user. In a dedicated link environment, a second equivalent or lower-speed dedicated link can be employed as a standby. As an alternative, two links of equal bandwidth can be used so that one can take over priority traffic in the event of a line failure. A third possibility is to use modems that can fall back to a speed supported by the public switched network on dial-up links. Public switched networks such as British Telecom's Tymnet also can provide backup link capabilities for either high-speed or low-speed environments.

For hard-wired networks such as LANs, dual rings or buses are widely used. Satellites and private microwave network users should either duplicate the equipment at both ends of the link, or have some number of dedicated back-up links available. You must exercise caution when considering backup capabilities so that the costs of backup links are justified. The creation of backup capabilities can be a major expense and may not be warranted for some networks. A key consideration is what the potential loss will be if backup capabilities do not exist and the network fails.

Hardware

Hardware components also may fail and require backup capabilities. As with line costs, the cost of the backup hardware must be evaluated against the amount of loss the company will experience if the hardware fails. The most common method for providing hard-

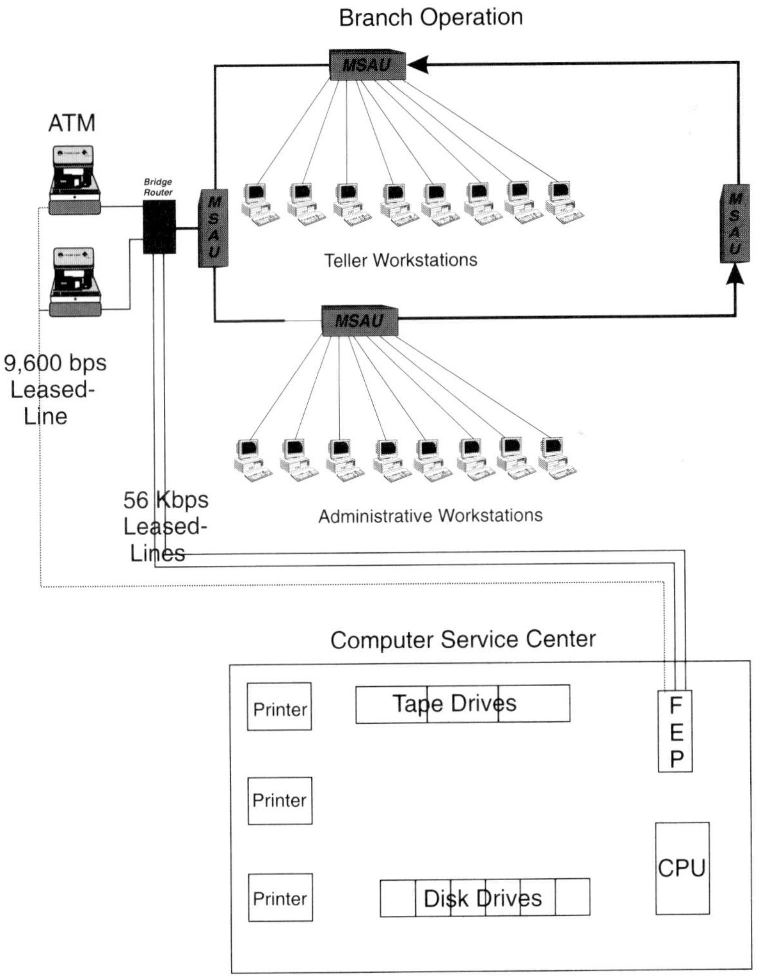

Figure 14.13 Connectivity at the Bank of Scotts Valley

ware backup is by having an identical component that can be used if the primary one fails. Because many modems and multiplexers are modular, spare boards may be all that are necessary. In some cases, a paired front-end processor arrangement is used with a line-switching capability. If one fails, the other one can assume all of the communications functions. In most cases, hardware failures are rare and can be kept to a minimum if time is allocated for preventive maintenance.

CASE STUDY (CONTINUED)

When considering the topological makeup of the Bank of Scotts Valley's network, it was important for the network planning group to identify the overall network requirement for redundancy, security, and backup. A review of these requirements helps to determine the overall network organization.

Within each of the branches, a double ring network is used for the connection of all administrative and teller workstations. The dual ring provides redundancy should the primary ring experience a failure. Each branch is connected using a star arrangement to its service center; the service center acts as the hub for the star. As described earlier, each branch is connected using a primary dedicated link, as well as a secondary backup link.

The service centers are connected to one another using dual high-speed point-to-point links. Additional redundancy is built into the service center connections because the bank uses the telephone company's network, which incorporates a mesh topology, so that service is 100 percent guaranteed. Figure 14.13 illustrates the types of topologies used by the Bank of Scotts Valley.

REVIEW QUESTIONS

1. What are the two basic types of line connections used in network communications?
2. True or false: Multipoint topologies can only use dedicated links.
3. Name the six basic network topologies.
4. What are the two main modes of polling in a multipoint line configuration?
5. Define the differences between polling and selection.
6. True or false: The mesh topology offers one of the better methods for backup and redundancy.
7. Are economic considerations important in the design of a communications network?
8. Name some of the factors that must be considered when designing a communications network.
9. Would a local network require different design criteria than a remote network? Why?
10. Define distributed processing.

CHAPTER 15

Network Architectures

INTRODUCTION

A network architecture identifies communications products and services so that the various components can operate together. Defining and implementing a network architecture was a necessity for many vendors because their product lines could not communicate with each other, let alone with other vendors' products. Several manufacturers have developed proprietary network architectures to address this internal incompatibility problem, including IBM with its Systems Network Architecture (SNA) and Digital Equipment with its DECnet architecture. It is important to note, however, that because they are proprietary, they cannot easily interface with one another.

Compatibility of processing techniques and transmission methods between different manufacturers historically has been a problem, and users have had to pay a large price for networking heterogeneous systems. Often, a company's technical support organization experiences great difficulty when attempting to link systems together that have dissimilar architectures. This was (and to some extent still is) due primarily to the reluctance of some manufacturers to address hardware and communications architectures other than their own.

To provide interconnection capabilities, many manufacturers have developed methods of interfacing their products with other vendors' systems. These interface methodologies generally provide a conversion activity using bridges or gateways (described in an earlier chapter) to manage the different formats. These interfaces make communications between different networks possible. Although this works reasonably well, it is considered a closed systems approach to internetworking and can be costly, because additional hardware and software must be acquired. Another problem with using bridges or gateways is that the conversion process may limit performance. Finally, conversion techniques are time-consuming to implement and to monitor.

A more functional approach to internetwork communications is an open systems architecture, which specifies communications using standard protocols. However, in the early stages of data communications growth, even if a manufacturer wanted to provide an open systems architecture, there was no agreement on what standards to implement. This was due, in part, to the multitude of incompatible national and international standards that existed. It also was due to some manufacturers' desires to sell their preferred networking solutions.

STANDARD ARCHITECTURES

When national and international standards groups began to take an interest in hardware and software compatibility, they reviewed various communications architectures. This review revealed that most hardware and software companies defined layers of functional responsibility for their networks. The functional responsibilities included electrical specifications, software handshaking, and others. These layers of responsibility provided different levels of service to the data being transported. Many architectures have been defined, such as the Xerox Networking System (XNS), DECnet, the Advanced Research Projects Agency Network (ARPANet), and the U.S. Department of Defense Network (DDN). But two architectures, IBM's SNA and the International Standards Organization's (ISO) OSI model, have emerged as the major network architectures.

Both of these architectures segregate functions into layers (levels). These layers form a protocol stack in that each layer's functions can be modified or enhanced without affecting the other layers. Modification can take place as long as a layer's basic functions and its method of interfacing to other layers remain intact. A protocol, simply put, is a "handshake" that can be implemented in hardware components (e.g., the EIA-232 interface) or in software (e.g., one of the layers in the OSI model or SNA). Each of these layered protocols addresses a different networking function.

Today, due to the emergence of international communications architectures, manufacturers provide hardware and software designed to interface with other vendors' products using international standards. Digital Equipment, for example, has stated that its proprietary DECnet architecture is fully compliant with the OSI model.

ELEMENTS OF LAYERED OPERATIONS

Network architectures are designed primarily to address the world of data communications; however, the top layers also support some data processing activities. These upper layers apply to the application process, the presentation format and the establishment of user sessions. The elements of a layered architecture consist of independent levels, each having specific responsibilities and functions. These functions include data transfer, flow control, segmentation and reassembly, sequence control, error detection, and notification. Each layer of the architecture also has a specific responsibility for the data.

Data Transfer

This function allows the transfer of data from one node in the network to another using the transport services of more than one layer. Several data transport options exist, including normal or expedited data flow, packet sizing, segmentation, and assembly.

Flow Control

Controlling the flow of packets throughout the network helps to reduce congestion and the degradation of network performance. Flow control is required because data traffic may at times be unpredictable, and congestion can occur when too many packets are placed on the network at the same time.

Segmentation and Reassembly

In large heterogeneous networks, it is often necessary to segment (take apart) the original packet and construct a number of smaller packets, because many networks do not have the same packet-sizing requirements. Specific layers of the communications architecture are responsible for this segmentation activity, as well as for the reassembly process on the receiving end.

Sequencing

One way to use the network efficiently is to keep all lines as active as possible. However, if multiple routes are available for moving the data to its destination, packets may arrive out of sequence. Packets segmented and sent across different routes take different amounts of time to arrive. To resolve this problem, each segmented packet is given an internal sequence number so that it can be reassembled in proper byte order by the receiver.

Error Detection

To provide guaranteed sequenced delivery of the data packets to their destinations, it is necessary to acknowledge receipt of the data back to the sender. It is equally important to inform the sender that one or more packets were received in error so that they can be resent. When several transmission and reception points exist within a large network environment, error detection and acknowledgment may be required within several of the operational layers of each node.

Notification

To maintain the data transfer activity, segmentation and reassembly, flow control and overall performance of the network, nodes must be able to communicate with each other. These communications use special management control packets and occur between network end points, internetwork gateway points, and from node to node.

IBM SYSTEMS NETWORK ARCHITECTURE (SNA)

In 1974, IBM developed SNA as a five-level architecture. SNA has evolved over a period of time into what is generally considered a seven-level architecture. (Some difference of opinion exists as to the actual number of levels.) SNA appears to closely follow the OSI model, but SNA does not map directly into the OSI model, and many aspects of it are not compatible with OSI.

SNA consists of software and hardware interfaces that permit various types of IBM equipment and software to communicate with each other. The architecture consists of network nodes, physical units, and logical units. Physical units manage the hardware and software needed for communications. Logical units manage the software for communications with end users. To understand how SNA functions, you need to understand the terms that IBM uses.

All physical devices in the network are identified as specific types of nodes. These represent strategic points in the network and consist of different types of hardware with specialized functions. A node can be a computer, communications processor, terminal controller, special processor, or terminal. The types of nodes defined for an SNA network are described below. Figure 15.1 shows the seven layers of the SNA architecture.

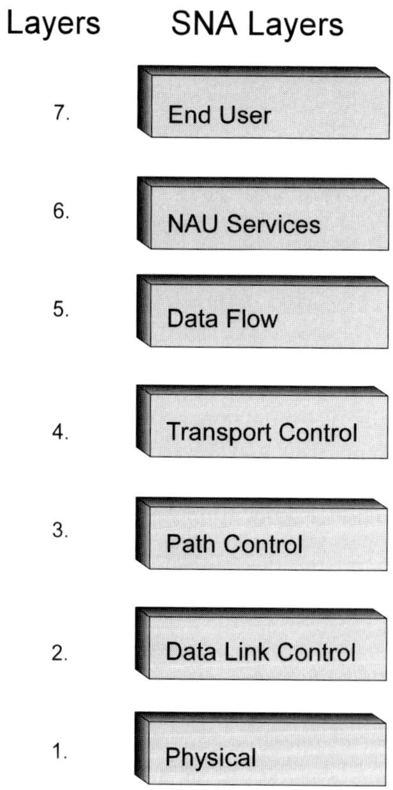

Figure 15.1 SNA Layers

Host Node. The CPU acts as the primary unit and is considered the host node. Each SNA network consists of one primary node and multiple secondary nodes that operate in a point-to-multipoint controlled mode.

CUCN. The CUCN (control unit communications node) controls the remote communications network to which it is attached. If this is a local node, the CUCN is a front-end processor such as the IBM 3745 communications processor. The CUCN also can be a remote device such as the IBM 3725 communications processor.

CCN. The CCN (cluster controller node) provides remote locations with access through the CUCNs to the host node for application services. It can be a programmable intelligent device and can support a wide variety of terminal devices. Typically, the IBM 3x74 is used as a CCN.

TN. The TN (terminal node) provides remote users with access to the host environment and its application programs.

Physical Units

All nodes in the SNA network can be classified as a physical unit (PU) type. A simple explanation of the physical unit concept is that it refers to a device that participates in controlling and routing network communications. Unique PU numbers are designated for different types of devices. PU types (illustrated in Figure 15.2) are as follows:

- PU Type 1 relates to non-SNA terminal devices. This PU type was a later addition to SNA and was implemented to recognize non-SNA/IBM devices and terminals.
- PU Type 2 devices are cluster or terminal controllers, such as IBM 3174, 3274, and 3276 types of controllers. PU Type 2.1 defines connectivity to IBM and non-IBM devices such as the IBM System/3x or the Tandem Nonstop system.
- PU Type 4 describes communications processors either remotely located or acting as front-end processors. It includes the IBM 3725, 3745, and 3765, as well as compatible processors from NCR and Amdahl Corp. running Network Control Program (NCP) software.
- PU Type 5 devices are host nodes such as the IBM S/390 family of computers and are responsible for managing one SNA domain.

Logical Units

Because end users are not identified within the SNA environment, a method was established for end-to-end communications called logical unit (LU) sessions. Logical units are the end-user terminals (e.g., CRTs and printers) or the originating application software located in the host. When an end user signs onto the SNA-controlled network, a pair of LUs are established at each end of the link, which creates a user session.

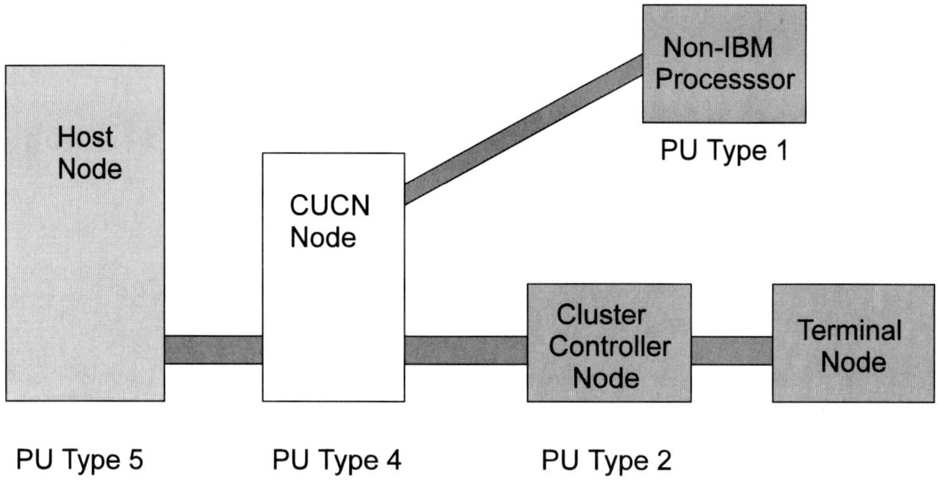

Figure 15.2 PU Types and Nodes

Systems Services Control Points

To establish a session between a user and an application process, all hardware must be identified, LUs must be established, and software must be present to establish and maintain the communications session. The Systems Services Control Point (SSCP) is a system software program resident in the host processor designed to manage user-host communications. In some respects, it can be likened to the systems communications software described in Chapter 3.

Domains

A typical domain is an independent SNA network linked to other SNA networks. A domain consists of a controlling host processor containing the SSCP software and user nodes. A typical SNA network can consist of multiple domains running multiple SSCP software.

SNA also supports intelligent nodes connected for peer-to-peer communications. This is known as LU 6.2 communications. It enables intelligent devices to function on a peer-to-peer level without requiring host activity for communications support functions. LU 6.2 is commonly referred to as Advanced Program-to-Program Communication (APPC) or Low-Entry Networking (LEN). APPC software enables a variety of processors to communicate on a peer basis.

LU 6.2 supports interface functions at the application, session, and presentation layers of the OSI model (described in a later section). Within SNA, the most important application layer facilities supported by LU 6.2 are SNA Distribution Services (SNADS), Distributed Office Support System (DISOSS), Document Content Architecture (DCA), and

Document Interchange Architecture (DIA). These application software systems support office automation and document handling/transfer services.

Many other companies that manufacture computers and data communications products have implemented SNA gateways and offer compatible equipment to provide interface capabilities for users of their products. Gateways are implemented using software and hardware to convert non-SNA data and communications formats into SNA-type formats.

Each layer of SNA's seven layer architecture provides specific responsibilities:

Layer seven, the application/end-user layer, is responsible for end-user activities such as those provided by CICS (communications monitor), TSO (time sharing), CMS (conversational monitor), and IMS (database/data communications) software.

Layer six, the presentation services layer in SNA, offers functions provided by ACF/VTAM (Advanced Communications Facility/Virtual Telecommunications Access Method) communications software.

Layer five, the data flow control layer, and layer four, the transmission control layer, currently handle ACF/NCP (Advanced Communications Facility/Network Control Program) software functions.

Layer three is the network/path control layer. In SNA, this layer is implemented in the front-end processor and remote communications processors in the form of NCP software.

Layer two, the data link control (DLC) layer, is implemented in SNA in the form of SDLC, which manages node-to-node communications.

Layer one provides the pathway for system and application programs to access the communications link. It is responsible for activating, maintaining, and deactivating the physical links. The physical layer also defines the electrical and mechanical aspects of interfacing to a physical medium for transmitting data.

THE OSI MODEL

The ISO OSI reference model is a layered architecture comprising a series of international networking standards known as the X.200 standards (see Figure 15.3). These standards were developed over a period of six years and were completed in 1983. They are published as OSI guideline IS #7498, which defines a common set of rules that specify the way participating networks and nodes must interact to exchange information. Although this standard was designed primarily for interconnection of different networks internationally, manufacturers of data processing and data communications equipment around the world have implemented the OSI standards as the foundation for their communications architectures.

As with IBM's SNA, the OSI specifications are a set of layered standards that provide, via the layers' specifications, communications relationships between the different software modules that can operate within a network node. The intent of the layered approach is to define specific levels of service that can be negotiated between two end

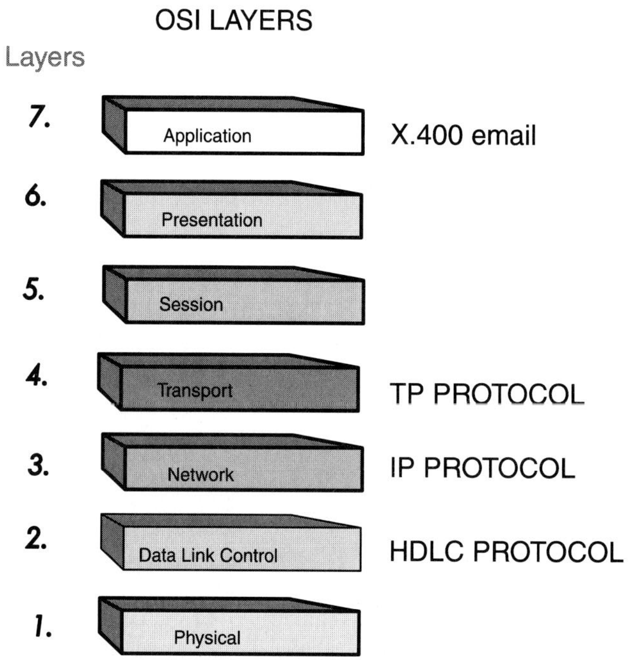

Figure 15.3 OSI Layers

points or nodes in a network. The supporting software modules within each node provide an interface between the higher and lower layers to satisfy service requests.

Starting with the top layer on the sending side, each layer takes data and information received from a higher level, places it into a packet, and provides a value-added service. This is then appended to the existing packet in the form of a header. Lower layers treat the header and data as a single entity referred to as a protocol data unit (PDU). Layers in the sending node negotiate levels of service with layers in the receiving node, establishing a common bond for communications. The receiving node takes the PDU, and then each subsequent layer strips off the header as it passes up the levels (see Figure 15.4).

Layer Seven: Application Control

This layer provides services that directly support user and application tasks, as well as overall system management. Network management, resource sharing, and database management services reside within this layer. Functions of this layer include file transfer, interpretation of graphic formats, and document processing. One activity that operates within layer seven is the X.400 electronic mail messaging service.

Layer Six: Presentation Control

This layer takes the transmitted data and translates it into a format that will enable it to be processed by the receiving system and displayed on terminal CRTs and printers. Func-

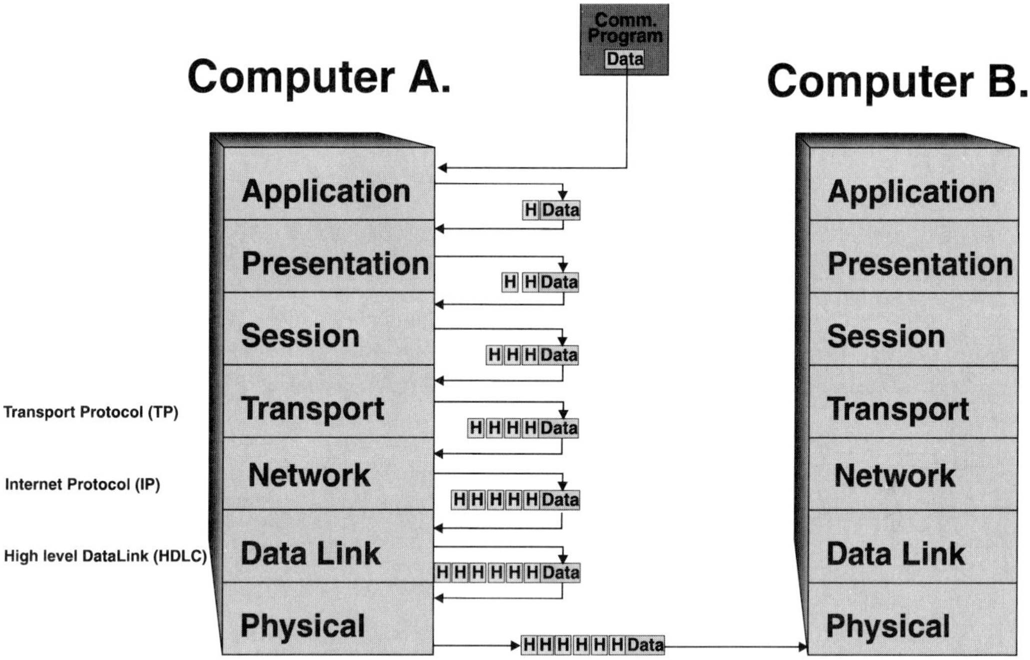

Figure 15.4 OSI Packet Construction

tions of this layer include data formatting, code set conversion (e.g., ASCII to EBCDIC), text compression for higher throughput efficiency, and encryption services.

Layer Five: Session Control

Layer five formats the data for transfer between end nodes of the network. This layer also provides session-to-session restart and recovery, and is responsible for maintaining the end-to-end connection until a termination request is received from the user or the system.

Layer Four: Transport Control

Layer four is responsible for end-to-end integrity and control of a communications session between network nodes. It accepts data from the next higher layer (session layer) and passes it to the network layer. Two protocols have been developed to provide transport services: transmission control protocol (TCP), which is not an official OSI protocol, and the four levels of the OSI Transport Protocol (TP).

Layer Three: Network Control

The network layer addresses the messages and transports them to their destinations by setting up the required paths to various communications nodes within the network. The network layer is responsible for message routing, error detection, and control of internodal traffic. The OSI Internetwork Protocol (IP) and other protocols provide services at this layer.

Layer Two: Data Link Control

The data link control layer establishes the communications link between devices over the physical channel. It is responsible for framing, error control, time-out levels, and data formatting. It also maintains sequence control for frames on the communications link. OSI uses the HDLC protocol for these services.

Layer One: Physical Control

Layer one is the physical link layer that defines the electrical and mechanical aspects of the interface to a physical medium for the transmission of data, as well as the initial setup. This layer includes the communications hardware and software drivers.

CCITT X.25 STANDARD

The CCITT X.25 standard defines a DTE-to-DCE interface and network. This standard closely follows the lower three levels of the OSI seven layer model. "Closely follows" means that it follows the concepts of OSI rather than maps directly onto the model (see Figure 15.5). In fact, X.25 was conceived prior to the development of the OSI model and had been implemented by a number of organizations in Europe before OSI became a standard.

X.25 is made up of three layers. Layer three is referred to as the packet layer. This layer is responsible for transporting data across the network to its destinations. It sets up the paths between various communications nodes within the network and is also responsible for packet routing and for the control of internodal traffic. The Internet protocol (IP) and other protocols provide services at this layer.

Layer two is the data link control layer. The DLC layer establishes the communications link between devices over the physical channel. It is responsible for framing, error control, time-out levels, and data formatting. It also maintains sequence control for frames on the communications link. X.25 uses the Link Access Protocol (LAP) and the Link Access Protocol-B (LAP-B), which are identical to the HDLC protocol for these services.

Layer one is the physical link layer that defines the electrical and mechanical aspects of the interface to a physical medium for the transmission of data, as well as the initial setup. This layer includes the hardware and the software driver for each communications device.

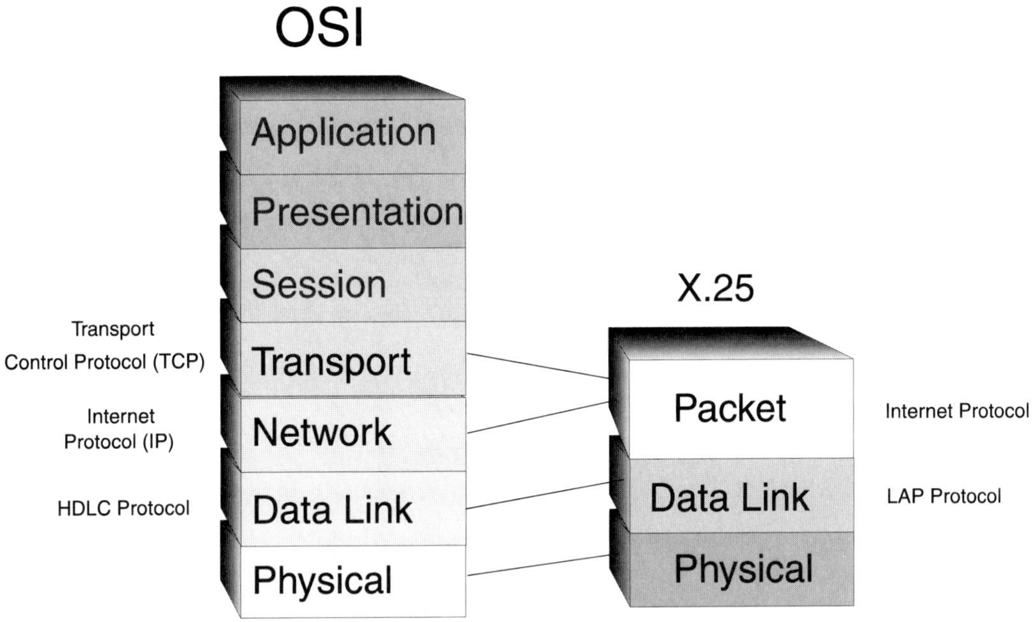

Figure 15.5 X.25 layers as they Relate to OSI Layers

ARPAnet / INTERNET

Before there was an Internet, there was the advanced research projects authority network (ARPAnet). ARPAnet was a network funded by the U.S. government and represented the beginning of communications between networks. This took place in 1975 when the Department of Defense funded the connection of four computer centers located at Stanford University, Cornell University, MIT, and the University of Illinois. The objective of this exercise was to allow the four computer centers to provide access to information at each other location to their users. Of course, it helped that all four centers were running the same software. ARPAnet was a valuable resource and grew at an accelerated rate because of the richness of the information base.

By the 1980s, the ARPAnet was a large connected network of government, military, manufacturing, and educational networks. In 1985, The ARPAnet name changed to reflect the diversity of the network. It became the defense advanced research authority network (DARPAnet). In 1992, the DARPAnet split into two separate networks because of the nature and its size. All military communications would migrate to the MILnet and all others to the Internet (see Figure 15.6).

Today the Internet is a very large network that is made up of hundreds of thousands of other large and small networks, each having one or more users and linked together electronically by the magic of the telephone companies. The Internet provides a vast net-

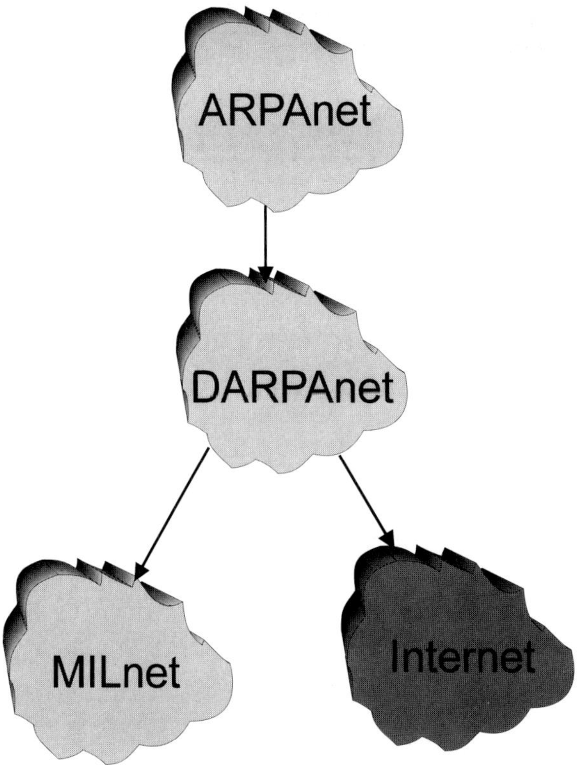

Figure 15.6 ARPAnet—Internet Migration

work for users who wish to make the investment and become a part of the network. Many resources are available from the Internet; it sometimes takes diligence and patience, but you probably will find what your looking for, be it trail software, information, or communications.

The Internet uses a common standard connection methodology that allows different types of computers on dissimilar networks to seemlessly connect to each other. At the present time, it is estimated that there is in excess of 30 million computers attached either full time or part time to the Internet. Many of these computers are part of other networks, LANs, Independent Service Providers (ISP) networks, and mainframe networks.

At one time, the National Science Foundation (NSF) established and maintained a high-speed backbone to several universities, however, the NSF no longer funds this backbone. Today telephone companies have assumed the responsibility of establishing, maintaining, and providing access to a shared Internet backbone. It is the responsibility of the user, the ISP, and companies to pay for their share of those services. This allows for a worldwide access capability that increases the value of the Internet because it is clearly a global network. Figure 15.7 illustrates cross connectivity between network providers.

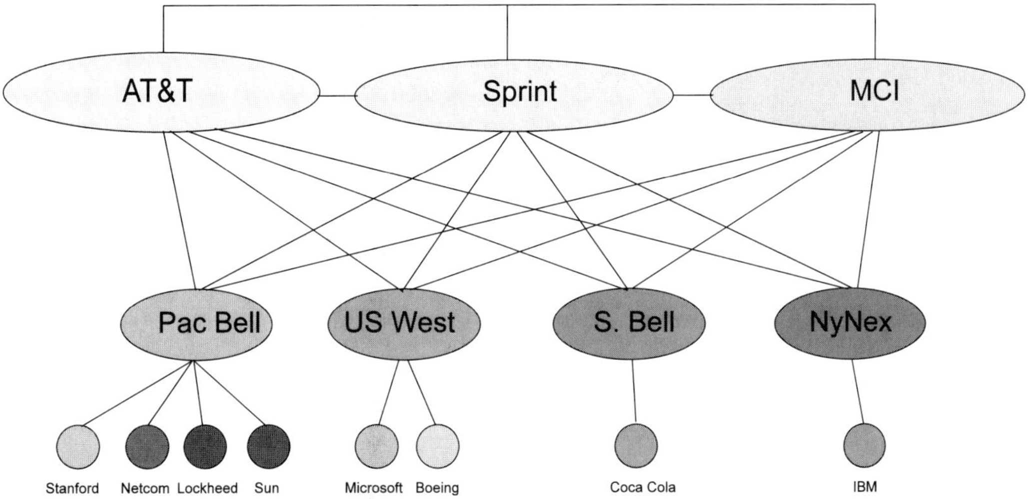

Figure 15.7 Internet Connectivity

An example of that might be a large university that wants to facilitate the sharing of research information. They would probably want direct high-speed access to the Internet backbone and would contract with the local exchange carrier (LEC) (local telephone company).

The ISP also can be considered a network. Their networks consist of many dial-in users, and if large enough, leased line users supporting speeds up to 1.5 Mbps. The ISP would then connect their network to the local telephone companies providing high-speed access to the backbone.

What of the Internet of tomorrow? If the Internet continues to grow as it has in the past, higher speed access will be required to keep up with the volume of users and the amount of information that can be delivered. The telephone companies and television cable providers are working on this right now. Cable providers are planning to implement cable modem service that will provide 10 to 37 Mbps of speed, and of course the telephone companies have something competitive as well that will run over standard telephone lines.

One final note is that many telephone companies are investing heavily in bringing fiber-optic cable to their curbside boxes and installing high capacity wire from those boxes to the final connection point. This will provide a whole host of new digital high-speed services they can offer. They include: high definition television, high-speed digital communications, interactive video phone service, and very high-speed computer links (billions of bits per second) which will be needed to support new Internet services. Some of the new Internet services include Internet phone, a telephone capability using a computer's microphone and sound board; three-dimensional virtual reality browsing capabilities; and interactive Internet video.

Internet Layers

Like the OSI model, the Internet also follows an architectural model and closely resembles OSI. As previously discussed, OSI is a model that is made up of definitions and protocol suites. Each layered protocol adds functionality to the communications activity and works closely with the other protocols in the architecture. This provides for isolation of functions and allows a great deal of flexibility. The Internet architecture is concerned with four primary layers—Figure 15.8 shows the four layers of the Internet Architecture.

The Application Layer. This layer supports several TCP/IP associated protocols, and while we can view them as being application based, they are not really considered application layer functions. They include: file transfer protocol (ftp), used for host-to-host file transfers; simple mail transfer protocol (SMTP), used for e-mail activities; telnet, used for virtual terminal access; Dynamic Name Service (DNS), used for converting names to addresses; Simple Network Management Protocol (SNMP) used to manage and control the network; and other application interfaces.

Transport Layer. The transport layer has the responsibility of managing end-to-end communications, and is not involved in the movement of the data. The Internet Trans-

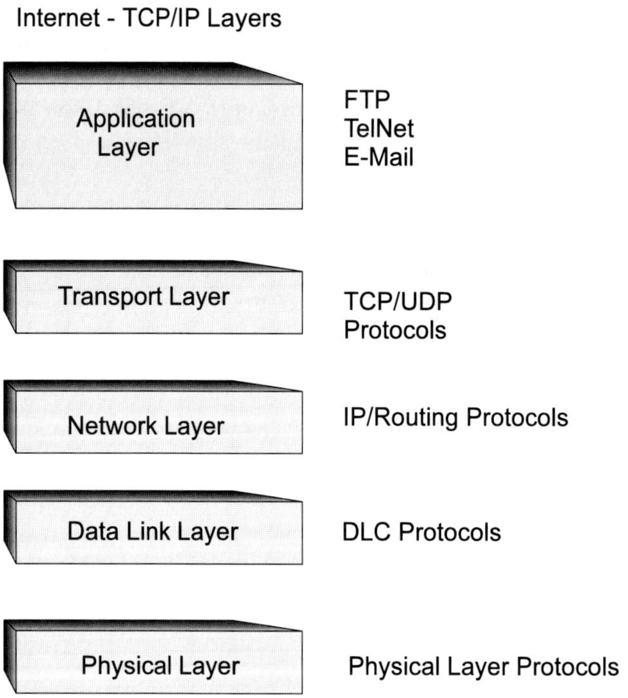

Figure 15.8 Internet Layers

port layer includes two end-to-end protocols. One is called UDP or unreliable datagram protocol and can be used for unreliable transport (best effort). The other is called TCP or transmission control protocol that can be used for reliable transport (guaranteed sequenced delivery).

Transmission Control Protocol (TCP). TCP provides application functions access to the network using a reliable connection-oriented transport layer service. This protocol is responsible for establishing sessions between user processes on the LAN or Internet and also ensures reliable communications between two or more processes. The functions that it provides are: to listen for incoming session establishment requests; to request a session with another network node; to send and receive data reliably; and to gracefully close a session.

User Datagram Protocol (UDP). UDP provides application functions access to the network using an unreliable connectionless transport layer service. It allows the transfer of data between the source and destination stations without having to establish a session before data is transferred, UDP does not support error checking. It is primarily used for those applications that do not require the robustness of the TCP protocol. For example, mail, broadcast messages, naming service and network management. These types of application services operate on a continuous or as needed basis and are designed to continually send data so if something becomes corrupt, it is either not important or will be resent again the next time around.

Network Layer. The responsibility of the network layer is to move and route the data from the sending computer to the receiving computer. The protocol designed for this activity is called the Internet protocol (IP); however, there are actually three protocols that can operate at this layer.

Internet Protocol (IP). The IP is designed to interconnect packet switched communications networks to form an intra- or inter-network. It transmits data records called datagrams, received from the upper layer protocols (TCP, UDP), to and from source and destination hosts. It provides a best effort or connectionless delivery service between the source and destination. Connectionless implies that it does not establish a session between the two end points before it transmits the data. IP also is responsible for the source and destination addresses.

Internet Control Message Protocol (ICMP). ICMP is not really an IP layer protocol, however, it does operate directly with the IP layer. Additionally, it is not used for the transport of user data, but rather for the transport of control and error information between routers and hosts on the network.

Address Resolution Protocol (ARP). This is another protocol that is not really part of the transport layer but rather provides translation services to convert 32-bit IP addresses to 48-bit LAN address to allow TCP/IP to be used across LANs.

Data Link Layer. The data link control layer establishes the communications link between nodes over the physical channel. It consists of activities for framing, error control, time-out levels, and data formatting. It also maintains sequence control for frames on the communications link. The Internet uses a variety of protocols at this layer that include Point-to-Point Protocol (PPP), Serial Line Interface Protocol (SLIP), Ethernet, FDDI, or token ring protocols.

Physical Layer. Layer one is the Physical Link layer that defines the electrical and mechanical aspects of the interface to a physical medium for the transmission of data as well as the initial setup. This layer includes the software driver for each communications device and the hardware.

MANUFACTURING AUTOMATION PROTOCOL (MAP)

The Manufacturing Automation Protocol (MAP) is a set of layered communications protocols (similar to those defined by the OSI model) that was developed by General Motors Corp. Its primary function is to manage factory automation activities. MAP uses elements of the IEEE 802.4 LAN standard, and interfaces to WANs via the OSI upper-layer protocols. Many hardware and software manufacturers offer products that incorporate MAP, enabling computerized equipment from different vendors to communicate with each other in a factory environment. MAP uses either broadband or baseband transmission over coaxial cable and operates similar to a token-passing ring network.

CASE STUDY (CONTINUED)

The Bank of Scotts Valley will implement IBM's SNA because of the large number of SNA networks and the level of support available. The bank's data communications network planners have decided to break the SNA network into two domains. Domain A, located in Northern California, will support branches located in the northern part of the state. Domain B will be located in Los Angeles and will support the southern part of the state. Each branch contains a control unit communications node and its workstations are terminal nodes. The ATM networks will not be part of the SNA but instead will be connected to a Tandem computer that is channel-attached to the IBM processor.

The service center located in Los Angeles will be able to support connectivity outside of the SNA network. It will support the CCITT X.25 communications standard and have a dedicated link to an X.25 public data carrier. This will allow internetwork communications to other computers that do not support SNA, but that have X.25 capabilities. It also will have OSI internetwork communications capabilities for communications with international networks. Figure 15.9 illustrates the Bank of Scotts Valley's SNA network.

Because each of the service centers maintains records for the accounts it supports, communications to the Northern California service center must be routed there from the Los Angeles service center. The computer in Los Angeles will take incoming requests

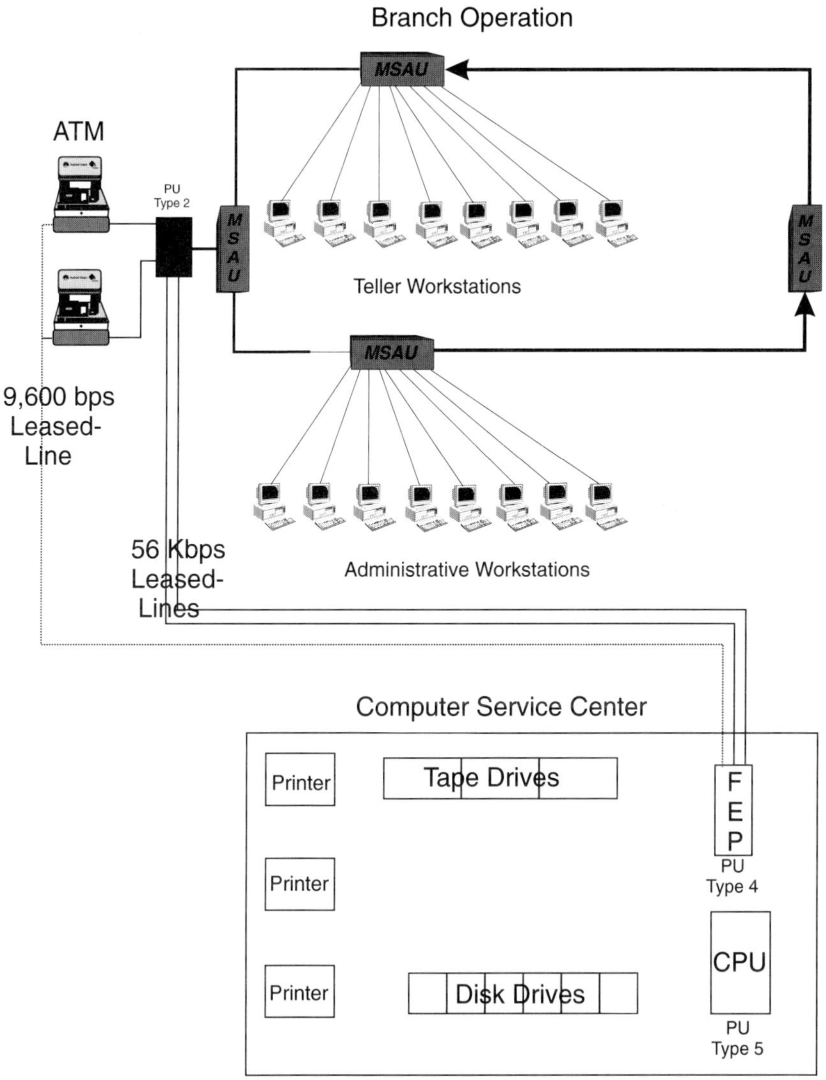

Figure 15.9 Bank of Scotts Valley SNA Network

from the OSI interface and convert them to an SNA format. At that point, the SNA-formatted request will be handled as if it were an interbranch activity. When either the Northern California or Los Angeles service center must communicate with OSI or X.25 networks, it will access an application program running in Los Angeles that is designed to reformat the communication into the correct format and transport it to the network interface.

REVIEW QUESTIONS

1. What does the ISO OSI model define?
2. What is SNA?
3. Are SNA and OSI architectures compatible?
4. SDLC is a protocol for what architecture?
5. What is layer two also known as?
6. True or false: The OSI model is designed to encompass both data processing and data communications.
7. For what is X.25 used?
8. What do the first three layers of X.25 define?
9. What protocol is used with the X.25 architecture?
10. What is MAP?

CHAPTER 16

Protocols

INTRODUCTION

Two intelligent devices that need to communicate with one another must use a protocol for controlling and exchanging information. A protocol can be defined as a set of rules that formats frames or packets with data and control information. The communications equipment at each end of a link uses the protocol's control information to determine the data's validity and its final destination within the network.

Another way of looking at a protocol is to compare it to the public mail system. The letter corresponds to the data, and it must be directed to a specific person. An envelope contains the letter and bears the addresses of the sender and receiver. The destination address also may contain routing information for speedy delivery. This method of exchanging information is conceptually similar to a protocol packet or framed data communications.

The function of a protocol is to provide a handshaking technique between two pieces of equipment, multiple nodes, multiple networks, or end-to-end sessions (user and application). Protocol functions are implemented as firmware or software processes, and reside in the various communications devices attached to the network. Protocols assist in the transport of data between two distal points in the communications network using physical media. The physical media typically provides connections between terminal controllers and front-end communications processors or host computers, or from one computer or communications node to another.

Every manufacturer of data processing and data communications equipment supports some type of protocol. This includes centralized terminal environments and networkwide communications activities.

Many companies originally provided only a protocol that was unique to their products. At that time, there was no need to provide communications with other vendors' products. As multivendor networking became a reality, new protocols were developed that, in many cases, were different from those used for network or remote communications. This multiprotocol environment often created incompatibilities between different manufacturers' networks, making it difficult for users to connect their networking hardware.

Today it is common for one manufacturer to adopt and use the protocol of another manufacturer for internodal communications. One example is the adaptation of IBM's BSC protocol by many other vendors. Adapting other manufacturers' protocols is meant to serve as a compatibility feature, allowing direct connection of multivendor equipment. Since the introduction of international standards, many companies have implemented new protocols that will allow for internetwork communications.

DATA COMMUNICATIONS PROTOCOL CONCEPTS

Transportation of data across a communications network mandates that certain activities take place. Specific control activities occur at each end of the link to ensure that the data arrives in a sequential fashion, with notification to the sender if an acknowledgment is desired. Such notification communicates a successful reception.

Bit- Versus Byte-Oriented Protocols

Two general types of protocols exist, byte-oriented and bit-oriented. In byte-oriented protocols, character or byte string is transferred in seven- or eight-bit formats. Control characters embedded within the header and trailer of the packet provide control indicators to the software. Byte-oriented protocols are generally found only at the second layer used with older protocols (Figure 16.1 shows a point-to-point, network-to-network, and endpoint-to-endpoint protocol environment).

In bit-oriented protocols such as SDLC and HDLC, bit streams are transported and only one bit pattern (01111110) is recognized (opening and closing flag). Bit-oriented protocols are generally considered more efficient, because they require less overhead for transmission control than do byte-oriented protocols. (Figure 16.2 presents a bit-oriented protocol.) Bit-oriented protocols are typically synchronous and full-duplex in operation.

Protocol Functions

All protocols follow a set of established procedures for the movement of data. Protocol software frames the data bytes or bit stream with communications codes or special bit patterns that indicate specific control functions. These control functions are used to indicate transmission status and to identify the beginning and end of the byte or bit stream transferred.

The protocol software also establishes and maintains some type of sequencing control to ensure that packets and frames are not lost and are received in the order in which they were sent. This usually involves placing a sequence number in a field within the packet.

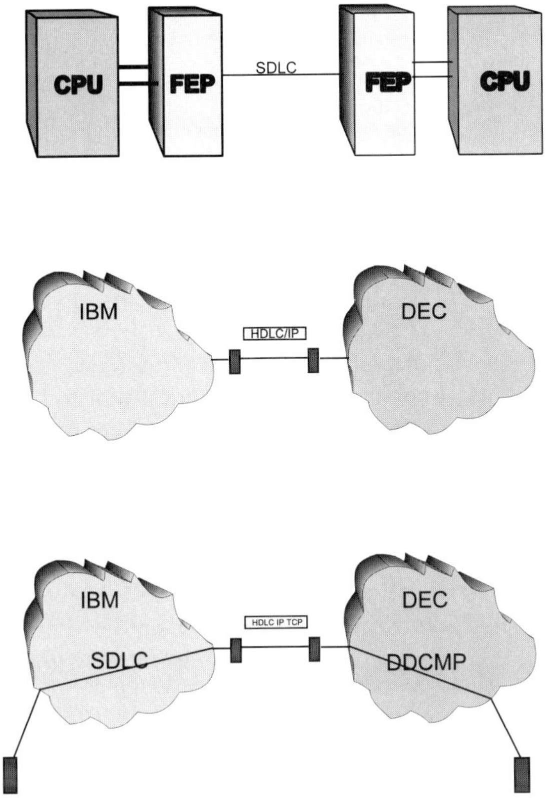

Figure 16.1 Protocols

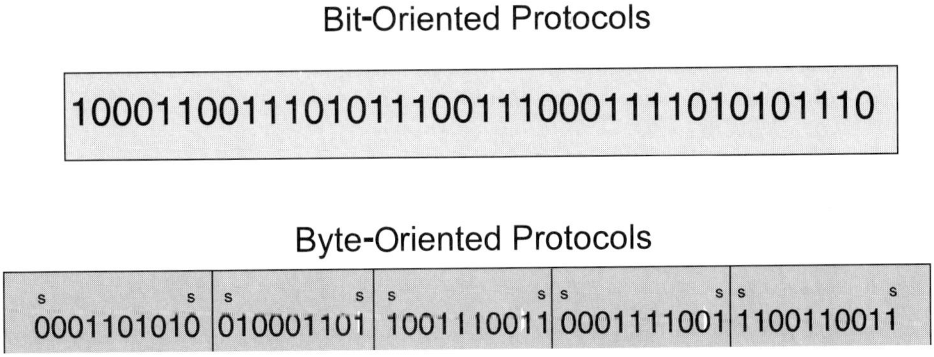

Figure 16.2 Bit-vs-Byte Oriented Protocols

Protocol software also must maintain direction control to ensure that when one node is transmitting, all other nodes are receiving. This is especially important in a half-duplex environment.

The protocol software also manages start and stop control. Start and stop control provides an orderly transition through all phases of data communications. Start and stop control indicates the beginning and end of the packet byte or bit stream. It also includes line turnaround in a half-duplex environment, indicators for the beginning and end of the packet flow, and proper linking and unlinking.

Protocol software must provide extensive error control to detect error conditions in the transmitted data. Upon discovery of an error in the received packet, the software requests retransmission of the data.

Protocol software must contain a function for time-out control, which is required to address deadlock situations that arise when a receiving node must wait to transmit because a response is expected from the sending node. Deadlock can occur when terminals and lines experience delay or are out of operation. If time-out control did not exist, a networkwide lockup could occur. Typically, time-out values are tunable in the software, based on the type of link used.

For certain protocols, transparency is needed to inform the protocol software that the data area begins starting at a certain byte within the packet. This control feature informs the protocol software to ignore any bit patterns that may indicate protocol functions. This is useful for transferring or downloading software across the link. Without transparency, the software that needs to be transferred could cause the protocol software to perform an unwanted function.

Depending on the protocol used, selected functions may not be required. Often, certain protocol functions may be implemented in such a fashion that they perform nothing more than basic activities.

Byte-Oriented Data Link Communications Codes

In byte-oriented protocols such as BSC, the standard code set used by the communications equipment often determines what control codes are available to the protocol software. The EBCDIC code set, for example, provides additional communications codes not available in ASCII. Specific data link control protocols selectively use the ASCII and EBCDIC standard communications codes differently. The protocol being used determines which codes are valid for that handshaking technique.

With bit-oriented protocols such as the SDLC protocol and the OSI HDLC protocol, ASCII and EBCDIC communications codes are not used because the code set employed is transparent to the protocols. Instead, HDLC and SDLC use bit groupings or binary counts to represent specific activities. Typical communications codes available in both ASCII and EBCDIC for byte-oriented DLC protocols include the following.

SOH: Start of Header. This is usually the first character after the synchronization characters and is used to indicate the start of the header information. The SOH is used to delineate the beginning of the frame, the control characters, and the data.

STX: Start of Text. The STX character is used to indicate the beginning of the data field, which immediately follows the STX character.

ETX: End of Text. The ETX character indicates that there are no more data and immediately follows the last character of data. This may be placed where additional blocks of data will appear in the frame.

DLE: Data Line Escape. The DLE character works in conjunction with the STX and the ETX to provide transparency for the data being sent. Transparency allows communications control characters to be sent as data and prevents the transmission software from responding to their functions. This is used primarily when downloading software to remote communications nodes.

ETB: End of Transmission Block. The ETB character is used to indicate the end of a transmission block. The ETB is the same as the ETX except that when the ETX is used, more data blocks may follow.

ITB: Intermediate Text Block. The ITB character is used to indicate an intermediate end of transmission block. It is used to separate the message into sections for error detection without causing a reversal of the transmission direction.

EOT: End of Transmission. The EOT character indicates an end of the transmission block. It is also used to indicate "nothing to respond," and for line turnaround in half-duplex environments.

WACK: Wait Before Transmit. A WACK response indicates that the previous frame was received without error (positive acknowledgment), but the receiving station is not ready to accept another frame. The transmitting station will respond with an enquiry until a character other than WACK is received.

ACK: Positive Acknowledgment. The ACK is used to acknowledge receipt of the data. There are two forms of the ACK, the ACK0 and the ACK1. ACK0 indicates that the data was received without error, and the receiver is ready to accept additional data. ACK0 is also used to acknowledge multipoint selection, point-to-point contention, and even-numbered blocks of data. ACK1 is used to acknowledge odd-numbered blocks of data received.

NAK: Negative Acknowledgment. NAK is a negative acknowledgment informing the sender that the previous data block was received in error. Receipt of the NAK indicates that a retransmission of the block received in error is required.

ENQ: Enquiry. An ENQ character is used to bid for use of the link when in a point-to-point, contention-based mode. It indicates the end of a polling or selection request. It also is used to request retransmission of a NAK or ACK response if the original response was garbled or not received when expected.

RVI: Reverse Interrupt. The RVI is similar to the ACK and NAK. However, the RVI is a priority interrupt request to the station controlling the link to turn the line around for an important transmission.

TTD: Temporary Text Delay. The transmitter uses the TTD to retain the line even though it is not ready to use the link. The receiving station will respond with a NAK.

Error Detection

Several types of error detection techniques are available, depending on the protocols in use. Often, multiple error detection schemes are used at the same time. As mentioned in Chapter 11, some additional error detection may be available from modem manufacturers, either in the form of standalone error correction devices attached to the link or built into each modem.

Vertical Redundancy Checking (VRC). This form of error detection adds an additional bit to each byte, known as the parity bit. This additional bit will be turned on or off at the sending side, depending on the parity of the machine (odd, even, or none) and the bit composition of the character sent. If the hardware operates in odd parity, then each byte that is composed of an even number of one bits will cause the hardware to set the parity bit to a one. This will make the number of one bits in the byte an odd amount. Also referred to as parity checking, VRC is not generally used, because it adds overhead to the data streams being transmitted. Figure 16.3 illustrates an example of VRC or parity checking.

Longitudinal Redundancy Checking (LRC). This is used primarily with byte-oriented protocols. LRC creates parity bits that form a character for each bit row in a block. By placing a number of characters side by side in the buffer, each bit row generates its own parity bit. These parity bits create characters referred to as block check characters, which are transmitted to the receiving modem and terminal as part of the protocol frame. When the data is received, the data is reconstructed in the buffer and each bit row is validated (see Figure 16.4). A problem with this form of error detection is that if a block experiences a multibit error, it may escape detection on the receiving side.

Cyclical Redundancy Checking (CRC). CRC is the most extensive and complicated error detection and correction scheme available. It uses polynomial arithmetic to calculate 8-, 16-, 32-, or 64-bit cyclical redundancy check characters. CRC operates by calculating multiple binary-encoded characters that represent the sum of the data stream within the protocol frame. CRC is used extensively in HDLC, SDLC, and other upper layer protocols. In SDLC, cyclic redundancy checking utilizes a generator polynomial. The CRC is usually the last field sent as part of the frame or packet.

Bit Position	Value
1	0
2	1
3	0
4	0
5	0
6	0
7	1
8	0
Odd Parity	1

Figure 16.3 VRC Error Detection

BIT POSITION	INFORMATION CHARACTERS					BLOCK PARITY CHARACTER
	1	2	3	4	5	
1	0	1	0	0	1	1
2	1	0	0	0	0	0
3	0	0	1	1	0	1
4	0	1	1	1	1	1
5	0	0	0	0	1	0
6	0	0	0	0	0	1
7	1	1	1	1	1	0
8	0	0	0	0	0	0
ODD PARITY	1	0	0	0	1	1

Figure 16.4 Example of VRC Error Checking

Several other error detection methods also are available, including frame check sequence (FCS) characters (similar in function to CRC error detection) and forward error correction.

LAYER TWO ASYNCHRONOUS PROTOCOLS

Many hardware and software manufacturers have developed asynchronous protocols for low-speed data communications. These protocols are used primarily within small computers and PCs.

While these protocols are usually simple, several of them include extensive error-checking procedures. Blast is perhaps one of the best-known PC-oriented protocols; it provides for recovery if transmission is interrupted. XMODEM is a PC-oriented protocol that creates 128-byte packets and uses CRC error detection. Kermit, another PC-oriented protocol similar to Blast and XMODEM, also supports a 128-byte data packet.

Many other enhanced asynchronous protocols are available that include pseudo-blocking and extensive error correction techniques. Enhancements differ from manufacturer to manufacturer, and are usually not compatible. In general, asynchronous protocols have the characteristics outlined in the following paragraphs.

Framing is at the character level and consists of a start bit before the data bits and a stop bit after the data bits. Because the asynchronous transmission method is used in telex transmission, an additional stop bit may be incorporated.

Although the data may be blocked in the application program, the data is sent one character at a time and not in blocks. Error correction is limited to character parity checking, except when error control has been implemented by the protocol developer under a particular manufacturer's specifications.

Asynchronous protocol speeds are usually limited to 56 Kbps or less, and are used primarily with dial-up links. One reason these protocols are used in low-speed environments is the large overhead associated with asynchronous transmission (in some cases up to 37 percent). Some modems do provide asynchronous transmission at speeds in excess of 50 Kbps using special modulation techniques and add-on error correction devices.

SLIP/PPP Protocols

The Serial Line Internet Protocol (SLIP) or Point-to-Point Protocol (PPP) are protocols primarily designed for asynchronous dial-up access to the Internet. SLIP provides the ability to transport TCP/IP traffic over serial lines, such as dial-up telephone lines, between two computers. This allows a home user to access the Internet from a PC with just a simple modem and a telephone line. The SLIP protocol is one of the most basic protocols ever developed. It performs no error checking and uses a special bit pattern to indicate the end of the packet. For all intents, SLIP has been replaced by the PPP protocol.

PPP also is considered to be an Internet protocol, and is designed to carry data packets created from other protocols. PPP can be viewed as a protocol that transports other protocols. PPP can be used in either dial-up or dedicated connections and consists of several sub-protocols that include: the link control protocol (LCP) which sets up, tests, and configures the link; the network control protocol (NCP) used to initialize, configure, and terminate use of a particular network protocol; and finally, a protocol for the exchange of actual information such as IP datagrams.

LAYER TWO SYNCHRONOUS PROTOCOLS

Binary Synchronous Protocol (BSC)

Developed in 1966, BSC (also known as bisync) was created by IBM to support its computer communications activities. BSC is a character-oriented (byte-oriented) protocol that frames its data with communications control codes (described at the beginning of this chapter). BSC has been adopted as a de facto standard by many manufacturers because of the widespread use of IBM equipment supporting bisync since the 1960s. Early versions of BSC were provided to users in software form, but today it can be implemented in the hardware's firmware. Figure 16.5 shows a BSC protocol frame.

BSC is a half-duplex protocol that sends a single frame at a time and requires that a response be received prior to the next transmission. It operates in either multipoint or point-to-point configurations in either controlled or contention mode. In the controlled mode, the protocol functions in a polled mode using master and secondary stations over point-to-point or multipoint configurations. In the contention mode, BSC uses a primary station connected to a secondary station in a point-to-point configuration.

The BSC protocol detects errors using a combination of VRC and LRC techniques. Parity bits are generated by the transmitting station for strings of characters and are used to create a Block Check Character (BCC). Each BCC is appended to the protocol frame transmitted. At the receiving side, the protocol software regenerates the BCCs and compares them to those received in the protocol frame. If the comparison determines that an error has occurred, a NAK frame is returned to the sender.

It is important to note that BSC is currently used in less than half of the IBM systems that perform data communications. Today, it is mainly used with systems that do not have high data throughput requirements or do not support SNA. BSC is most commonly used on low-end systems. It also may be found on systems that have not been upgraded to IBM's newer protocol, SDLC.

Synchronous Data Link Control (SDLC)

SDLC is the layer two DLC protocol currently considered to be the heart of SNA. (Figure 16.6 illustrates an SNA SDLC header and frame.) It was developed in the mid-1970s, and provides increased capability over the BSC protocol. IBM has stated that SDLC and the SNA architecture were developed as an overall strategy for its data processing and data communications environments.

HDLC, the layer two protocol of the OSI model, is a superset of SDLC and ADCCP, which is a U.S. NBS protocol. HDLC is considered a superset because SDLC was developed by IBM prior to OSI development and was used as a model for HDLC.

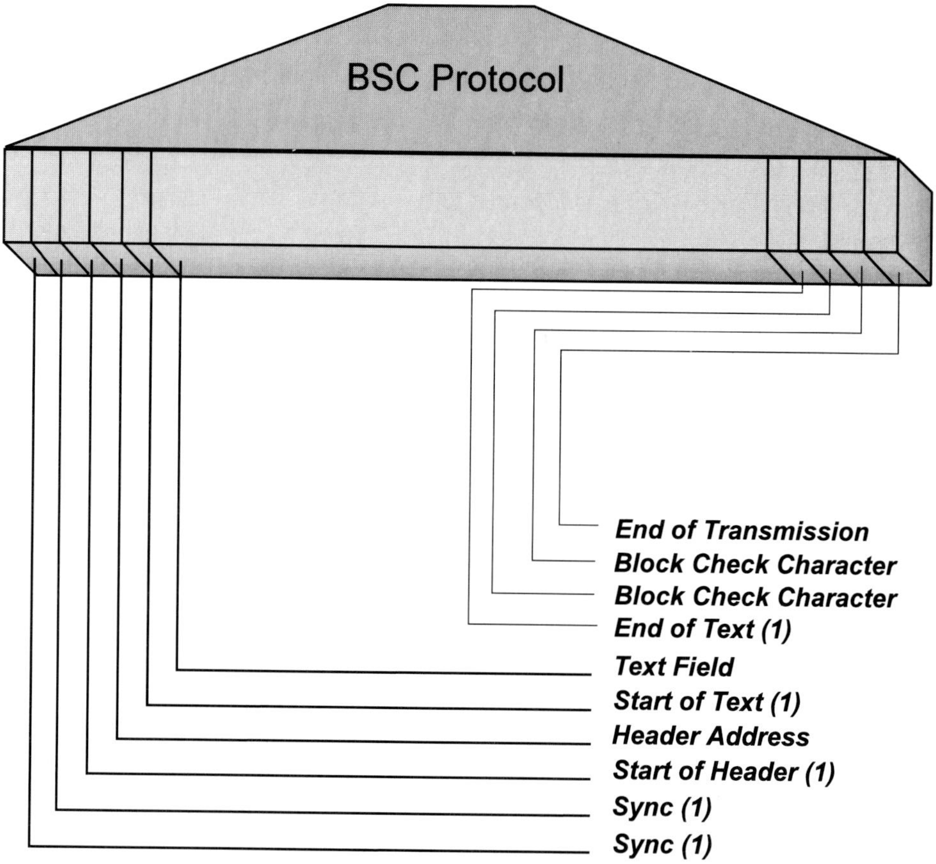

Figure 16.5 BSC Protocol Frame

The SDLC protocol operates either in a full- or half-duplex environment in point-to-point or multipoint configurations. It may support peer-to-peer or primary-to-secondary nodes; however, it is designed primarily as a host control protocol. Multiple data blocks can be embedded in frames, and up to seven frames can be transmitted without acknowledgments between each transmitted frame.

SDLC is a bit-oriented protocol that uses bit strings to represent text. Bit patterns are transparent to SDLC, with the exception of a one-byte flag indicating the start or end of the frame. The flag consists of a 01111110 bit pattern. To prevent a data bit pattern from indicating a flag, a technique called zero bit stuffing is employed. After the opening flag is generated, if the sending side encounters a sequence of five one bits in a row, the protocol will insert a zero bit between the fifth bit and the bits that follow. The receiving side will detect the string of one bits and remove the inserted zero bit when it is encountered.

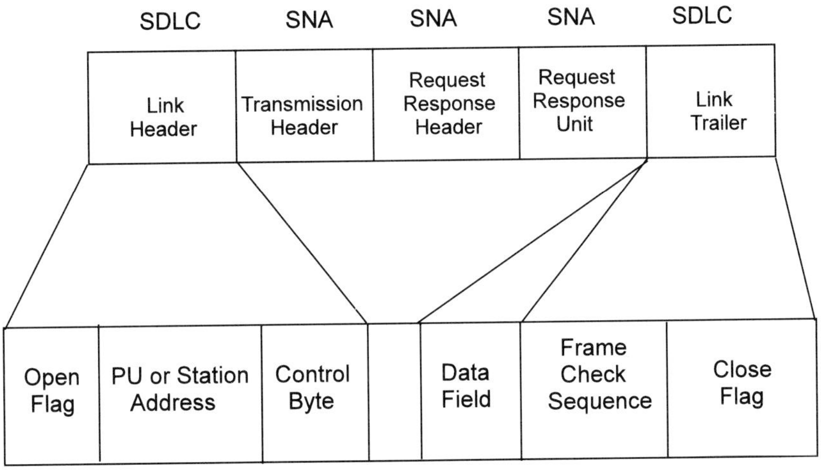

Figure 16.6 SDLC Frame

Communications functions exist for many types of frame formats, including supervisory, unnumbered, and informational frames. Unnumbered and informational frames transport data, and supervisory frames are used for control. The control field in each frame identifies the type of frame. Each frame also carries error-checking information in the form of a CRC field. One problem with SDLC is in its error correction. If an error is detected in the transmission, all contiguous frames after the error is encountered must be resent. This requires that a large buffer be established to hold all frames until the receiver acknowledges them as error free.

SDLC is designed for high transmission rates and supports satellite data links. Whereas BSC operates on a frame-at-a-time basis, SDLC can be configured to use extended formats. This extension supports transmission of up to 128 contiguous frames. Figure 16.7 illustrates the frame formats and control codes associated with the three types of SDLC frames.

High-Level Data Link Control (HDLC)

This protocol was developed by the ISO (ISO 4335) and is a superset of the SDLC protocol used in IBM's SNA. It is also used by many manufacturers as their synchronous protocol and is specified as the OSI layer two protocol. A version of HDLC referred to as Link Access Procedure-Balanced (LAP-B) is in widespread use as the CCITT protocol that supports the X.25 standard. Many LAN systems also use HDLC for data link control. Figure 16.8 shows a basic HDLC frame format.

HDLC has many of the same characteristics of SDLC, because it is a superset implementation. Normally, HDLC and SDLC are not compatible, however, depending on the implementation of the HDLC protocol, frame conversion from one format to another can be a simple process. Because HDLC is a standard definition, many computer manu-

Unnumbered Format

000	F	1111	Disconnect Mode
010	P	0011	Disconnect
011	F	0011	Unnumbered Ack
100	P	0011	Set Normal Response Mode
100	F	0111	Frame Reject
101	P/F	1111	Exchange Station ID
101	P/F	0011	Test

Supervisory Format

RRR	P/F	0001	RR	Receive Ready
RRR	P/F	0101	RNR	Receive Not ready
RRR	P/F	1001	REJ	Reject

Information Format

RRR	P/F	INF	Information frame

Legend

P	Poll Bit
F	Final Bit
RRR	Next Receive Count

Figure 16.7 SDLC Control Byte Information

facturers have adopted it. The HDLC standard defines types of stations, link configurations, and data transfer modes.

Types of Stations. Three types of stations can be identified:

- Primary stations that issue commands.
- Secondary stations that issue responses.
- Combined stations. Combined stations can operate as either a primary or secondary station, issuing both requests and responses.

Link Configurations. Two types of link configurations can be used with HDLC: unbalanced and balanced. In the unbalanced mode, one primary station and multiple secondary stations operate over point-to-point or multipoint links. Balanced mode uses a point-to-point configuration with the combined station concept in a peer-to-peer arrangement.

Data Transfer Modes. The data transfer modes indicate how a station will participate in a data transfer operation. The three data transfer modes are:

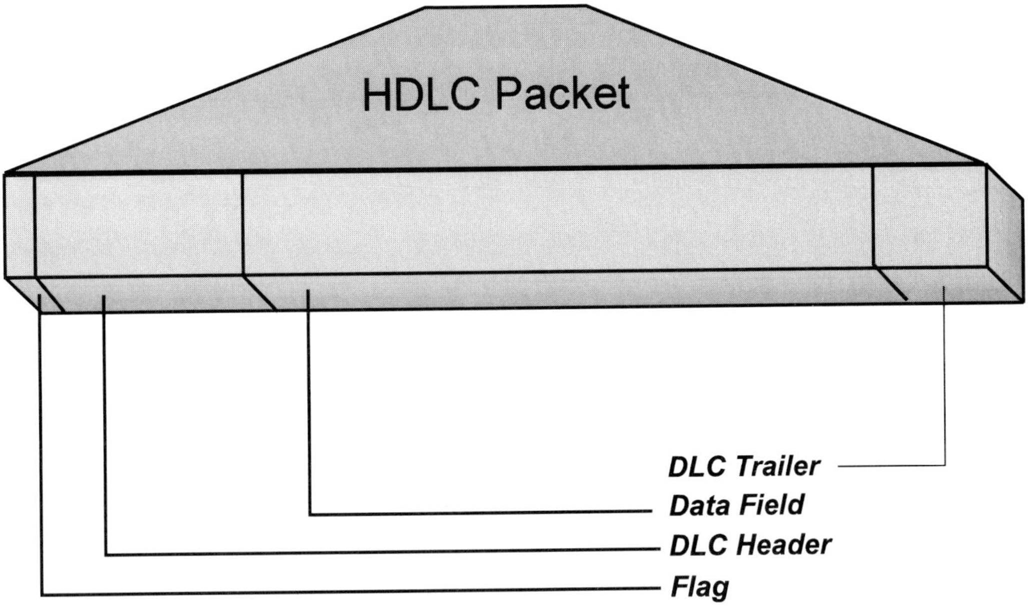

- o Frames contain data link control information (header and trailer) and may also contain higher-level control information
- o Data link header contains sequence numbers, addresses, and control information. Data link trailer contains error dectection information

Figure 16.8 HDLC Frame Format

- Normal response mode (NRM)
- Asynchronous balanced mode (ABM)
- Asynchronous response mode (ARM)

NRM uses an unbalanced link configuration with a single primary and multiple secondary stations. The primary station controls data transfer to the secondary stations, but secondary stations can transfer data only when they are polled by the primary station. ABM uses a balanced link configuration with combined stations. Either station can initiate data transfer operations at any time. ARM uses an unbalanced configuration with a single primary and multiple secondary stations. Unlike NRM, ARM allows secondary stations to initiate data transfer without a poll from the primary station.

HDLC uses a frame format concept similar to SDLC that defines supervisory (S), information frames (I), and unnumbered (U) frames for control and data transfer. Figure 16.9 shows the basic format for the three types of HDLC frames.

Frame Types

- o Information (I) Frames always contain information
- o Supervisory (S) Frames do not carry information
- o Unnumbered ((U) Frames may carry some information

Basic Control Field Format

Frame Type	Bit Position							
	8	7	6	5	4	3	2	1
Information	N/R			P/F	N/S			0
Supervisory	N/R			P/F	S		0	1
Unnumbered	M(2)			P/F	M(1)		1	1

Legend

N(S) Sequence number of this I frame
N(R) Sequence number of next expected I frame
P/F Poll/final bit
S Supervisory Frame
M(1) M(2) Unnumbered frame type

Figure 16.9 HDLC frame types and control byte information

The format of the frame fields includes the flag, address, and control fields.

Flag Field. The flag field is used to indicate the beginning and end of a frame. Similar to SDLC, the flag consists of a 01111110 bit pattern. To prevent a data bit pattern from being mistaken for a flag, the same technique as used in SDLC, zero bit stuffing is employed. After the opening flag is generated, if the sending side encounters a byte containing five one bits in a row, the protocol will insert a zero bit between the fifth bit and the bits that follow. The receiving side will detect the string of one bits and remove the inserted zero bit when it is encountered.

Address Field. The address field is normally an eight-bit field, but it can be longer if the number of devices in the network exceeds 127. The first seven bits of the address field are used for device addresses, and the eighth bit is used to indicate extended addressing, if implemented. The eighth bit indicates that an additional eight-bit address field is present. When an address field consists entirely of one bits, it is considered to be a broadcast address and all devices will receive the frames.

Control Field. The control field is an eight- or 16-bit field that consists of subfields for different protocol activities. The first subfield is a one- or two-bit field that identifies the type of frame (I, S, or U) it resides in. Depending upon the type of frame, the next field may differ. For I types, the next subfield is a three-bit (seven-bit in extended format) sequence number used for the "next send" value. This field controls the number of frames that can be transferred contiguously without acknowledgments. This portion of the control field is also used for negative or positive acknowledgments to and from the receiver. If an out-of-sequence condition is detected in the sequence number, an error has been detected and frames represented by the overlapped numbers must be resent. For U and S packets, the second subfield is a one-bit field used to indicate the function.

The next subfield is a one-bit field indicating poll or final status. If a poll bit is indicated, the packet is a command type from the primary to the secondary station. The secondary station uses the final bit indicator to indicate that the last I type data frame is being transmitted. The last subfield in the unnumbered format is used for protocol functions and, for I and S frames, is the next received field.

The next field in the I frame and also in some unnumbered frames contains the data to be transported. This may be variable in length but is usually set to some fixed multiple of eight bytes. Following the data field is the frame check sequence (FCS) field, which uses the standard CRC-16 or CRC-32 format. Table 16.1 lists the various command and response codes and frame formats that can be used with the HDLC protocol.

LINK ACCESS PROCEDURE-D (LAP-D)

The LAP-D protocol (ISDN Standard I.440/I.441) is designed to support ISDN basic and primary rate D channel activities. LAP-D is a variation of the LAP-B protocol used in X.25 networks. LAP-D's primary function is to communicate control information in a

Table 16.1 SNA Commands

	Information Frames
INF:	Describes Data Carrying Frames
	Supervisory Frames
RR:	Receive Ready Indicator
RNR:	Receive Not ready Indicator
REJ:	Frame Reject Command
SREJ:	Selective Reject
	Unnumbered Frames
DM:	Disconnect Mode Request
DISC:	Disconnect
UA:	Unnumbered Acknowledgment
SNRM:	Set Normal Response Mode
SNRME:	Set Normal Response Mode Extended
SADM:	Set Asynchronous Balanced Mode
SABME:	Set Asynchronous Balanced Mode Extended
SIM:	Set Initialization Mode
RD:	Request Disconnect
RIM:	Request Initialization Mode
UI:	Unnumbered Information Frame
UP:	Unnumbered Poll
RESET:	Reset
FRMR	Frame Reject
XID:	Exchange Station ID
Test:	Test

multiplexed environment between the user nodes and the central office. In an ISDN network, user nodes can be considered operationally equivalent to OSI layer two. Central offices are functionally equivalent to OSI layer three (network layer).

LAP-D provides two types of service: unacknowledged information transfer service and acknowledged information transfer service.

In unacknowledged transfers, no acknowledgment is required from the receiver, and neither flow control nor error detection services are supported. This operates like X.25 datagram services and is used for rapid information exchange. With acknowledged transfers, a virtual connection is made and the protocol operates like HDLC with the same types of services, including error control, congestion control, and acknowledged delivery. Both types of service can be supported concurrently on the D channel. LAP-D frame structures are similar to HDLC frame structures and consist of the flag, address, and control fields, described below.

Flag Field. The flag field indicates the start and end of a frame. Like the HDLC flag, the LAP-D flag consists of a 01111110 bit pattern. To prevent a data bit pattern from being mistaken for a flag, zero bit stuffing is used. After the opening flag, the sending side encounters a sequence of five one bits in a row. The LAP-D protocol will insert a zero bit between the fifth bit and the bits that follow. The receiving side will detect the string of one bits and remove the inserted zero bit when it is encountered.

Address Field. The address field in LAP-D (identical to HDLC) can be up to 16 bits in length. This is because the protocol must provide two levels of multiplexing to the devices it supports. At user sites, several pieces of equipment can share the ISDN interface and each device can support multiple types of traffic. A two-part addressing scheme is used to provide the multiplexing capability. It consists of a field identifier, command/response subfield, service access point identifier (SAPI), and a terminal end-point identifier (TEI) subfield. The SAPI field identifies the type of traffic and the TEI subfield identifies specific user devices.

Control Field. The control field defines three types of ISDN frames and, like HDLC, has supervisory, information, and unnumbered formats. Additionally, subfields are used for sequence number control, command status, and poll final activity, just as in HDLC.

The I-type frames are the only ones allowed to carry data. However, again like HDLC, some unnumbered types of frames (response types) also can carry data. The control and data fields are limited to 260 bytes.

LAYER THREE (NETWORK-LAYER) PROTOCOLS

There are several protocols for layer three (network-layer) operations. Layer three activities manage internetwork communications between similar or dissimilar networks. Layer three protocol functions are considered to be the most complex because they must be designed to support many types of networks. In X.25 networks, layer three protocols operate in what is referred to as the packet layer.

The network layer supports many types of functions, including data relay, addressing, congestion/flow control, and error detection. Data relay activities support transport of the data packets from one network to another and are routed through to the receiving network. Address resolution for packets includes network-to-network addressing and dynamic addressing generation to assist in relay activities. Congestion and flow control are supported at layer three to avoid degrading the network's performance. Error control to manage traffic and resolve error conditions within layer three also may be provided, but error detection is not a specification of this layer. Layer three protocols include IP and the Internet Control Message Protocol (ICMP). Figures 16.10 and 16.11 shows network layer connectivity.

Internet Protocol (IP)

IP was the result of a joint development effort by Xerox, DARPA, and Stanford University. IP was designed to provide services between TCP and the Data Link Control proto-

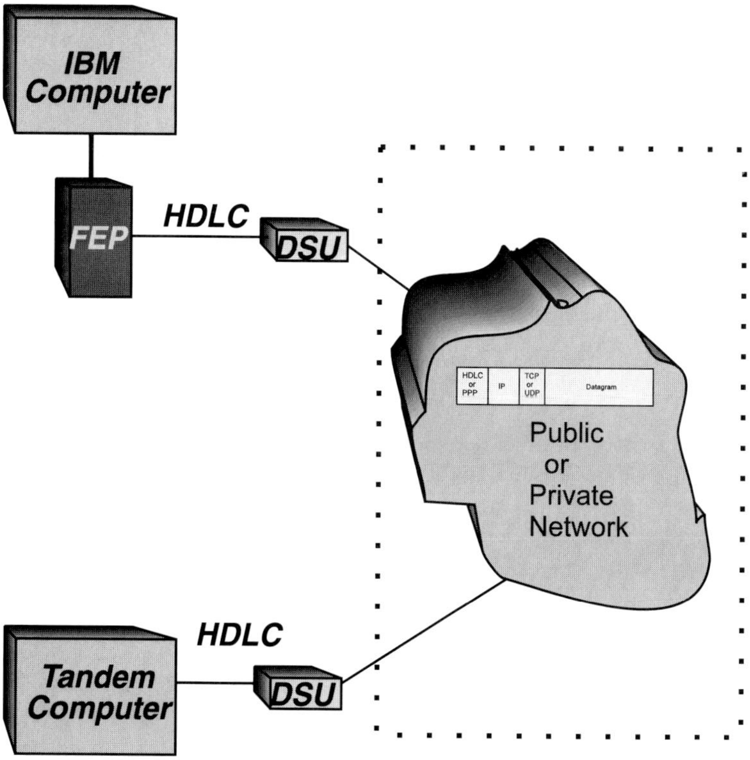

Figure 16.10 Network Layer Communications

col in the data link layer. Its function is to connect networks with minimal impact on each network. IP operates as a sub-network-layer protocol and can connect networks with different internal protocols and performance. Figure 16.12 shows the IP PDU frame format.

IP components within a given network define the format of Internet packets they will be responsible for transporting. The protocol software in each node supporting IP must establish rules (via negotiations) for protocol and PDU management functions.

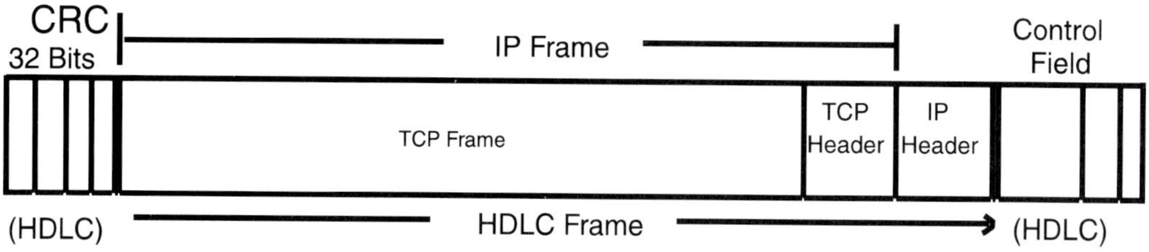

Figure 16.11 Layer Three Packet Format

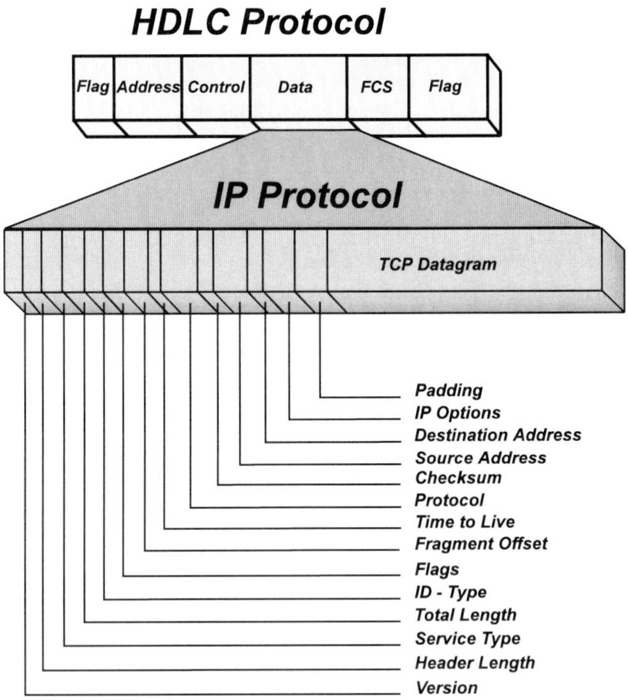

Figure 16.12 Internet Protocol Format

These functions are based on control information passed between IP nodes contained within the IP header portion of the packet. The Internet protocol supports delivery of datagram packets from source location to destination; however, it does not provide acknowledgment, data error checking, or retransmission. These functions are provided from the transport layer via TP or TCP. IP provides limited flow control to avoid congestion that slows network operation.

Values can be set via bit switches in the control portion of the IP header to provide information about the actions desired and agreed upon. IP supports two types of frames, data frames that consist of four parts and error-reporting PDU frames that contain three parts. The PDU consists of the fixed, address, and option fields, described below.

Fixed Field. This contains the protocol identifier, which is not used if the sending and receiving nodes are in the same network. A length indicator defines the number of bytes in the IP header. A version number ensures compatibility with other IP implementations.

A PDU time-to-live (TTL) value can be used to ensure that the PDU will not continue to move through the network if it is not deliverable. The TTL is usually calculated in multiples of 500 milliseconds on entry into a network, and is based on the number of nodes that the packet must traverse within the network. Each node decrements the value by one as it transports the packet. If the value becomes negative, the node discards the packet.

A segmentation flag is used to indicate that segmenting of a packet into smaller packets is allowed. Another flag, called the more flag, indicates that segmentation has or has not occurred. A third flag, the error flag, indicates that an error packet is needed. This occurs if the IP packet has to be discarded because an error is detected.

The last three subfields in the fixed part include the type of packet (data or error), segment length of the PDU (including the header information), and the checksum calculation.

Address Field. The address part of the PDU consists of subfields for the source and destination addresses. This is because they may be of different lengths depending on the source and destination of the packet. Addressing is handled across networks by establishing classes that define the type of address being used. This provides flexibility and allows a large number of small, intermediate, and large networks to be interconnected. Five address classes have been established:

- Class A addresses are used for large networks. They consist of an 8-bit network address and a 24-bit local address. Class A addresses use decimal values between 0 and 127 for the network portion.
- Class B addresses are used for medium-sized networks. They consist of a 16-bit network address and a 16-bit local address. Class B addresses use decimal values between 128 and 191 for the network portion.
- Class C addresses are used for small-sized networks. They consist of a 24-bit network address and a 8-bit local address. Class C addresses use decimal values between 192 and 223 for the network portion.
- Class D addresses are used for broadcast multicast activities. Class D uses decimal values between 224 and 239 for the network portion.
- Class E addresses are reserved for experimental use. Class E addresses use decimal values between 204 and 255 for the network portion.

Options Field. An options field may be included to communicate additional functions, such as:

- User-defined security
- Source routing that identifies the networks to be used
- Route information that keeps track of the nodes visited
- Time stamping
- Priority for normal or expedited data transfers
- Quality of service, which specifies degrees of reliability and delay

The final part of the network PDU includes fields for the data and padding, if necessary, to provide a fixed-length packet. The TCP datagram field can be up to 65,535 bytes in length.

Error PDU Format

An error PDU is created when a data packet cannot be delivered and must be discarded. This will occur only if the data packet contains an indicator requesting the creation of the error PDU. Formats of error PDUs are identical to that of the data PDU; however, they provide an additional two-byte field that contains a reason for discard rather than data. Twenty-two different discard classifications currently exist.

Internet Protocol (Next Generation)

As of this writing, Internet Protocol, Version 6 (IPv6) is being proposed as a successor to the current version of IP, Version 4. A new version of IP is needed because under the existing addressing method, all IP numbers have been assigned. The changes to IPv6 are designed to address the following issues:

Expanded Addressing Capabilities. IPv6 increases the IP address size from 32 bits to 128 bits, and will have the ability to support 665,570,793,348,866,943,898,599 unique addresses. This increase is designed to support more levels of the addressing hierarchy, a much greater number of addressable nodes, and simpler auto-configuration of addresses. The scalability of multicast routing is improved by adding a "scope" field to multicast addresses. And a new type of address called an "anycast address" is defined, used to send a packet to any one of a group of nodes.

Three types of IPng addresses exist:

unicast address—a single node.

cluster address—a group of nodes that share a common prefix such that a packet sent to the address will be delivered to one of the nodes.

multicast address—represents a group address methodology in that a packet sent to a multicast address will be delivered to all nodes in the group.

Header Format Simplification. Some IPv4 header fields have been dropped or moved to optional extension headers, to reduce the common-case processing cost of packet handling, and to limit the added bandwidth cost of the IPv6 header (beyond the long addresses). Fragmentation and reassembly are limited to the source and destination.

Improved Support for Extensions and Options. Changes in the way IP header options are encoded allow for more efficient forwarding, less stringent limits on the length of options, and greater flexibility for introducing new options in the future. Among the improvements is a flexible format for the transport of specific routing information.

Internet Control Message Protocol (ICMP)

A companion to IP is ICMP. It is required in networks such as the U.S. DDN, and it must be implemented by every IP node. Figure 16.13 illustrates the ICMP header format.

ICMP is used by a gateway or destination host for error notification to a source host. ICMP uses the basic support of the IP packet or frame. This includes the IP headers as if the ICMP frame was data within the higher-level IP. ICMP is actually an integral part of the IP.

ICMP provides feedback to each of the gateways about network communications problems occurring within that specific network. Typically, message packets are created in a gateway node that detects the problem, then sent to other gateway nodes by the ICMP software. This occurs when a datagram cannot reach its destination or a problem exists with a gateway (e.g., no buffer space). It also can occur when a shorter route is available or when there are fragmentation problems.

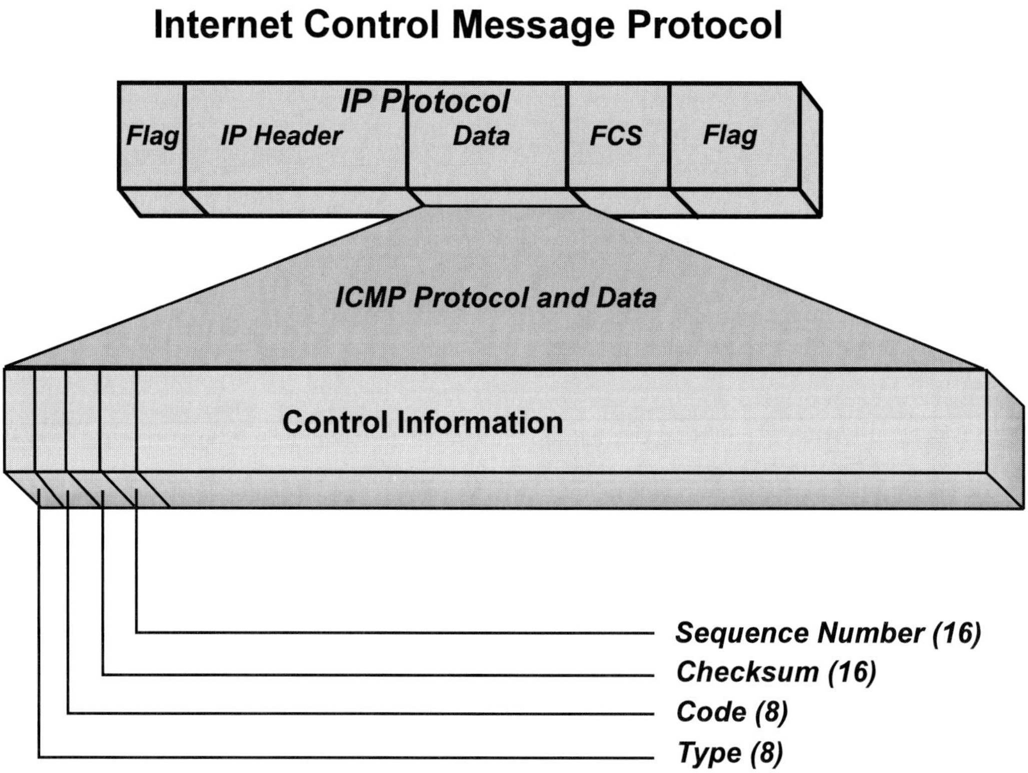

Figure 16.13 ICMP Protocol Format

ICMP message types include the following:

- Redirect—Used when a gateway has determined that greater efficiency can be achieved by redirecting the packet using a different link than the one specified
- Source quench—Indicates to a node or station that it needs to reduce its flow because of a congestion problem
- Time exceeded—Used when the time parameter of the datagram packet has been exceeded and the packet must be discarded
- Parameter problem—Indicates that a syntactical error has occurred and the parameter value in the IP header is unrecognizable
- Destination unreachable—Used when a network or node is unreachable for a variety of reasons
- Echo/echo Reply—Used for testing the links and connections between nodes
- Time stamp/time stamp reply—Used for monitoring the performance of the network

LAYER FOUR (TRANSPORT LAYER) PROTOCOLS

OSI layer four protocols provide end-to-end integrity and guarantee delivery of packets across the network. One function of layer four protocols is the segmentation of packets into smaller pieces. Segmentation applies to packets considered too large for transport across a particular network. Reassembly of those segmented packets into larger packets in the proper sequence, and sequenced delivery of the packet to its destination point, must also be guaranteed.

The two basic protocols at this level are the OSI Transport Protocol (TP), which is broken into five functional levels, depending on user requirements, and the Transmission Control Protocol (TCP), which offers the same capabilities as TP, but is not an ISO-defined protocol.

Despite the fact that TCP is not an ISO protocol, it seems to be more popular than TP. TP, on the other hand, is used extensively in Europe and within the U.S. government.

Transmission Control Protocol (TCP)

TCP operates within the boundaries of the transport layer of the OSI model. It provides reliable interprocess communications between pairs of processes in distant hosts. TCP was designed for communications among dissimilar networks and was developed to support IP. As discussed, IP does not guarantee the delivery of packets or acknowledge a successful transfer.

TCP interfaces with and manages the user's application data flowing from the upper-layer protocols such as the file transfer protocol (ftp) and telnet protocol. TCP interfaces with the lower-layer protocols and provides a translation service to convert appli-

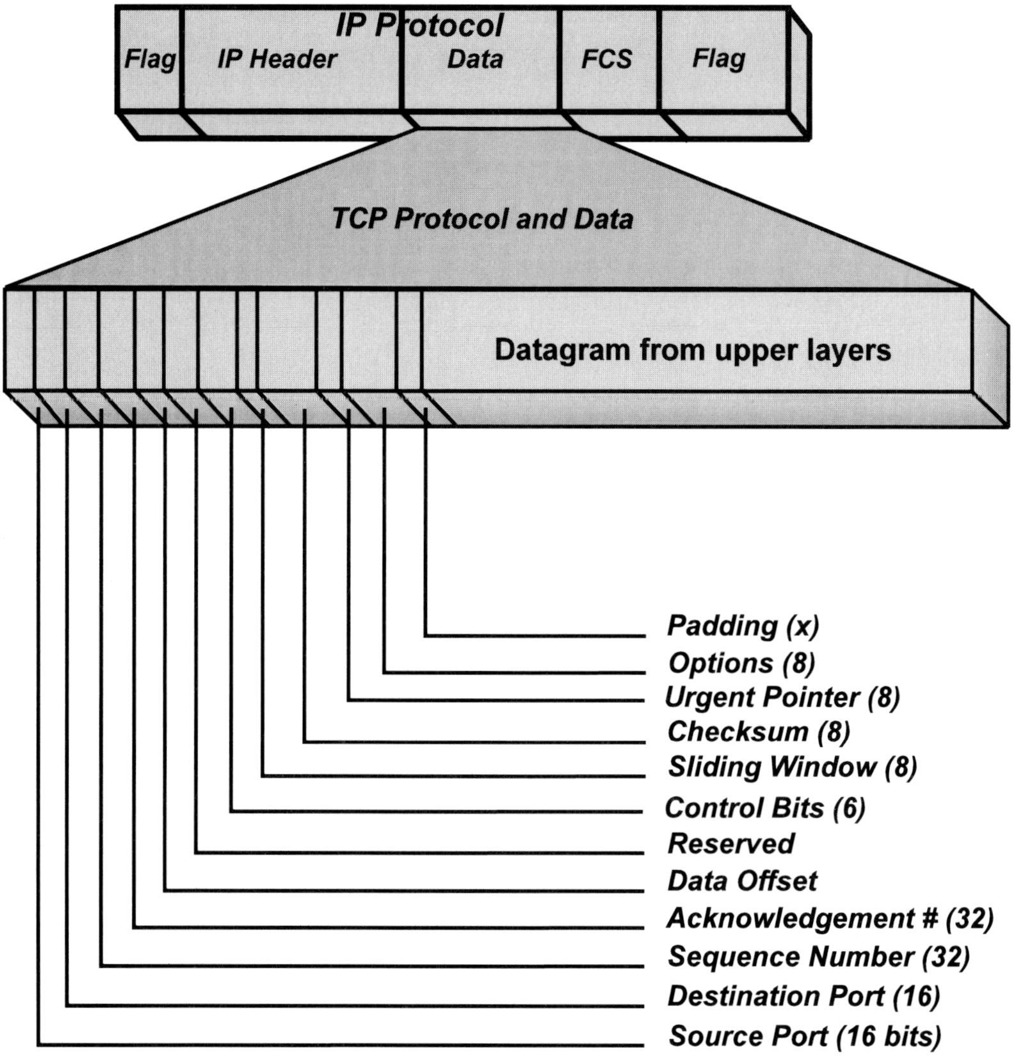

Figure 16.14 TCP Protocol

cation names to lower-level addresses for routing through the network. Figure 16.14 illustrates the header portion of the TCP data unit.

TCP is a virtual circuit protocol in that it establishes end-to-end connections using the concept of port and socket addressing. This establishes unique sets of addresses (ports) within the hosts for each pair of concurrent processes requiring communications.

Ports and sockets consist of unique network and node addresses and a generated circuit ID, which are used for packet routing and delivery. Like other protocols already discussed, TCP operates by taking protocol data units from upper layers of the network and establishing levels of service based on the negotiated end-to-end connection. TCP offers two basic data transfer services and provides several options that guarantee packet delivery and protection against duplicate packets.

Data to be transferred via TCP is broken into segments within the node's send buffer area (allocated in the memory of the node). These segments, known as transport protocol data units (TPDUs), are carried by the IP service for delivery to their destinations. The TPDU consists of TCP header information, control information, and user data.

TCP provides many functions, including connection establishment, data transfer, segmentation and reassembly services, error control, and flow control.

Connection Establishment. Before data can be transferred using TCP, the two end points must establish a connection and agree on the level of services required. Agreements must be reached on activities such as security, sequence numbering used for acknowledgment, window sizing and other options that will be required during connection establishment. The activities involved in the transport of TPDUs are discussed below.

Data Transfer. Transfer of the TPDU across the network is accomplished via a mechanism that ensures the acknowledgment of each byte of the packet received. Embedded byte sequence numbers accomplish this activity by indicating starting and ending data byte offsets within the TCP frame.

Segmentation and Reassembly Services. Different networks often require packets to be a specific size, which can sometimes create a frame size incompatibility between networks and nodes. To address this issue, TCP performs packet segmentation and reassembly. A segment (frame) size is negotiated between the sender and the receiver during connection establishment. This may require that large packets be segmented (broken into smaller units) and then reassembled at the receiving node.

Error Control. A major activity of TCP is error control, which is required when packets either do not arrive or are damaged in transit. Acknowledgment packets containing sequence numbers of the last byte string received ensure that all packets are successfully transported. An incorrect sequence number returned to the sender by the receiver indicates that an error has occurred in transit and the packet needs to be resent. Timers are used by both the sender and the receiver. When the sender does not receive an acknowledgment in a specified period of time, it means that the packet was not received and should be resent.

Flow Control. Flow control logic is resident in each node of a TCP-oriented network and allows the receiver to regulate the arrival speed of the transported packets. Flow control between nodes is also managed using timer mechanisms that indicate congestion on the network. This is managed and controlled by using a negotiated window size between nodes. The result of this negotiation establishes the number of bytes that can be transported before an acknowledgment is needed.

Note: TCP is not recommended by the ISO. The ISO supports the Transport Protocol, Classes 0 through 4. TCP was developed primarily to support U.S. DoD networks. However, over the past few years, TCP has become available from a variety of vendors and is now widely used in the commercial sector.

TCP Formats. The transport packet consists of a 28-byte header and a variable-sized packet depending on the specific network. The header includes fields for addressing, sequencing, acknowledgment, service indicators and pointers, a flow control window for congestion management, checksum, options, padding, and user data. The user data also may be part of an encapsulated upper-layer protocol frame that is treated as data by TCP.

Addressing. TCP addressing includes source port and destination port addressing. Each address is 16 bits long.

Sequence Number. Sequence numbers are contained within a 32-bit field and apply to the first byte of data within this segment. The sequence number corresponds to the byte offset of this packet as it relates to other segments for the packet. This can be illustrated by using an 800-byte packet and breaking it into 100-byte segments. In segment number 1, the sequence number begins at 0; in segment number 2, the sequence number starts at 100. This is piggybacked with the acknowledgment field and used for error and sequence control.

Acknowledgment Number. The receiver uses this field to tell the sending node the first byte of the next segment it expects. If an out-of-sequence number is received, the sender will be able to take corrective action and resend the missing or damaged segment.

Data Offset. This field is a four-bit pointer to the location within the segment of the first byte of data.

Pointers and Reserved Area. A reserved field follows the data offset and is not used at the present time. This field is followed by six one-bit flag fields indicating special activities, such as urgent data, acknowledgment, push data, connection on reset, synchronization of the sequence numbers, and an end of data.

Window. This 16-bit field is used for flow and congestion control. It indicates how many bytes (segments) the receiver is willing to accept without requiring an acknowledgment.

Checksum Field. The checksum field is 16 bits long and is used for validating the entire segment (header and data).

The remainder of the segment includes a 16-bit urgent pointer for urgent data, any future options, padding, and the user data.

User Datagram Protocol

Many applications have data that will fit into a single datagram, and do not need the complexity of TCP and its reliable delivery service. With this type of data, we will resend it if it does not get delivered the first time. This is where the UDP protocol can be best used.

User Datagram (UDP) is designed for applications that do not require sequenced delivery of the datagrams. It is encapsulated into the IP protocol just like TCP. The packet consists of a UDP header that is placed in front of your data. UDP's protocol number is placed in the protocol field instead of TCP's protocol number. UDP does not do as much as TCP does. It performs no fragmentation or assembly, nor can it perform error control. Note that the UDP header is shorter than a TCP header, however, it contains many of the same fields. Figure 16.15 shows the header format of a UDP frame.

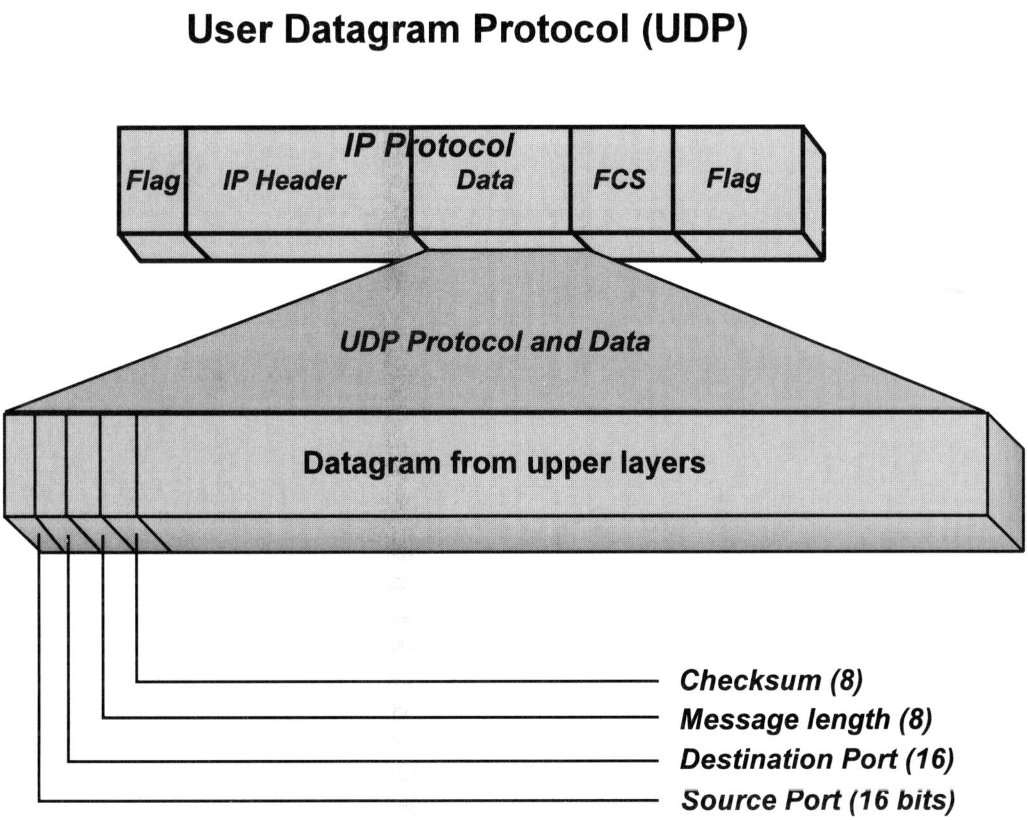

Figure 16.15 UDP Protocol Frame

OSI Transport Protocol (TP)

TP is really a suite of protocols supported by ISO standard 8073. TP can be viewed as supporting three types of networks with five classes of service. The three types of networks are:

- Type A networks—Provide services with an acceptable error rate and an acceptable rate of signal failures. Type A networks provide a reliable level of service to the user.
- Type B networks—Provide an acceptable error rate and an unacceptable rate for signal failures.
- Type C networks—Provide network connections with an error rate not acceptable to the transport service users.

TP Class 0: Simple Protocol (Type A Network). The CCITT developed Class 0 service to support the basic telex/teletype user. This is the simplest protocol class available and provides an end-to-end connection. TP Class 0 uses type A network-level services for flow control management.

TP Class 1: Basic Error Recovery (Type B Network). Class 1 TP provides basic error recovery using sequence numbers contained within the protocol data unit. It was designed by CCITT to operate over X.25 type B networks. TP Class 1 is used to synchronize the network in the event of a failure.

TP Class 2: Multiplexing (Type A Network). This class operates over a type A network and provides multiplexing (multiple transport connections) across a single connection. It uses basic flow control with a credit allocation management technique.

TP Class 3: Error Recovery and Multiplexing (Type B Network). Class three combines the services of Classes 1 and 2. Additional capabilities are provided to manage networks with unacceptable levels of errors.

TP Class 4: Error Detection and Recovery (Type C Network). TP Class 4 is designed to operate across all types of networks and provides the greatest degree of control and recoverability. All services provided in Classes 0 through 3 are incorporated into TP Class 4 networks.

TP is designed to provide reliable internetwork communications between pairs of processes operating in physically distant nodes. TP operates by taking protocol data units from upper layers of the network, providing value-added services, and transferring the PDU to the network layer for transport. Reliable network service, unreliable network service, data transfer, error control, and flow control functions are incorporated into TP, as in TCP.

TP Formats. Ten types of TP formats exist, and each has three parts: a fixed part, a variable part, and the data area. The fixed part contains parameters for header length, flow control, source and destination references, the protocol class, options, connect or disconnect reason, segmentation, and sequencing. The variable part contains the called and calling transport IDs, PDU size, version number, security options, and a checksum. It also includes additional options, fall-back protocol class, timers, requested packet transport rate, and sequence numbers.

UPPER-LAYER (APPLICATION SERVICE) PROTOCOLS

Many types of protocols fall within the definition of upper-layer protocols in the OSI model. But many of these protocols are in fact application services that provide end-user functions. They include telnet, ftp, SNMP, and other protocols for electronic mail and directory services.

Telnet Protocol

The telnet protocol originally was designed for the Advanced Research Projects Authority Network (ARPAnet). Telnet is an application protocol service that allows remote asynchronous terminals and terminal-oriented processes to communicate with each other across the network. Telnet's design is based on network virtual terminal concepts that allow host access to be transparent to all end users. This allows the user to log on to a remote host using any defined terminal type and appear as a locally attached terminal. Like several other upper-layer protocols, telnet uses negotiations for establishing levels of service between hosts and users. It also uses the services of TCP/IP for transport.

Telnet permits the specification of pseudo-terminals, enabling the operating system on the host computer to treat the requesting remote terminal as a locally attached device. Telnet also permits the definition of I/O routines that enable an operating system to handle all I/O terminal activity in the same way, regardless of the type of device.

Telnet provides virtual terminal services that allow a common language to be used when performing connection activities. Virtual terminals are pseudo-device definitions that can be overlaid onto a real terminal, allowing the physical device to appear as another type.

Telnet uses a transfer protocol that defines the form of the data when transferred. In addition, it provides the control signals required for communications between the host and the physical terminal. It also describes the mode of data transfer (half or full duplex) and the data delivery method.

Telnet makes available a series of control functions for status and communications between the terminal and the host. Like other protocols already described, three major activities must occur within telnet: initialization, data transfer, and orderly termination. During the initialization phase (terminal setup), negotiations take place to agree on what format will be used in the communications activity. Data transfer is the actual communications activity and orderly termination deactivates the virtual terminal protocol activity in an orderly manner.

File Transfer Protocol (ftp)

Ftp is another Internet protocol that falls within the definition for upper-layer activities. Ftp was developed because different computer manufacturers use different file formats for data storage.

Ftp is a set of protocol services that allows file transfer between a host computer and the end user. Ftp is similar in design to the telnet protocol. One major difference is that ftp was designed specifically for file transfer rather than data access in a terminal environment. Ftp specifies how files should be transferred from one computer to another over a data communications network. Like other Internet and OSI protocols, ftp is designed to support computers from different manufacturers. One additional point is that similar to telnet, ftp is also a set of services that uses TCP/IP for data transfer.

Early in the evolution of data processing/data communications, a significant problem was how to transport files from one computer to another, regardless of computer type. Within the U.S. federal government, this problem was resolved with ftp, which was designed to shield users from file and storage variations between different types of hosts. Today, ftp is extensively used for all types of data communications and is one of the foundation protocols of the Internet. Ftp uses the concept of a source and destination file, and provides for three types of data transfers: stream, block, and compressed format.

File transfer using the stream mode is the simplest to use and is the default for all transfer activity. No options exist for special processing and no restrictions exist on different file types.

Block mode provides capabilities for restarting a failed transfer. This is accomplished by encapsulating the data records into ftp frames, and using an error detection procedure embedded into each record to provide a file retransmission capability should a failure occur.

The compressed format mode can be used on top of the other formats to improve efficiency. The source host compresses the file data to be transported and the receiving host expands it on receipt.

Four different data types can be transferred using ftp: ASCII, EBCDIC, image, and logical byte size. ASCII and EBCDIC use an eight-bit byte format and transfer each byte in an unmodified format. Image transfer provides no modification and delivers each byte in the exact form it was transmitted. Image file transfers are used primarily for object code (a program translated into a computer machine language) between similar machines. The final data type, logical byte, allows for the creation of a byte greater than eight bits in length. This may be used when sending program codes from one type of machine to another that use different byte sizes because of their architectures. For example, the 12-bit "slab" was at one time used in NCR computers, versus the eight-bit byte used by IBM.

Trivial File Transfer Protocol (tftp)

Tftp is another protocol designed to transfer files between host computers. Because tftp is a minimal file transfer protocol, tftp does not provide all of the features supported by ftp. As an example, tftp does not provide subcommands to list remote files or change directories at the remote host; additionally it only provides limited file access privileges on the remote tftp machine.

The Tftp protocol has two formats: interactive and command line. With the interactive format, tftp is issued alone or with a host argument that specifies the default host to use for file transfers during this session. A port number can also specified for the tftp connection. For interactive tftp, the tftp prompt appears and you are then allowed to specify command arguments. The tftp command can overwrite a file, but prompts before doing so.

When transferring data to a remote host, the transferred data is placed in the directory specified by the remote file argument. The remote file must be a fully specified file name, and the remote file must both exist and have write permission set for others. Tftp attempts to write the data to the specified file. However, if the remote tftp machine does not have the appropriate privileges to write the remote file or if the file does not already exist, the transfer fails.

Simple Network Management Protocol (SNMP)

While they can provide cost-controlled, efficient communications, multivendor networks also create an interesting challenge. With all of the different equipment, how do you monitor and maintain the network? Several methods and products have been developed to address this problem. One such method is the Simple Network Management Protocol. SNMP is not an OSI protocol, and it uses the services of TCP/IP for its transport activities.

SNMP is referred to as a protocol, but is really a protocol subsystem consisting of many components. SNMP was developed to provide network management capabilities across a wide range of interconnected products. Introduced in 1988, SNMP is supported by many communications hardware and software vendors, including IBM and Digital Equipment.

Three key elements make up the SNMP protocol environment: the structure of management information (SIM), the management information base (MIB), and the network management station (NMS). These components, working together, provide visibility into the operation, and performance of a heterogeneous network.

Structure of Management Information. This component defines how each piece of equipment in the network is described in the management information base. The description includes information such as the device's line speed, bit rate, protocol, and address.

Management Information Base. The MIB contains information about the overall structure of the network, including such items as test points and variable alarm information that the SNMP is expected to support.

Network Management Station. The NMS is usually a dedicated console that provides access to a database containing information on all network-defined devices.

While SNMP is defined as an application-layer process of the OSI model, it requires both TCP and IP. SNMP operates using datagram packets to carry status information between SNMP software resident in each network node. SNMP also provides interface capabilities to support network elements not designed to SNMP specifications.

Two problems that exist with the protocol are the large number of packets needed to support the network and its connectionless orientation. A large number of packets are required to support the network because SNMP must send a packet out in order to receive one. Connectionless service could cause a problem in that packets carrying important information might not be delivered. Consequently, this type of service does not guarantee delivery.

Exterior Gateway Protocol (EGP)

The exterior gateway protocol is used to connect an organization's networks to other networks outside the organization. To connect to external networks, a router must be designated as the external gateway through which traffic to external networks is sent and received. This router informs the external networks that it is the access point for its internal networks, and it receives information from external networks on how to reach the external networks. Figure 16.16 shows the format of the EGP protocol.

EGP is a interdomain reachability (routing) protocol transported by IP and used on the Internet. EGP enables networks to exchange information on how to reach other networks. Internetworking and the Internet typically uses internal and external gateways. (Gateways are the same as routers in this context). An internal gateway routes data within an organization's networks. Normally an internal routing protocol (such as the Routing Information Protocol (RIP)), is used to maintain routing information between workstations on such networks. Unfortunately, the weaknesses of exterior gateway protocols have become more visible as internetworking has grown and matured. Because of these weaknesses, EGP is currently being phased out of the Internet, and is being replaced by other exterior gateway protocols such as the Border Gateway Protocol (BGP) and the Interdomain Routing Protocol (IDRP).

EGP was originally designed to communicate reachability to and from the ARPAnet backbone routers. Information was passed from individual source nodes in distinct Internet administrative domains called *autonomous systems* (AS) up to the backbone routers, which passed the information through the backbone until it could be passed down to the destination network within another AS.

Although EGP is a dynamic routing protocol, it uses a very simple design and therefore cannot make intelligent routing decisions. EGP routing updates contain network reachability information. In other words, they specify that certain networks are reachable through certain routers.

EGP has three primary functions.

1. Routers running EGP establish a set of neighbors. These neighbors are also routers running EGP that need to share reachability information; there is no implication of geographic proximity.

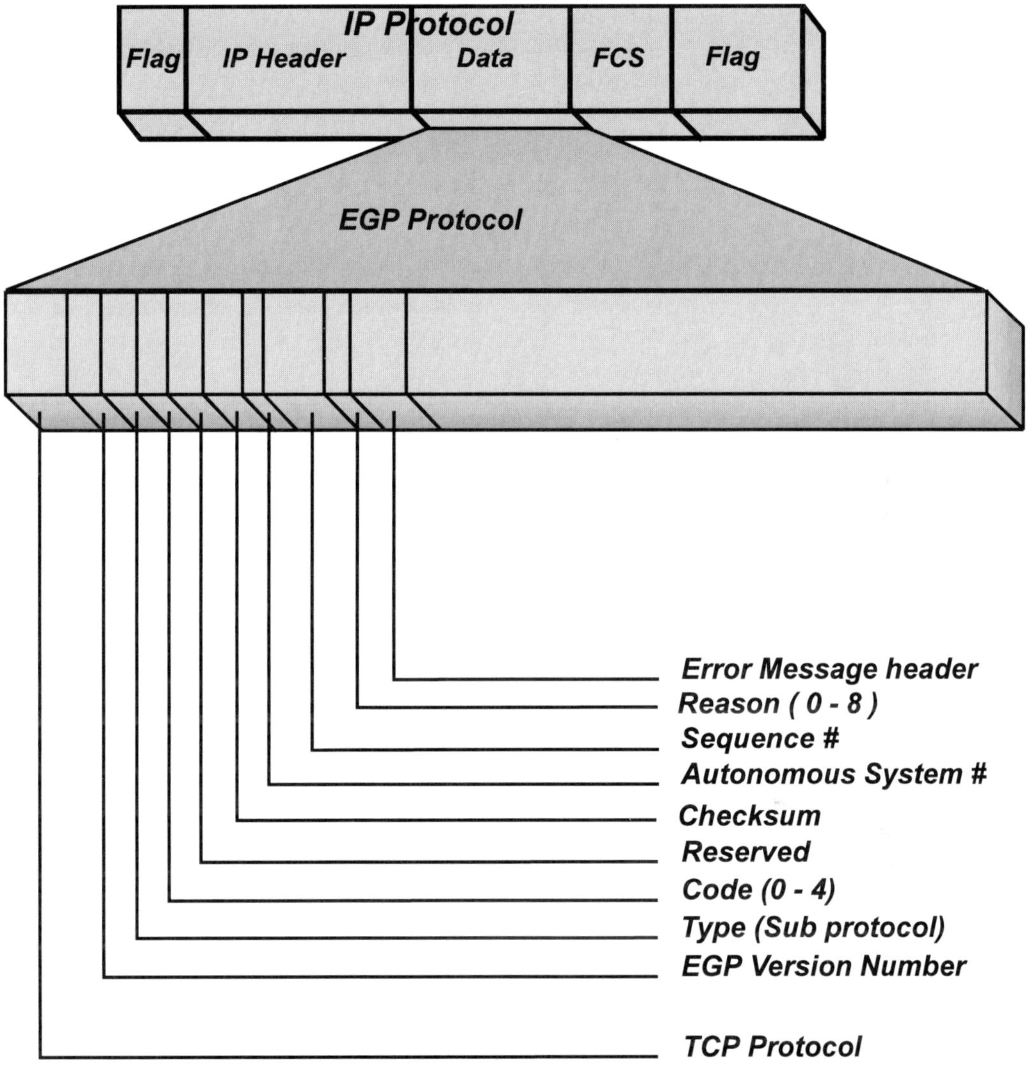

Figure 16.16 Exterior Gateway Protocol Frame Format

2. EGP routers poll their neighbors to see if they are alive.
3. EGP routers send update messages containing information about the reachability of networks within their ASs.

EGP Packet Format

The fields of the EGP packet are as follows:

EGP Version Number. Identifies the current EGP version and is checked by recipients to determine whether there is a match between the sender and recipient version numbers.

Type. Identifies the message type. EGP defines five separate message types.

Type 1. Neighbor acquisition. The neighbor acquisition message includes a hello interval field and a poll interval field. The hello interval field specifies the interval period for testing whether neighbors are alive. The poll interval field specifies the routing update frequency.

Type 2. Neighbor reachability. The neighbor reachability message adds no extra fields to the EGP header. These messages use the code field to indicate whether the message is a *"Hello packet"* message or a response to a hello message. Separating the reachability assessment function from the routing update function reduces network traffic because network reachability changes usually occur more often than routing parameter changes. Only after a specified percentage of reachability messages have not been received does an EGP node declare a neighbor to be down.

Type 3. Poll. Determines reachability of a particular network; to provide correct routing between ASs, EGP must know the relative location of remote hosts. The poll message allows EGP routers to acquire reachability information about the networks on which these hosts reside. These messages only have one field beyond the common header, the IP source network field. This field specifies the network to be used as a reference point for the request.

Type 4. Routing update. Routing update provides a way for EGP routers to indicate the locations of various networks within their ASs. In addition to the common header, these messages include many additional fields. The number of interior gateways field indicates the number of interior gateways appearing in the message. The number of exterior gateways field indicates the number of exterior gateways appearing in the message. The IP source network field provides the IP address of the network from which reachability is measured. Following this field is a series of gateway blocks. Each gateway block provides the IP address of a gateway and a list of networks and distances associated with reaching those networks.

Within the gateway block, EGP lists networks by distances. In other words, at distance three, there may be four networks. These networks are then listed by address. The next group of networks may be those that are distance four away, and so on.

EGP does not interpret the distance metrics that are contained within the routing update messages. In essence, EGP uses the distance field to indicate whether a path exists; the distance value can only be used to compare paths, if those paths exist wholly within a particular AS. For this reason, EGP is more of a reachability protocol than a routing protocol. This restriction also places topology limitations on the structure of the Internet. Specifically, an EGP portion of the Internet must be a tree structure in which a core gateway is the root, and there are no loops among other ASs within the tree. This restriction is a primary limitation of EGP, and provides an impetus for its gradual replacement by other, more capable, exterior gateway protocols.

Type 5. Error. Error messages identify various EGP error conditions. In addition to the common EGP header, EGP error messages provide a reason field, followed by an error message header. Typical EGP errors (reasons) include bad EGP header format, bad EGP data field format, excessive polling rate, and the unavailability of reachability information. The error message header consists of the first three 32-bit words of the EGP header.

Code. Determines the specific packet type.

Status. Contains message-dependent status information. Status codes include insufficient resources, parameter problems, protocol violations, as well as other types of information.

Checksum. Used to detect possible problems that may have developed with the packet in transit.

AS Sys #. Identifies the AS to which the sending router belongs.

Sequence #. Allows two EGP routers exchanging messages to match requests with replies. The sequence number is initialized to zero when a neighbor is established, and incremented by one with each request-response transaction.

Border Gateway Protocol (BGP)

Border Gateway Protocol (BGP) is an exterior gateway protocol, it is designed to connect exterior networks together. In some cases EGP and even static routes can also be used to connect to the Internet backbone. BGP has many dynamic characteristics that make it an excellent choice as an interface between interior networks using RIP and OSPF and the Internet backbone that is considered an exterior network.

For routing purposes, large networks are partitioned into separate autonomous regions called *Autonomous Systems*. For example, a large division of a company might control its own autonomous system. The system managers will be responsible for setting up the routers and selecting the protocol. A routing protocol used within an autonomous system is call an *Interior Gateway Protocol.* An *Exterior Gateway Protocol* is used to connect *Autonomous Systems.* Routers that connect areas are called *border or boundary routers.*

IP packets follow a path made up of sequences of hops calculated at each router location. Nodes adjacent to each other are considered one hop from each other. Routers that are separated by one hop are called neighbors. Because the IP packet carries the source and destination addresses, it is routed by selecting the next hop destination at each router along its path.

BGP offers optimized routing table procedures in which the first route update includes all routes and all additional routes are incremental. Also, the distance vector algorithm recognizes Autonomous Systems with BGP. Furthermore, BGP detects failures on routes and Internet hosts using 30 second "keep alive" messages. BGP supports CIDR (classless inter-domain routing) where IP addresses can be summarized to a single route, routing decisions made within BGP can be based on the CIDR protocol 32-bit addresses and masks, and the 32-bit addresses and masks are included in BGP routing information.

Although BGP is not required in stub or multihomed configurations, it is required in transit configurations. In a transit configuration, there is a central site with at least two connections to remote locations. Each location is an autonomous aystem, one to a remote location with an Internet connection, the others to an additional Internet connection. Figure 16.17 illustrates the format of the BGP protocol frame.

Routing Information Protocol (RIP)

RIP is a protocol that automatically creates and maintains network routes among routers that support RIP. Without RIP, you would need to manually update routing tables each time a routing change is made. This would be complicated if many routers were involved. With RIP, when one router learns about changes in its routes from one of its neighbors, the router gives this information to other neighboring routers so they also update their routing tables. As a result of RIP, if a network component fails (such as a router or phone line), the other network components calculate how to work around the failure if possible. When the faulty component is repaired, the network changes back to the previous (superior) configuration.

Static (or fixed) routes can be set by command. If a router running RIP also has static routes, the static routes take precedence over automatically selected routes. Mixing static and dynamic routes is not recommended.

One downside of the RIP protocol is its 15-metric constraint. In large networks, such as university and corporate environments, many routers are required to connect multiple network segments together. Each router increments the RIP count by one as it passes

Border Gateway Protocol

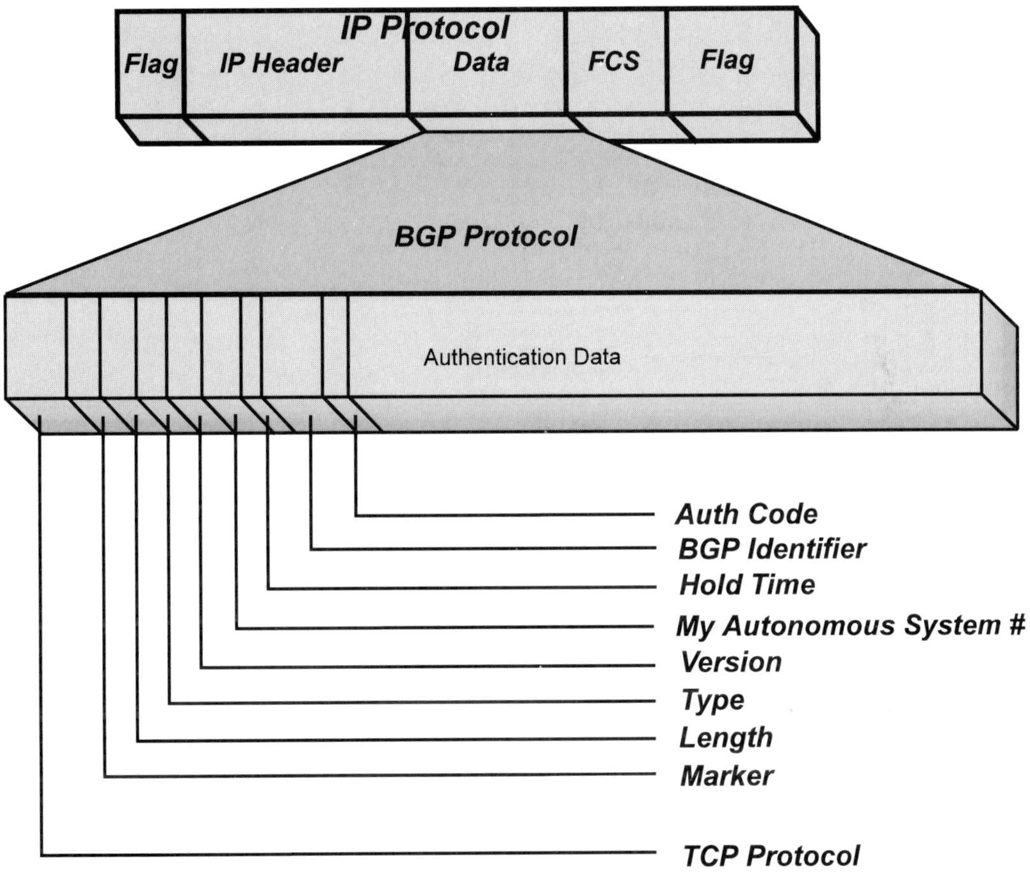

Figure 16.17 Border Gateway Protocol Packet Format

through a RIP table update. In large hierarchical networks, it is often a problem for network administrators to guarantee that the 15-metric barrier will not be exceeded.

RIP does not factor the speed of a link or "circuit cost" into route computation. This lack of information could result in RIP making sub-optimal routing decisions due to lack of information. The most notable examples are illustrated in remote routing environments where a mix of T1 and fractional T1 links are available. In such cases, a RIP router will always choose the "shortest" route in terms of hop count, not the "shortest" route with regard to network delay. Figure 16.18 shows an example of a RIP protocol frame.

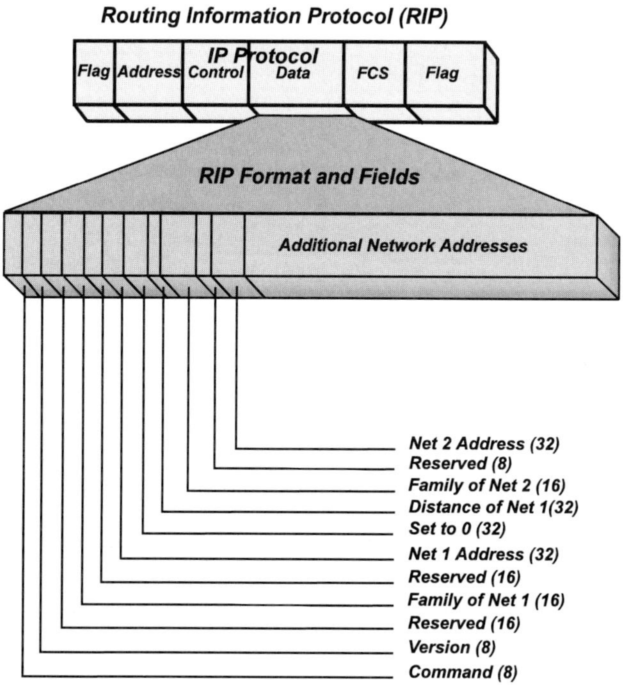

Figure 16.18 Routing information Protocol Packet Format

TCP/IP networks have used many routing schemes to move data through Internets. Schemes range from static tables to proprietary dynamic routing protocols. The RIP protocol can be considered one of the major dynamic routing protocols employed in the Internet; however, it has several shortcomings. They include:

- Inability to understand the speed of a particular interface
- Restrictive metric (or hop count) field that does not allow for routes beyond 15 routers
- Delay in responding to network changes
- Lack of security.

To overcome these and other shortcomings in RIP, the Internet Engineering Task Force proposed a new routing protocol, OSPF, which is described in the next section.

With RIP, knowledge of a change in the network is included with the table updates and can take a considerable amount of time to propagate through the network. As the changes are being processed by the closest neighbor, erroneous routing tables remain in the downstream routers. The opportunity for path congestion and data loss is exponentially increased during a RIP update.

RIP's method of table updating also contributes to delays in responding to network changes. A RIP router forwards its entire routing table to its neighboring router(s). The receiving router must process that table by comparing each entry with entries it currently has. In large networks this task is CPU and memory intensive. This computation is done on every routing table exchange every 30 seconds, whether the tables have changed or not. The receiving router cannot know if a change in the network has occurred until it has completed the table comparison.

Open Shortest Path First (OSPF) Protocol

Open Shortest Path First (OSPF) is a routing protocol for TCP/IP networks based on the DARPA Internet Protocol (IP) network layer. The basic algorithm of OSPF is called the Shortest Path First Algorithm, or Dijkstra's algorithm (named for its inventor). It is a protocol designed to be used inside autonomous systems. It is not designed to route between Autonomous Systems. OSPF is an Interior Gateway Protocol, and is intended to be used within an IP network under common administration, such as a campus, corporate, or regional network.

OSPF enables routers to exchange information about their interfaces (whether they are up or down, how fast they go and to whom they connect) and maintain a complete, synchronized, link state database in every router. OSPF was designed to meet a number of specific goals:

- Provide a single protocol that all vendors could implement interoperably
- Provide rapid, deterministic calculation of Internet routes
- Facilitate separate administration of differing parts of the Internet
- Facilitate hiding of detailed information about the network
- Use detailed information about the network to optimize route calculations
- Isolate misconfigured or malfunctioning routers in the network and route around them
- Provide for the effective use of information derived from other routing protocols, even those that use metrics that OSPF does not understand.

The OSPF protocol consists of three sub-protocols, and runs directly on top of IP (protocol type 89). They are: hello, exchange, and flooding.

OSPF sends only table changes. The messages between OSPF routers contain either a change or no change statement. OSPF routers exchange information about their interfaces and each router maintains a complete, synchronized link state database. When a router's table changes, only the changed information is propagated through the network. Then, every router can calculate new optimal routes. Once a failure is detected in the network, recovery from the failure takes slightly longer than the time required to send a message through the network, and the result is deterministic. Likewise, taking full advantage of a new network path takes little longer than the time required to notify others of its presence.

OSPF packets contain 64 bits of authentication information in the header. Administrators can set these bits and restrict unauthorized routers from communicating with an OSPF router. Route authentication has two distinct advantages; it can prevent accidental misconfiguration of the network; and multiple independent OSPF networks can coexist in the same AS by using different authentication keys. OSPF requires a great deal of processing power to keep the link state information up to date. But in today's routers, fast CPUs and ample memory are prevalent allowing the OSPF protocol to function in an efficient manner. Figure 16.19 illustrates an example of the OSPF protocol frame.

Address Resolution Protocol (ARP)

The ARP protocol is a companion protocol designed to help determine an Ethernet address when using IP-oriented Ethernet. ARP requests must be sent as broadcasts, because there is no way that an ARP request can be sent directly to the right system. When the

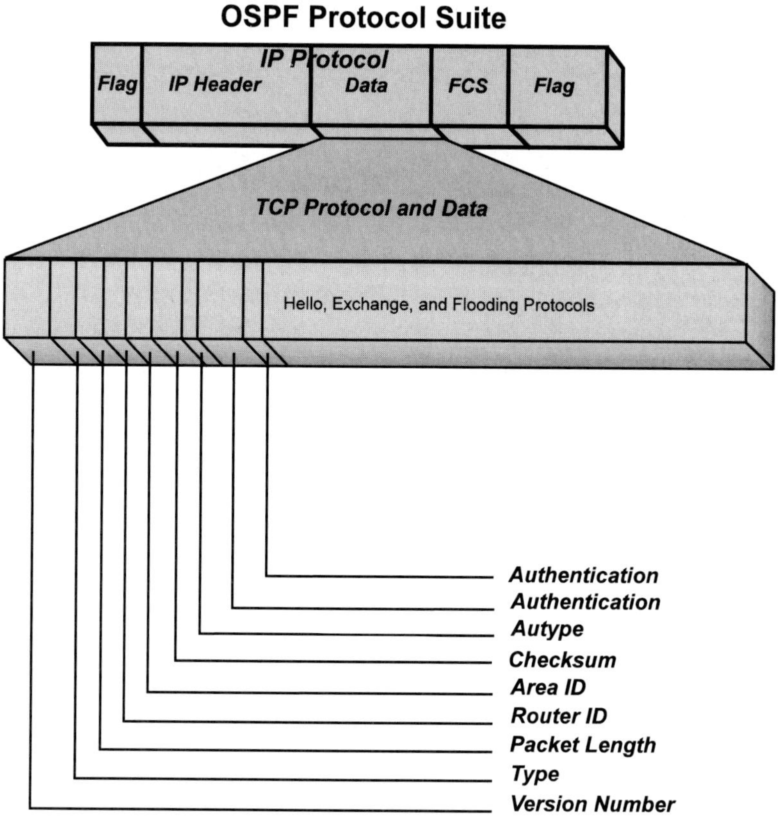

Figure 16.19 Open Shortest Path First Routing Protocol

ARP protocol is used, it functions by establishing an ARP table containing Ethernet NIC addresses and IP addresses.

As an example, a connection is needed to IP address 125.10.15.2 from IP source address 125.10.15.7. The source node will first verify that 125.10.15.2 is on the same network. This will allow it to talk directly via Ethernet. Once it has completed the verification, it will look up 125.10.15.2 in its ARP table to see if it already has the Ethernet address. If so, it will reformat the IP packet with an Ethernet header, and transport the packet to its Ethernet destination.

If the destination IP address is not in the ARP table, it uses the ARP protocol to send an ARP request. An ARP request simply asks for the Ethernet address assigned to

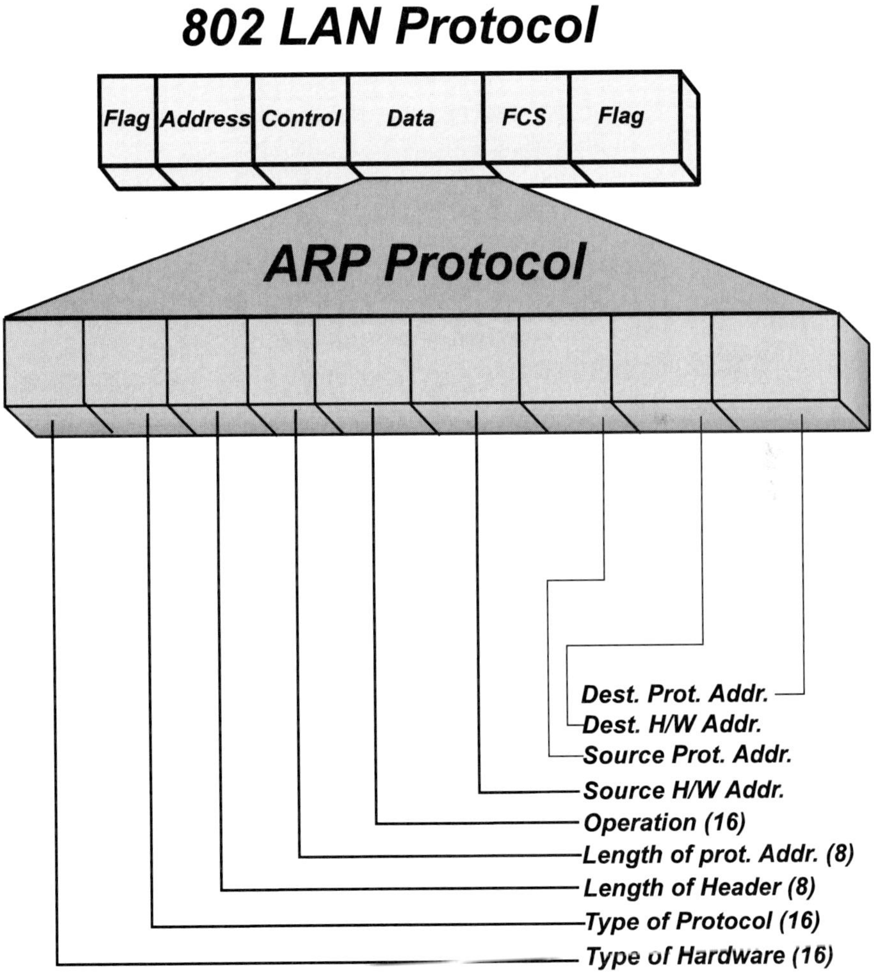

Figure 16.20 Address Resolution Protocol Format

125.10.15.2. All connected networks listen to ARP requests. When a system sees an ARP request for itself, it is required to respond. The node-assigned IP address125.10.15.2 will see the request, and will respond with an ARP reply providing the Ethernet NIC address. The system will save the information in its ARP table, so it can be used again if necessary. Figure 16.20 illustrates an example of the ARP protocol frame.

MAIL PROTOCOLS

Mail protocols are considered to be application layer protocols and are designed to allow you to send messages to users on other computers. These services are typically supported with most implementations of TCP/IP.

Simple Mail Transfer Protocol (SMTP)

SMTP is an electronic mail system that is robust enough to run on the entire Internet. This protocol allows for the exchange of electronic mail between two or more systems and allows users to transmit mail messages between other users. It is one of the most widely used applications of the TCP/IP protocol suite. SMTP in reality is a server-to-server application that consists of two components and a series of commands. The two components are used for sending and receiving mail messages to each other. The sender establishes a TCP session with the receiver and mail messages are exchanged.

Address Field Format. The address has the general format of local-part@ domain-name. By this, you should recognize the domain name format. As an example, an address at the SMTP header could be j_smith@ix.netcom.com. This indicates that the

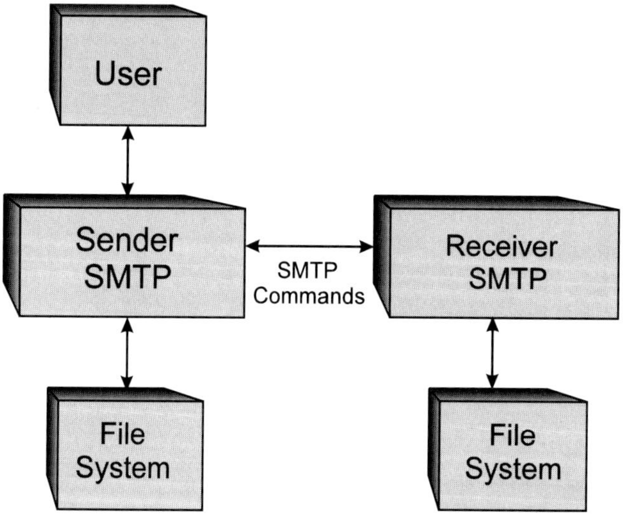

Figure 16.21 SMTP Flow

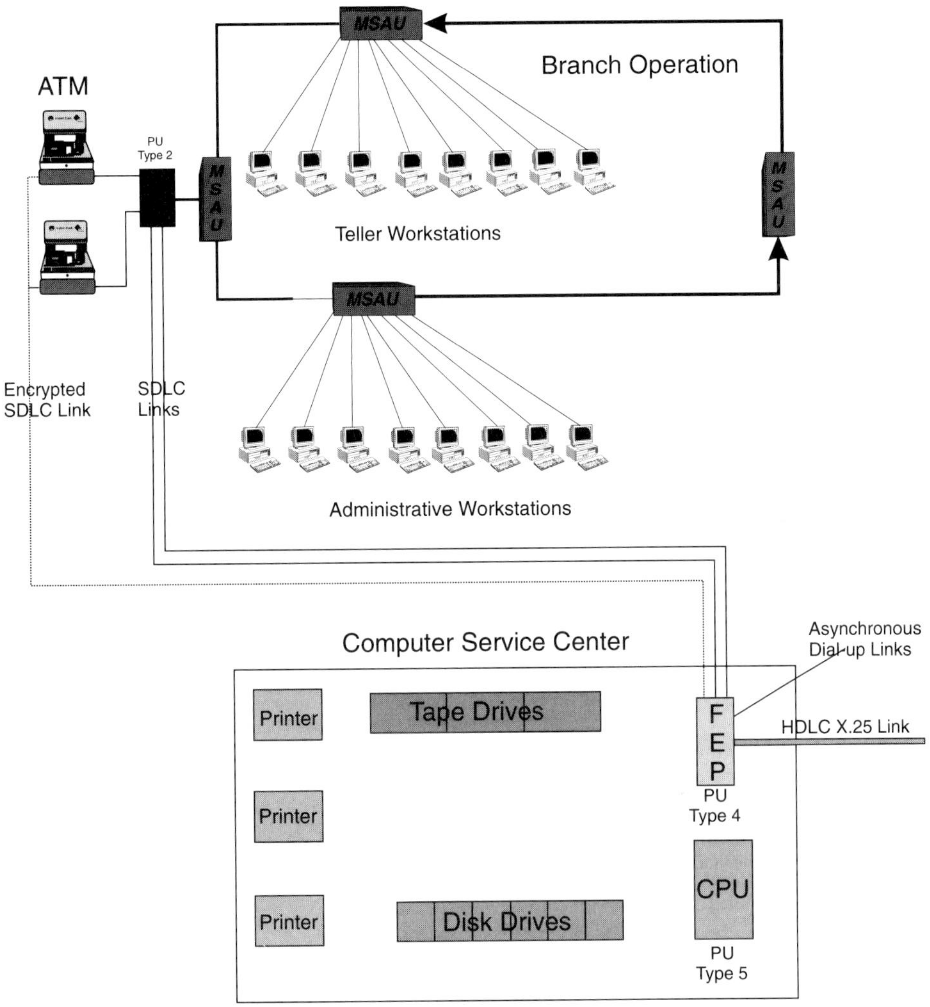

Figure 16.22 Bank of Scotts Valley Link Protocols

message is addressed to a user named j_smith at the commercial entity known as Netcom.com, specifically on a computer named ix.

Post Office Protocols (POP)

The post office protocol is used to transfer mail between desktop stations and a mail server running SMTP. A users outgoing mail is transferred to a server using SMTP. For incoming mail, users can view a list of waiting items and their sizes and can selectively retrieve and delete mail items. Figure 16.21 shows an SMTP flow between servers and clients.

CASE STUDY (CONTINUED)

Communications between each of the workstations at the branch level, as well as from the branch to the service center, will use SNA SDLC in a full-duplex environment. To guarantee security, the bank has decided that data will be encrypted to scramble the EBCDIC code set whenever it is transmitted. One of the major features of SDLC that is important to the bank is the protocol's ability to provide a high level of error detection. SNA will be the transport mechanism between the branch and the service center. Within the branch, a token ring LAN will carry the SDLC frames between the server and the workstations.

The Los Angeles service center will be responsible for converting data intended for international communications. The computer at that center will convert the SDLC frames to either an X.25 LAP-B datagram-type packet or an OSI HDLC virtual circuit-type packet. The computer also will be programmed to recognize and decode TCP/IP-type packets and convert them into SDLC frames. Figure 16.22 illustrates the Bank of Scotts Valley's protocol links.

REVIEW QUESTIONS

1. Define the term "protocol."
2. What functions does data link control accomplish?
3. What layer of the OSI or SNA protocol stack is the DLC protocol?
4. What older transmission method sends one byte at a time?
5. What are the functions of control bits in bit-oriented protocols?
6. What control codes in byte-oriented protocols accomplish line turnaround?
7. What does a NAK do?
8. Name the three types of error detection in use.
9. What is the major difference between asynchronous and synchronous protocols?
10. What are IBM's current synchronous protocols?
11. What BSC response does a secondary station send when it is not ready to receive the next frame or packet?
12. For which architecture are TP and IP defined?
13. What is SNMP?
14. What is RIP?

PART VI

LANS, INTER- AND INTRANETS, TROUBLESHOOTING, NETWORK MANAGEMENT, AND FUTURE TRENDS

CHAPTER 17

Local Area Networks and Client-Server Networks

INTRODUCTION

A LAN can be defined as a network of hardware and software that provides connectivity for directly attached intelligent devices (computers) distributed over a limited physical area. With this arrangement, users can access and manipulate data in a simple, efficient, and inexpensive manner on a local level, rather than via a host or mainframe computer.

Local area networks represent the culmination of 25 years of technology development. LAN technology was first used by the cable television industry (CATV) for the transport of multiple television channels over coaxial cable to locations that could not receive broadcast signals. Because of CATV industry requirements for very wide bandwidth, a system of transmission was developed using transmitters, receivers (transceivers), coaxial cable, and connectors. This transmission system can carry dozens of television signals (multiple channels) over a single transmission medium. Xerox first employed CATV technology for local area networking in its Ethernet system. (Ethernet will be discussed later in this chapter.)

Today, LANs provide connectivity between all types of intelligent devices over short distances. Physical LANs usually are restricted to 500 meters or less; however, they can be extended for much greater distances via the use of optical fiber and/or digital repeaters. Connections are accomplished using a variety of transmission methods and communications links. The cable and wire available includes coaxial cable, twisted pairs, and fiber-optic cable. Wireless LANs are also becoming popular for locations that do not support wire. Throughput rates (bits per second or packets per second) and user services vary, depending on the LAN technology and transmission technology used.

IEEE PROJECT 802

Local area networking got its official start when the IEEE established Project 802 in February 1980, at the request of several of the organization's members who wanted to develop a standard to specify how computers and intelligent devices could communicate over a limited distance. Project 802 continued for more than two years, resolving a number of technical issues, and defining and documenting a set of technical specifications to be submitted to the main body of the organization. The results of these activities led to the development of the IEEE 802 and ISO 8802 international standards.

At the formation of Project 802, a large segment of the group wanted the IEEE membership to accept Xerox's Ethernet LAN as a standard. Designed and implemented at the Xerox Research Center in the early 1970s, Ethernet was developed to provide communications between Xerox's Palo Alto Research Center (PARC) and other Xerox facilities. The original Ethernet operated at a speed of 2.96 Mbps. By the end of 1980, Xerox had installed more than 30 Ethernet systems supporting more than 1,000 attached devices. It was in the best interests of Xerox and its installed customer base to try to have the Ethernet specifications adopted as standard.

The IEEE 802 group agreed to look at several different implementations because other manufacturers were also developing LAN technologies. Project 802 established a number of working groups to evaluate the different technical specifications. This evaluation led to recommendations for the creation of the IEEE LAN standards.

Project 802 Activity Log

February 1980. The IEEE 802 Committee was formed to pursue the specifications for local area networks and produce a working draft (Draft A).

December 1980. The committees decided to pursue multiple access methods based on submission of a joint publication on Ethernet from Digital Equipment, Xerox, and Intel.

November 1981. Draft B was circulated for voting and suggested revisions among IEEE membership.

May 1982. Draft C was circulated for voting and suggested revisions among IEEE membership.

December 1982. The 802 final draft was accepted by the IEEE. Immediately after acceptance by its membership, the IEEE, U.S. National Bureau of Standards, and ECMA endorsed three LAN standards, which became the ISO and IEEE LAN standards. IEEE 802 and ISO 8802 LAN standards are listed in Appendix A. Figure 17.1 lists the major IEEE 802 standards.

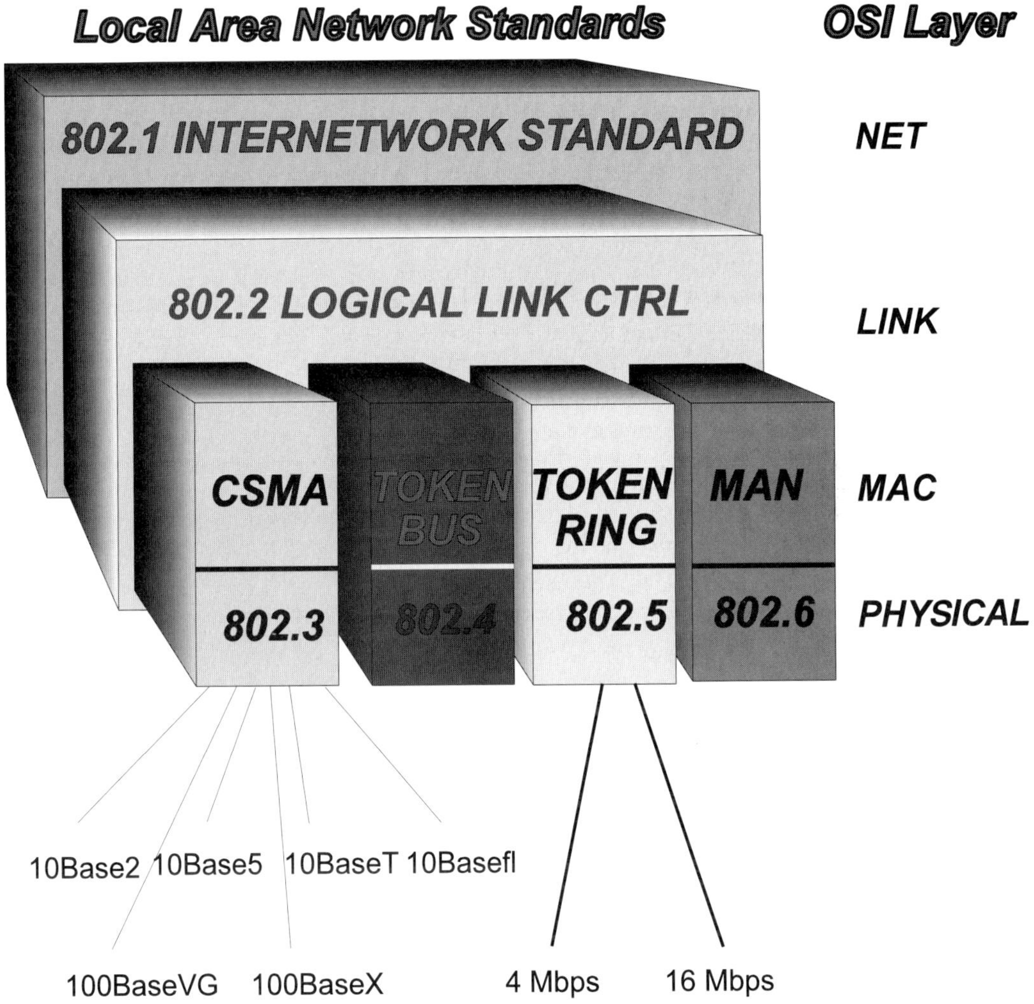

Figure 17.1 IEEE LAN Standards

LAN DESIGN

The objectives for developing LAN technology were determined by user requirements for locally shared access to common resources. Over the years, these objectives have become the overall guidelines for LAN design and are discussed below.

Media Connectivity

Media connectivity addresses the ability to link together the different devices that make up the network. Network users need to access and connect many types of devices regardless of their communications protocols and operating speeds. A flexible connectivity capability permits each workstation to access many network resources. The connection method must be cost-effective and provide the ability to extend and enlarge the network quickly when required. An important point to remember is that a cost-effective network should take advantage of any on-site wiring systems. The existing wiring must be able to support a bandwidth wide enough for all planned devices over the life of the system.

Internetwork Connectivity

Internetwork connectivity is required so LANs can access wide area networks and other network resources such as the Internet if necessary. Users often need connectivity to IBM SNA networks, TCP/IP networks, or X.25 packet networks. The connection may be either direct or via routers, gateways, and bridges.

Network Management

Network control and management are also design considerations. Regardless of the size of the LAN, network control and management may be required to maintain acceptable uptime and performance. When networks go down or grow rapidly and change configuration, the need for network control and management becomes evident.

Cost Control

Cost control may require that end users share the expense for incorporating and maintaining a LAN. Frequently, users manage and track costs associated with the implementation and maintenance of networks. It is common for management to require expense breakdowns not only for the network components, but for usage as well. This information is required to justify the acquisition of new software or hardware when the network needs to be expanded. If needed, the capabilities for managing this function should be provided by the LAN product vendors.

Security

In light of computer hackers' ability to obtain network access, security should be integral to the LAN. Security ensures that network and data integrity are maintained. Security activities include log-on and password access control, workstation access control, and public and private file access control using passwords, data compression, and encryption. If the LAN is to be connected to the Internet, additional security such as firewalls need to also be considered.

Integration

Local area networks should incorporate the technology necessary for easy integration of other networking services such as the world wide web interface, video, or audio. The integration of data, imaging, and process control systems into a single data communications network may be needed in the future. The LAN should provide capabilities for future integration of company operations without impacting the user base.

LAN Organization

Computers attached to a LAN can be configured as either clients and/or servers. A client is a workstation such as a PC that provides end-user access to application processes. A server can be a PC, a midrange machine such as an IBM AS400 or a mainframe computer. The role of the server is to provide shared access and services to the client. Using special software (network operating systems), and hardware (repeaters, routers, bridges, gateways), LANs can be logically organized into different types of networks depending upon the user needs and requirements.

One type of network is called a *peer-to-peer LAN*, where each node in the network is considered a peer level device. Peer level simply means that all devices control their own environments and, upon validation, provide permission to others when access is requested. Peer-to-peer LANs are also called combined client-server LANs, all workstations on the LAN act as both clients and servers. Figure 17.2 illustrates a peer-to-peer LAN environment.

Another type of network can be referred to as a managed *client-server network*. This type of organization defines specific roles for both client and server hardware, as well as how application programs will process information, security be performed, and internetwork communications accomplished. Figure 17.3 shows a client-server environment. In addition to the term client-server defining a hardware relationship, it is also important to recognize that client-server additionally defines how application processes relate to each other.

Often times a hierarchy will be created using multilevel servers. An example would be a network of PCs connected to a PC server that is connected to a high-end server (for example, a multiprocessor superserver such as the IBM AS 400, Digital VAX clusters or Tandem processors). The superserver in turn may be connected to a host processor such as an IBM S/390 platform. Figure 17.4 shows a multi-tiered LAN environment.

LANS AND CLIENT SERVER COMPUTING

Using LAN technologies, a client-server computing environment splits applications into tasks, and is designed to provide platform independence. Presentation of the application is designed for the client system, while data management and storage is handled on a server. Data processing is split between the client machine and the server and, by using LAN WAN technology, seemlessly connects functionality. Client-server environments can be defined as fat clients and thin servers, or thin clients and fat servers.

A fat client and thin server environment is one in which the client either downloads the application processes from the server or has them stored on their own hard drives. The client runs the entire process locally, and only uses the server for support activities such as communications and shared storage, or printing functions.

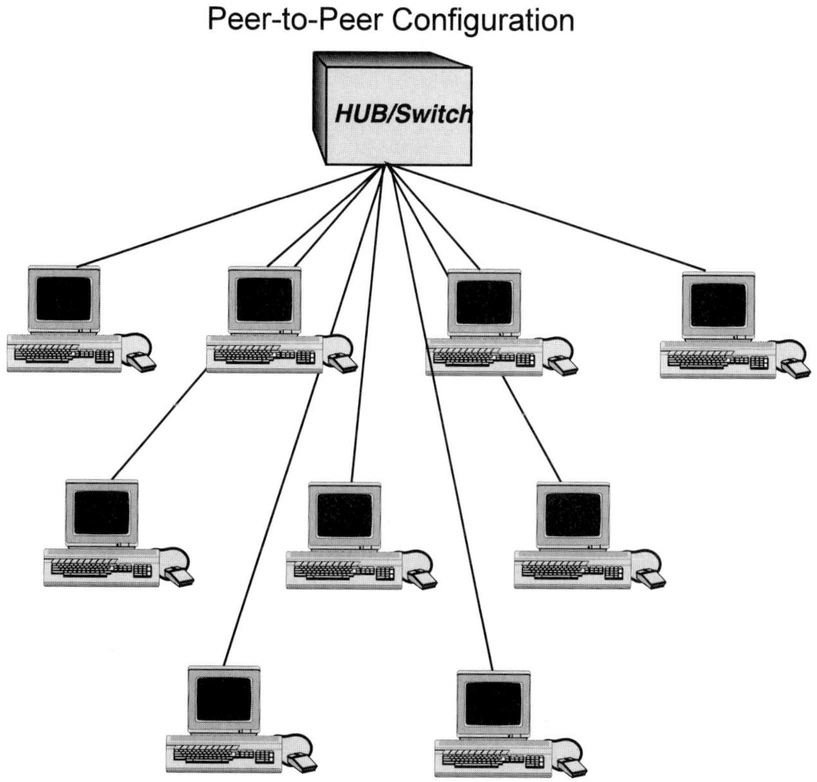

Figure 17.2 Example of a Peer-to Peer LAN

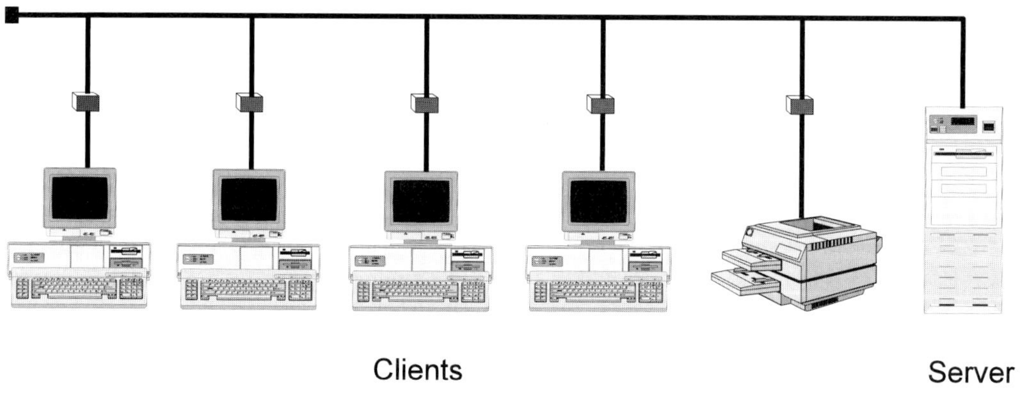

Figure 17.3 Client-Server Environment

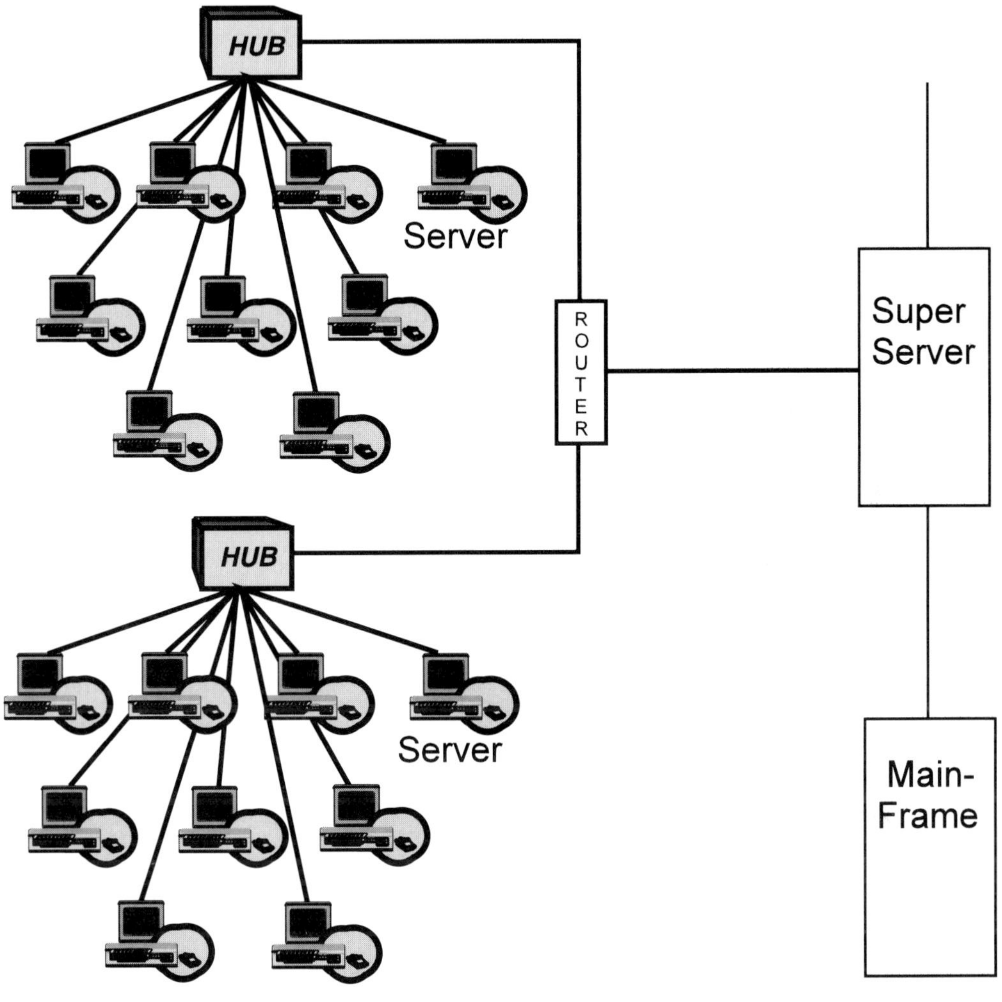

Figure 17.4 Three-Tiered Client-Server Environment

The fat server thin client model is one in which the server provides most of the application processing to the clients that need it. The client contains the interface software (GUI, application front-end, etc.) and relies on the server for most of its processing needs.

Emerging business environments for client-server computing include the following: a national and global business environment; a diversity of data types and uses; a variety of hardware platforms; a variety of software environments; and accelerated business growth.

Physical Organization

Local area networking is the emerging processing technology for tomorrow. LANs are also the fastest growing methodology for information processing and communications within the data processing/data communications industry. In its simplest form, a LAN is a hardwired self-contained network providing shared services between a group of users. For quite some time, LANs have been replacing traditional terminal networks because of their flexibility and power to provide processing at the desktop level in a shared environment. We now commonly see LANs used for internetwork communications, voice, video, and local processing.

LAN interconnection on a corporate wide basis is rapidly becoming a major objective for many organizations that need connectivity on other than a local level. In fact, many organizations have adopted Internet concepts using both Internet application software processes, such as the World Wide Web and file transfer, and the TCP/IP protocol suite to form what are known as intranets. The typical LAN environment of today provides services at a variety of levels. These include: workgroup LANs, departmental LANs, building LANs, campus LANs, and enterprise LANs.

Workgroup LANs. Workgroups provide connectivity for desktop computers within a specific department or functional area. An example of this might be a LAN designed to support billing and accounts receivable functions. This type of LAN can be considered to be either an "open" LAN or a "closed" LAN. An open LAN is one that requires connectivity for communications to other LANs within the organization. A closed LAN is a standalone network, and will not routinely be used for internetwork communications. In cases where a closed LAN needs internetwork communications, a modem will be typically used for connection to a wide area network. Figure 17.5 shows a workgroup LAN environment.

Departmental LANs. Departmental LANs can either span an entire department, or if the department is large, consist of multiple workgroup LANs interconnected together. A departmental LAN might consist of several workgroup LANs with each of the workgroups having specific responsibilities within the department. As an example using the accounting department, several workgroup LANs supporting accounts receivable, accounts payable, shipping, and vendor control might be linked together to form the departmental LAN. Figure 17.6 illustrates a departmental LAN that consists of several workgroup LANs.

Building LANs. Multiple departmental LANs linked together within a building are referred to as building LANs. Building LANs are often referred to as building backbones or backbone LANs. Simply put, a building LAN consists of multiple departmental and workgroup LANs connected together by bridges, routers or gateways, or by another LAN. Figure 17.7 illustrates a building LAN connected to department and workgroup LANs using routers.

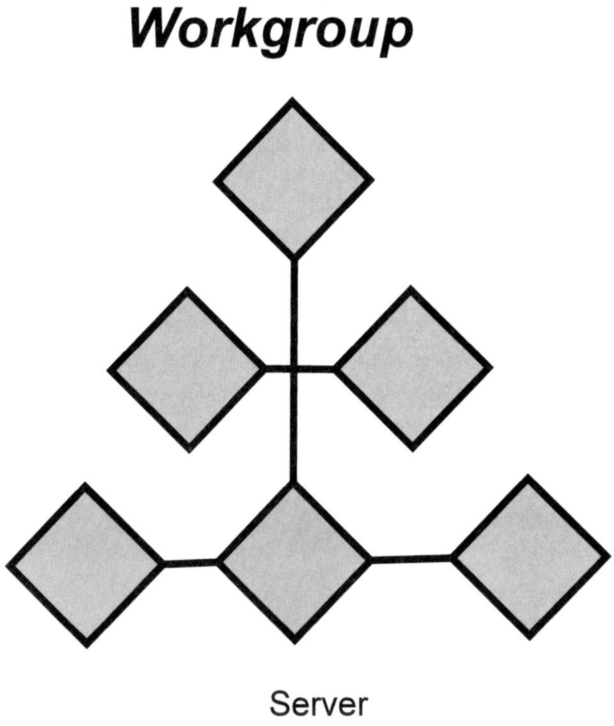

Figure 17.5 Workgroup LAN

Campus LANs. A campus LAN is a LAN connecting multiple buildings, multiple departmental LANs, and multiple workgroup LANs together to form a linked network between physical locations. Figure 17.8 shows a campus LAN backbone connecting multiple building LANs.

BASEBAND AND BROADBAND LANS

A LAN can be described as baseband- or broadband-oriented, depending on whether it was designed to use digital or analog signaling techniques. Some vendors offer LANs that support both types of transmission; however, the majority of installed LANs are of the baseband variety.

Baseband Transmission

The term "baseband" defines a single-channel transmission methodology that is mainly used for digital communications. The workstation uses digital signals to produce electrical signals (voltage) corresponding to the data it wishes to transmit. These electrical sig-

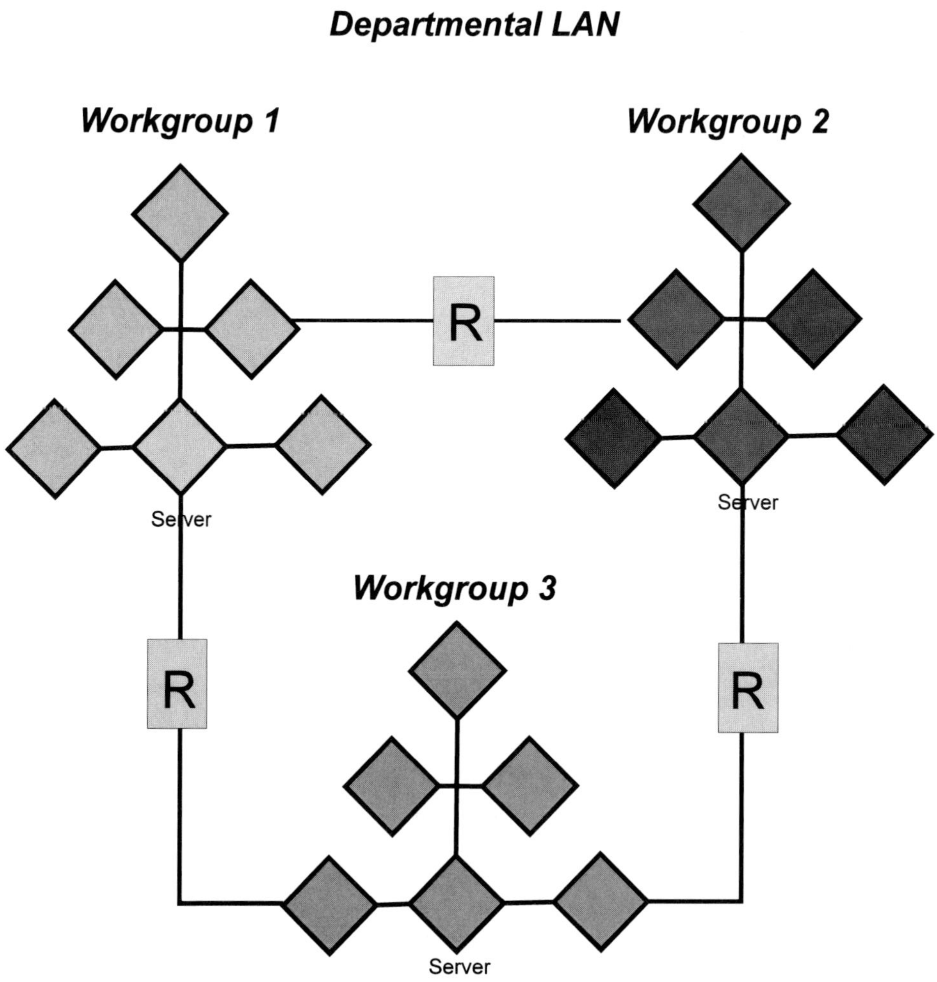

Figure 17.6 Departmental LAN

nals are placed directly onto the baseband cable. Baseband transmission allows the transmission of only one device's signal on the wire or cable because the device requires the entire bandwidth for digital transmission.

Baseband is the popular transmission method because the technology is quite mature and costs are higher with the equipment needed for broadband analog conversion. In addition, baseband systems are usually less complex than broadband systems. Most vendors have stopped producing broadband systems in favor of baseband transmission.

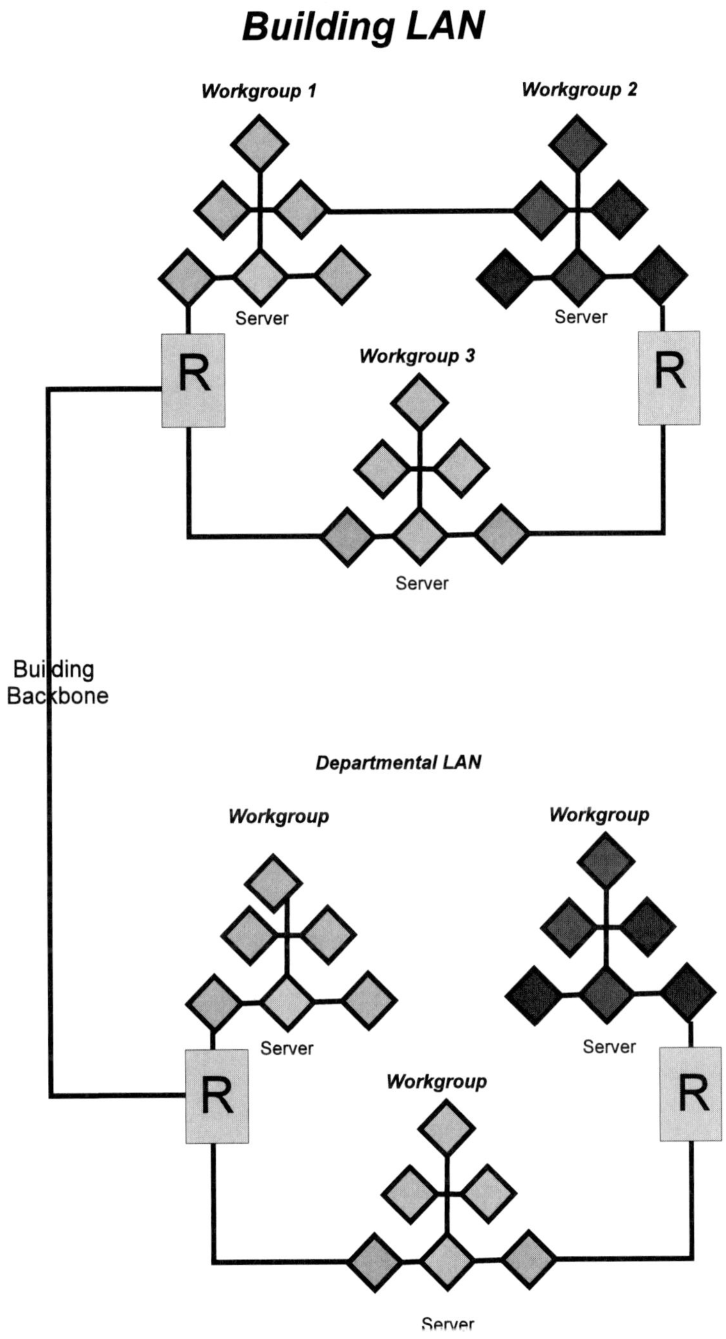

Figure 17.7 Building Backbone

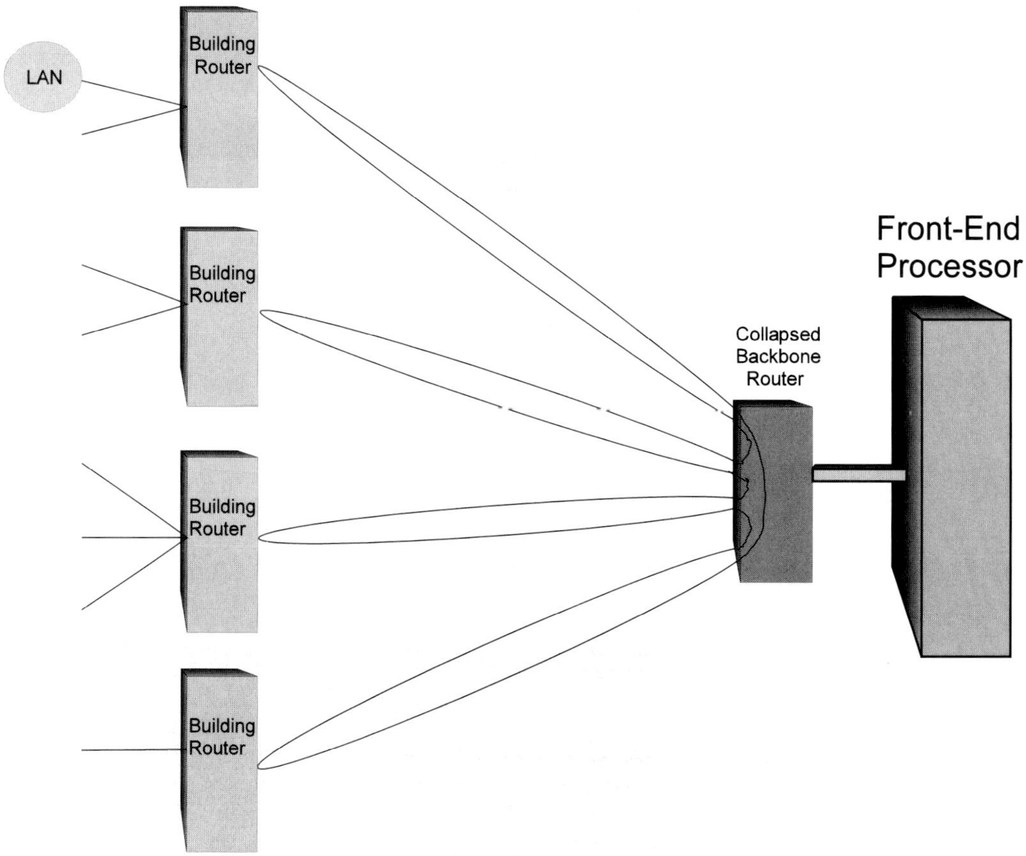

Figure 17.8 Campus Backbone LAN

Broadband Transmission

Although broadband LANs have more capabilities than baseband LANs, they are more difficult to design, implement, and maintain. They also are more expensive, primarily due to the analog components required for modulation. Differences between baseband and broadband systems depend upon the characteristics of the transmission medium (normally coaxial cable for broadband), its length, and the associated components being used.

Broadband transmission provides multiple channel capabilities dispersed across the entire bandwidth of the medium. In addition, the various channel carrier waves are modulated to represent the information they carry.

An advantage of broadband transmission is its ability to accommodate several independent transmission sources simultaneously, unlike baseband systems, which can handle only a single transmission path. Through the division and allocation of its available bandwidth, a broadband system can service data, text, voice, and video applications at the same time by allocating channels to each function.

LAN CONTROL

In order to transmit messages across a network, a communications path must be established using circuit switching or packet switching. Data is placed onto the media by the LAN node and is then transmitted to its destination using centralized or decentralized control.

With *centralized control*, a central node receives all inbound data and processes it for routing to its destination. Centralized control costs are initially high, but the incremental costs are lower. Centrally controlled environments require redundancy to ensure reliability, because a controller failure will disrupt the LAN operation. Traffic is easy to control from a central node, but routing software for handling and switching traffic can be quite complex. This type of LAN is referred to as an intelligent hub-oriented LAN, and has been implemented by both of the leading LAN technologies, token-passing rings and Ethernet.

With *decentralized control*, the medium is shared by a large number of users. As packets travel through the network, each node receives all packets, but selects only those messages intended for itself, ignoring or passing all other packets to the next node. Each node incorporates software and hardware to perform this packet selection process, as well as the routing functions necessary to forward other packets to their proper destinations.

Decentralized control is more reliable than centralized control, because a node failure does not necessarily halt network operation; rather, the failed node is simply bypassed. Software updates are more difficult in a decentralized environment, however, because each node requires the new software, and a method must be developed for its distribution.

MEDIA ACCESS CONTROL

Media access control (MAC) is the technology used to access the media. It is responsible for the access methods required for managing and controlling network traffic in a local area network. If media access control is centralized, a central controller determines when access and transmission by each station can occur. It accomplishes this by using polling techniques similar to those used with wide area networks. Stations transmit when requested to do so, or when a service request is honored by the central controller. Overhead associated with polling does, however, decrease the throughput of the network because it requires additional packet traffic for control.

Failure of the central controller disrupts the entire network. However, network node failures have minimal impact on overall operations. In this type of access, the failed node is taken off-line. Fault diagnosis and hardware/software updates are relatively easy in a centralized environment. Most centrally controlled LANs use deterministic access control to avoid contention and collision.

Decentralized media access control permits each node to be responsible for its own access to the network facilities, and for resolving contention and collision. Most decentralized environments use nondeterministic access.

Deterministic Access

When deterministic techniques are used, one method for controlling access is by token passing. A token is a specific bit pattern that indicates whether the media is available for use. Typically, only one token per network is allowed and passed from node to node. The node in current possession of the token acts as the master for all other nodes on the network, and it may require them to respond to its request for data. After transmitting, the node passes the token and associated control to its successor in a predefined order, and within a prescribed amount of time. Token control and recovery procedures are complex, but this type of access control is reliable and efficient.

Nondeterministic Access

In a nondeterministic environment, network nodes must contend for access and control of the media. If two or more devices contend for the media simultaneously, collisions will cause the transmitted data to be invalid and require data retransmission. Collision detection and retransmission control are straightforward and are based primarily on timing algorithms. This type of access control is efficient in low-traffic conditions, but is not as efficient as deterministic access for high-traffic conditions.

ACCESS METHODS

LAN media access methods are either contention-based or noncontention-based. The most common standard methods are token passing (noncontention) and CSMA (contention). Many differences exist between the two access methods and their specific implementations. The IEEE provides a standard for both token passing and CSMA. Each method has its strengths and weaknesses, and the selection of a particular method should be based on the requirements of the applications.

NONCONTENTION-BASED ACCESS

Token-Passing Ring

Token-passing LANs most often are associated with ring topologies, but may also use the bus topology. Token-passing rings use special bit patterns referred to as tokens, which are passed around the ring from node to node. In this type of LAN, no central authority controls the token (see Figure 17.9).

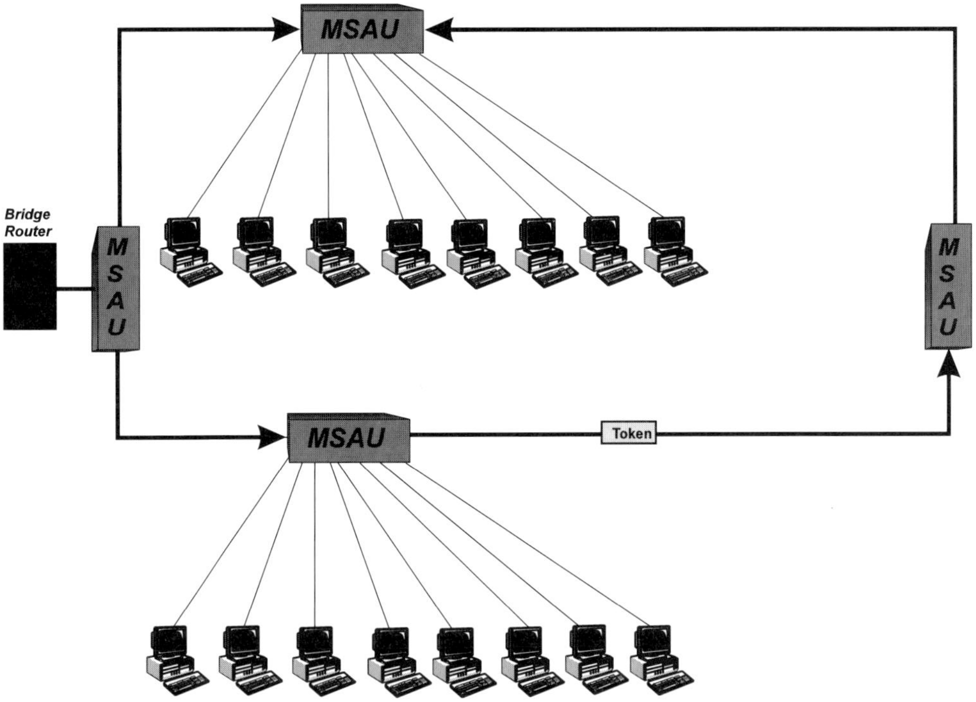

Figure 17.9 Token-passing Ring

The token normally consists of eight bits (see Figure 17.10). Token availability is represented by either a binary 1 or 0 in one of the bit positions. A zero bit indicates that the token is available and a node wishing to transmit may take control of the token as it passes by. A binary 1 is used to indicate a busy token (information has been appended to the token). The remaining seven bits represent different activities such as monitoring, reservation schemes, and prioritizing. The functions of these bits vary according to specific products. The monitor bit, which provides control capabilities, is used for the welfare of the ring operation.

As an illustration, if two tokens happen to be generated onto the ring at the same time, one of the workstation control interfaces would have to correct this condition. The interface, using a timing mechanism and the monitor bit, can remove damaged tokens from the ring and create a good token to restore operation. Lost or duplicate tokens can be caused by a damaged interface or a hit on the line.

A node gains use of the ring when in possession of an empty token. The workstation marks the token as active and appends the data behind it. Because only one device at a time may transmit in a token ring LAN, packet data byte sizes are determined by the length and number of nodes on the ring. If a station has more information to transmit than a data packet can handle, segmentation and reassembly will take place in the node's net-

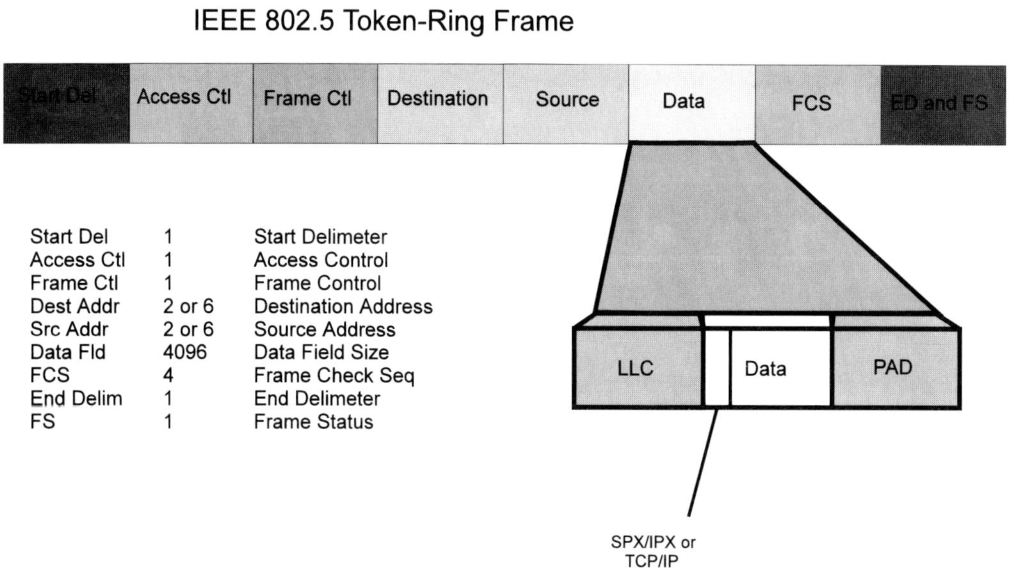

Figure 17.10 Token-ring Frame Format

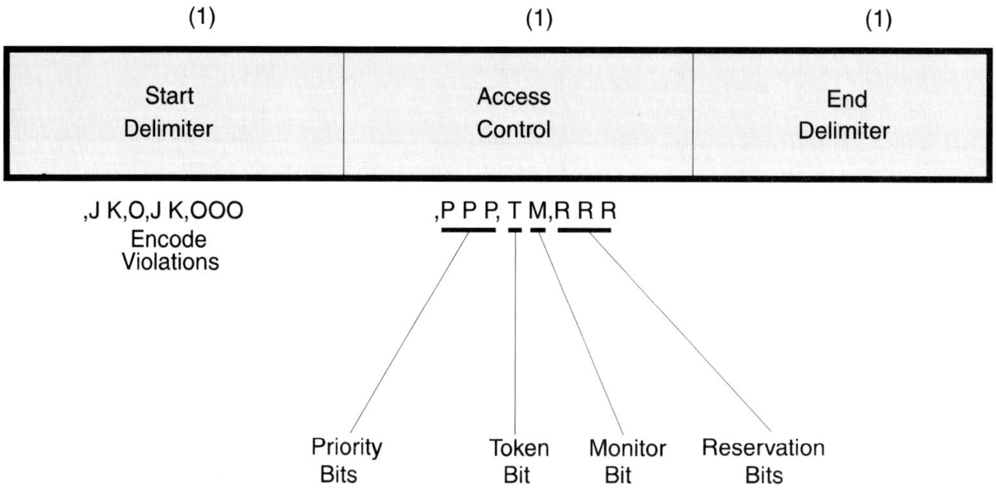

Figure 17.11 IEEE 802.5 Token Format

work interface. The sending device's interface stores the data in a buffer and transmits it in separate packets using a sequence number. Figure 17.11 illustrates the frame structure of an IBM Token-Ring network.

Token-Passing Bus

A token-passing bus establishes a logical ring structure (see Figure 17.12). Because each interface unit is directly attached to the bus, the logical ring token is passed from device to device using an addressing scheme established during network implementation. The device in possession of the token has the right to transmit. If it has nothing to communicate, it passes the token on to the next device in the logical ring. Each node maintains the address of the next node to which the token must be passed.

A node possessing the token may use the bus for a specified maximum hold time. When this time expires and the transmission is completed, or the device has no data to send, the token is passed to the next device. A specific set of algorithms governs interface

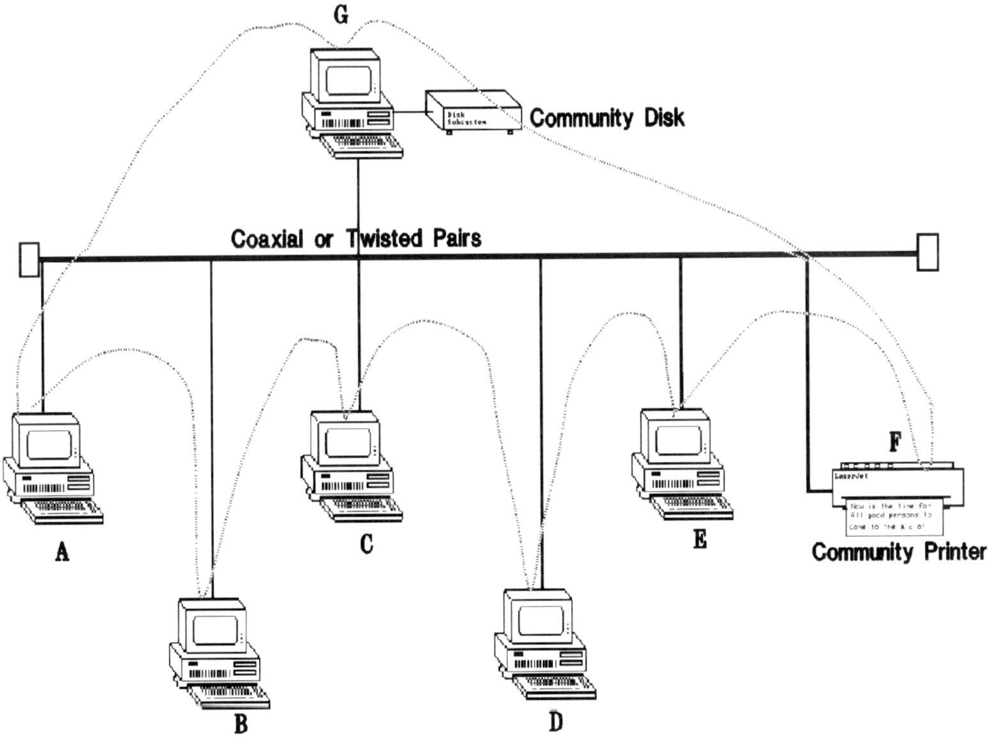

Figure 17.12 Token-passing Bus

operation. All activities on the bus are synchronized to a slot time equal to twice the maximum propagation delay plus the interface's internal response time.

Error conditions are handled by controlled contention mechanisms. As an example, device A passes the token to device B, but B fails to respond. Device A, via a time-out mechanism, recognizes this failure. Device A is responsible for re-establishing linkage for the logical ring. Device A must reissue the token, which must then be followed by carrier-free frames or quiet windows, to be broadcast over the channel.

Devices with data to send could, at this point, respond to these windows by demanding membership to the ring via a set-next-interface frame. This demand would indicate the new member's address. Device A then would respond by issuing a token link-up to the new member's address, restoring operation to the ring.

With token passing, either bus or ring, the size (number of bytes of data) and number of devices have a great effect on the throughput. Of the token-based topologies, token-passing rings are the most common. Token-passing buses are the second most frequently used.

Token-Passing Fiber Ring

Token-passing fiber ring networks operate identically to token-ring networks that use coaxial cable or twisted-pair wire. The two major advantages of fiber are its speed and its ability to filter out electromagnetic interference. At the present time, throughput rates are in the 100 Mbps range.

Several methods have been developed to interface LAN drivers to the fiber-optic (FO) media. LED or laser emitters are the most common devices used. Interfaces include RS-232-to-FO interface drivers, RS-422-to-FO drivers, and baseband FO drivers.

CONTENTION-BASED ACCESS

Contention access methods most often used with LANs are based on a modification of the ALOHANet network approach, named for the University of Hawaii's system. The ALOHANet provides a large number of terminal users with inexpensive and easy access to central computing facilities using radio links. The ALOHA system is a broadcast radio system covering the Hawaiian Islands; the central processor is located in Honolulu.

Several hundred terminals have access to the ALOHANet using a time-sharing system. In the original ALOHA system, stations transmitted without attempting to determine if the channel was available. This allowed data packets from different sources to be transmitted at the same time, creating the possibility of a collision as they arrived at the central processing unit. A collision caused the destruction of the entire data packet. Extensive efforts have been made to develop variants of the ALOHA method that can be used in local area networks. Variations of the ALOHA version include the CSMA techniques described in the following sections. Figure 17.13 shows a bus-oriented CSMA (Ethernet) LAN.

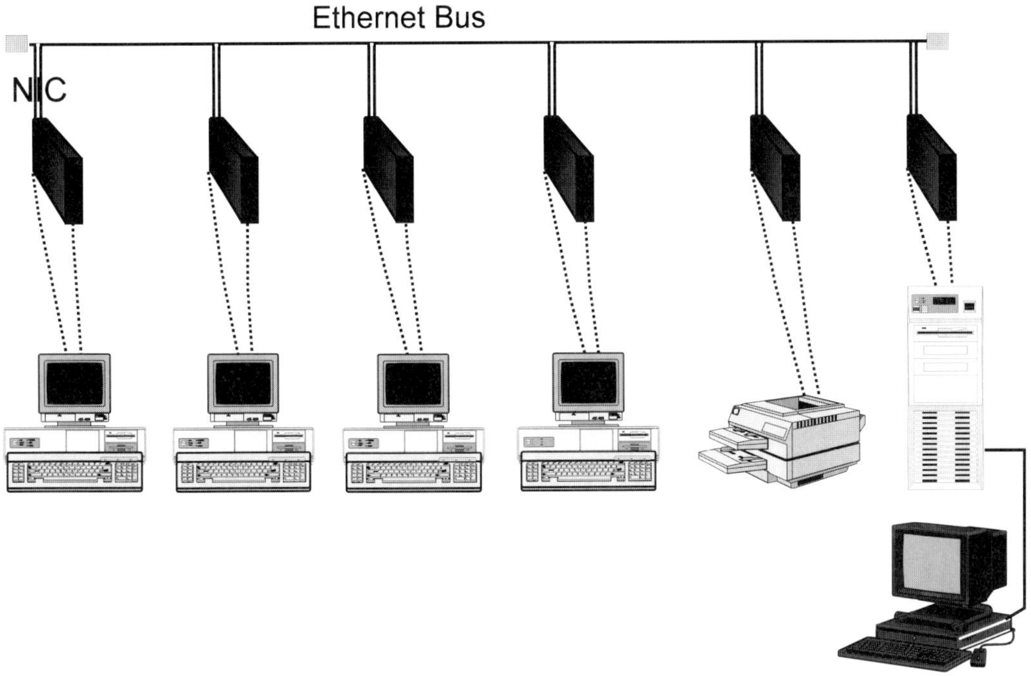

Figure 17.13 Bus-Oriented CSMA (Ethernet) LAN

CSMA

CSMA is a contention-based access method used in many bus-oriented LANs, as well as in Ethernet. With CSMA, devices must listen to the channel before sending in order to determine if the carrier has data traffic. If the bus is busy, all other stations are able to sense the traffic and hold transmission until the bus is free. Figure 17.14 depicts a CSMA frame structure.

With CSMA, a node must listen to the medium to determine if another node is transmitting. This access method is often referred to as Listen Before Talking (LBT). If the bus is free, a station may proceed. If two or more devices attempt to use the medium at the same time, a collision will occur and the data from both devices will be destroyed. CSMA resolves this problem by specifying what a station should do if the channel is busy. Three methods of resolving contention are in use: Nonpersistent, 1-Persistent, and P-Persistent

Nonpersistent CSMA requires that a station wishing to send data listen to the medium, and if the medium is idle, it may transmit. If the medium is busy, it waits a specified amount of time calculated from a probability algorithm. This is accomplished using

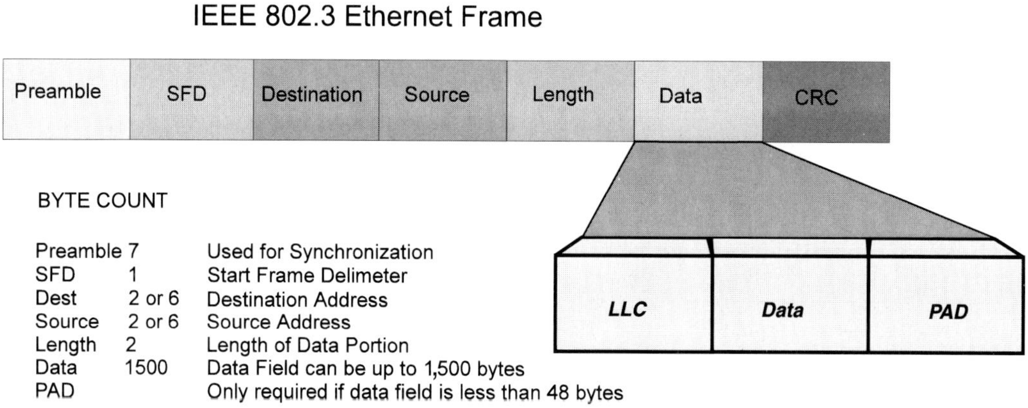

Figure 17.14 CSMA (Ethernet) Frame Format

logic built into the interface that provides a random access to the medium. When an interface encounters a busy condition while attempting to access the channel, a random wait time is invoked. If, on the next attempt, the same condition is met, the cycle is repeated until access is accomplished. The use of random timing reduces the possibility of a collision. If several stations have data to send, there may be wasted idle time following a transmission. To avoid channel idle time, the 1-Persistent access method can be used.

1-Persistent CSMA. This CSMA technique enables a station that needs to transmit to listen to the medium and determine if it is idle. If it is idle, the station may transmit. If it is busy, the device will continue to listen until the channel is idle, then it will transmit immediately. If a collision is sensed, due to lack of acknowledgment, the cycle is repeated. Although this method reduces idle time, a collision is more likely with 1-Persistent, because transmission is immediate rather than random, as with Nonpersistent CSMA. A compromise that attempts to reduce collisions and idle time is referred to as P-Persistent CSMA.

P-Persistent CSMA. Using the P-Persistent technique, a device first determines if the bus is free, then it transmits with the probability of a variable (P) plus one time unit equal to the propagation delay time. If the bus is busy, the device will continue to listen until the channel is idle, then it will repeat the first step. To make this access method feasible, different P values are established for each device to retry. These retry intervals involve spacing retransmission attempts further and further apart to maximize channel use at the cost of message delay.

A major problem with CSMA is that when two frames do collide, the medium is unstable for the duration of the transmission. For long data strings, the amount of time wasted can be considerable. This problem can be reduced if a station continues to listen to the channel after transmission has begun, thereby enabling the device to detect the collision. This access method is referred to as CSMA with Collision Detection (CD).

CSMA/CD

A CSMA/CD-based network station continuously monitors the channel after it initiates transmission. Because local area networks operate in a limited geographic area, feedback is not a problem. If a collision is detected during transmission, the device immediately ceases transmitting the data and sends a brief jamming signal to ensure that all stations are informed that a collision has occurred. After transmitting the jamming signal, the station waits a random amount of time, then attempts to transmit again using CSMA.

Collision detection saves transmission time. However, a comparison of delay between token passing and CSMA/CD indicates that the delay factor is less of a problem with token passing, because token-based system nodes do not have to contend for the link. A number of factors do affect the performance of both token passing and CSMA/CD. Currently, the most common contention-based access method is CSMA/CD, but another contention-based access method is available, CSMA with Collision Avoidance (CA).

CSMA/CA

CSMA/CA employs some type of priority scheme to safeguard the data communications of high-priority devices. The needs of these devices are catered to before the needs of the lower-priority devices.

CSMA/CA is implemented by using a delay after each data transmission that is proportional to the priority of the node. In other words, high-priority devices experience short delays while low-priority devices experience long delays. Using a priority technique of this type, packet collisions are usually avoided but still may occur on occasion. A solution is still required for handling collisions; this may be either collision detection or retransmission-based on a negative acknowledgment.

LAN HARDWARE

A variety of hardware devices have been developed that support local area networks. Many different types of components are available for use from many different manufacturers. Compatibility between different manufacturers hardware is common when they are compliant to the various standards that govern local area networks.

Because the end user generally assumes the responsibility for purchasing, installation, and maintenance of the network, care must be taken with the selection of these components. This is primarily due to the fact that the different technologies (CSMA-, token-ring-, baseband-, or broadband-oriented LANs) require different types of hardware and software.

A variety of local area network hardware is available for use. Many components from a multitude of manufacturers have been designed to support each others systems and are interchangeable providing a rich field of equipment from which to select. The following pages identify major LAN components.

Workstations

In general, workstations are usually of the personal computer variety. Some manufacturers such as 3Com and AST manufacture a diskless workstation (Intel-based 80386/486) for connection to a network. In most cases, a standard IBM or compatible personal computer may be used for workstation functions. Several companies have also introduced a device called the network computer. This is basically a medialess workstation and adds very little to the existing hardware base. Figure 17.15 illustrates a medialess workstation.

Servers

Some installed local area networks require the use of a network server or network manager to manage and control the network and its shared resources. Regardless of whether a dedicated server is required, or each node operates on a peer basis, network management functions need to be present. Often, network vendors will use the term server to identify a node designed to strictly provide shared print, disk storage, and communications capabilities to the workstations on the network.

These devices, in most cases a personal computer, may be dedicated to server functions. As an alternative, it may be a personal computer that acts as a workstation as well as performing server functions. The user must be aware that in the second case, perfor-

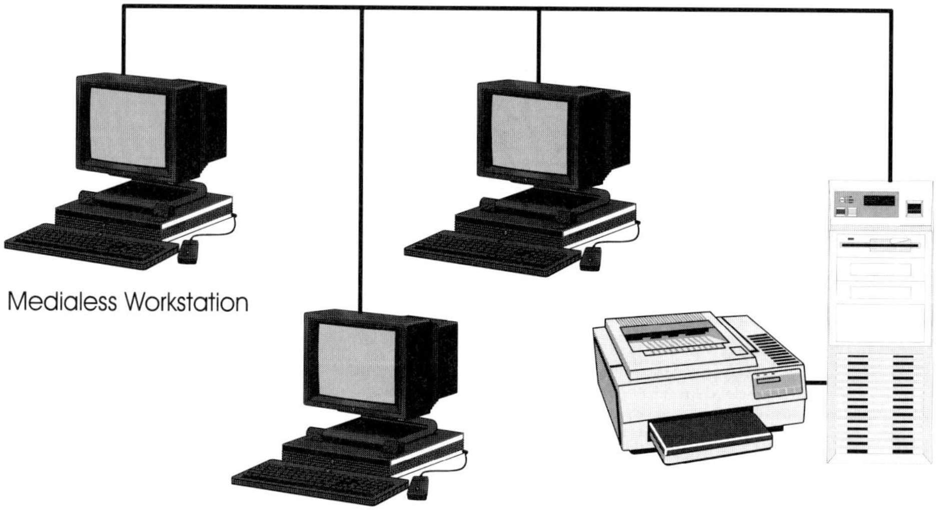

Figure 17.15 Medialess Workstations

mance will be significantly impaired unless the server/workstation is designed for higher performance.

It is quite common to find multiple or remote servers on a local area network. In order to improve performance, you might find a server allocated for gateway communications functions, a separate print server, and a separate file server for the storage of public and private files. Figure 17.16 illustrates an example of a server.

Network Printers

A network printer is normally a device, such as a laser printer, which can be a standalone device or attached to a machine designated as the print server. This network printer provides print access to all workstations on the network and, usually in association with the network server, provides spooling and queuing capabilities. Several manufacturers provide LAN-ready printers that have built-in interfaces, and can attach directly to the LAN.

Shared File Systems

One of the major advantages provided by local area networks is that of shared disk storage. One commonly finds large capacity disk storage attached to a server device that provides private, as well as public, access to all workstations on the network. Password protection and security are some of the features that may be incorporated into the file server software environment. Because file servers are the most vulnerable part of a LAN, many

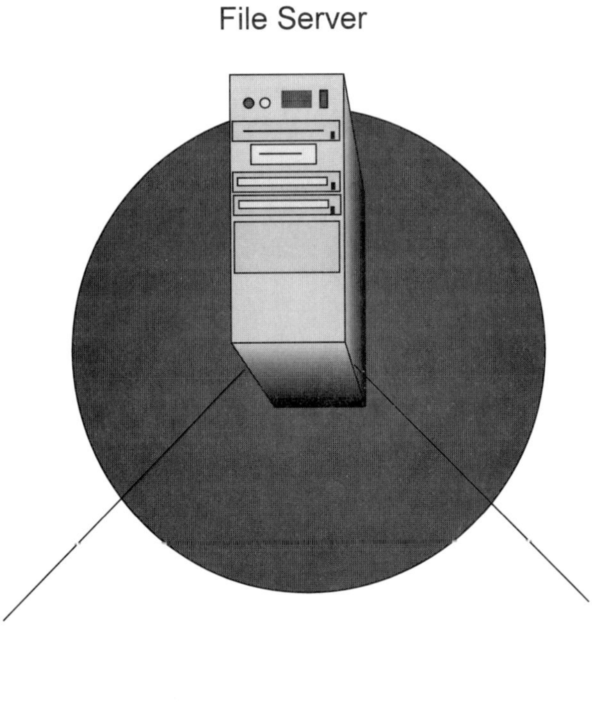

Figure 17.16 Example of a File Server

vendors are providing fault tolerant capabilities in the form of mirrored disk storage systems and recovery restore software.

Network Interface Devices

A tremendous variety of network interface devices are available for use by the multitude of local area network products on the market today.

Network Interface Cards

The network interface card (also referred to as a NIC, Ethernet or token-ring/bus card) provides a connection between the user workstation/personal computer and the network media. The Ethernet or token-ring card contains a microprocessor such as an Intel 80186, as well as memory, so it can operate independently of the workstation. This is used for transmitting high-level protocol frames while the PC continues application program processing. Most boards today interface with the PC using a 16-bit data path which provides greater throughput than eight-bit boards. The cards attach to the media using either an AUI 15-pin connector, a BNC type coax connector, a DB-9 connector, or a RJ-45 twisted-pairs telephone cable. Figure 17.17 shows an example of a NIC card.

Cables. For analog based coaxial CSMA and 10Base5 thicknet cable, the coaxial AUI transceiver cable connects the CSMA/Ethernet board to the media using one of the

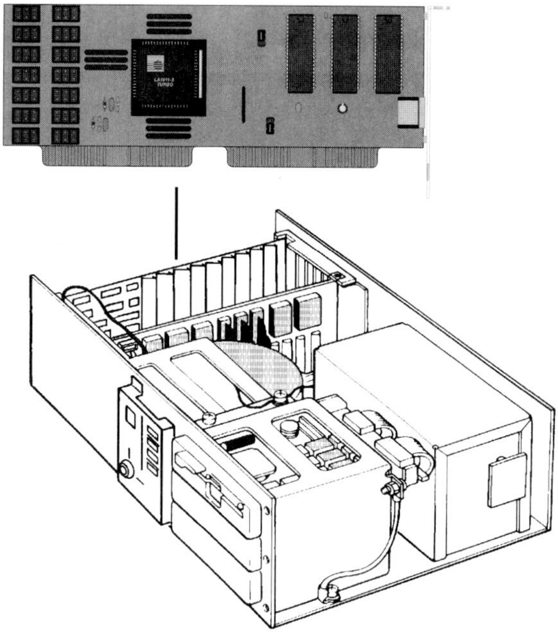

Figure 17.17 Network Interface Card

connectors mentioned above. 10Base5 Ethernets use 52 ohm coaxial BNC and T-connectors. Token ring boards can be connected using the DB-9 connector or the RJ-45 modular telephone connector. Cable lengths for CSMA/Ethernet type networks can be up to 50 meters between the workstation of the LAN media. For token rings this can be 100 meters.

Multi-Station Access Units (MSAU). An MSAU (also referred to as a MAU) acts as a hub on a token-ring network and allows for the attachment of up to eight workstations using the connectors and cable already described. The MSAU contains all of the electronics to send, receive, and manage the encoded data moving between the PC and the media. This device can be viewed as a type of modem with some variations. The difference is in its use of digital transmission techniques rather than modulation techniques. A standard MSAU has little intelligence; however, a number of vendors have enhanced the standard MSAU with intelligence so that remote monitoring and control can be implemented. Figure 17.18 shows placement of a MSAU in a token-ring network.

Transceivers. Transceivers are primarily designed to provide connectivity to CSMA analog broadband types of networks. Two types of transceivers are used, a single user type that is contained on the Ethernet board and connects the PC to the coaxial media, and a hub type of transceiver for clustering workstations together.

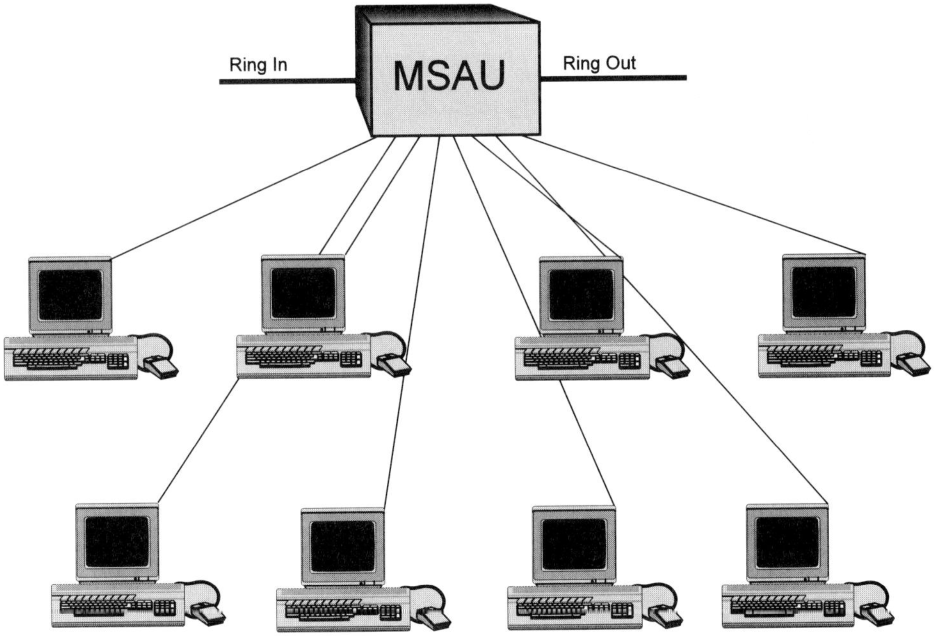

Figure 17.18 Example of an MSAU used in a token-ring Network

Frequency Agile Modems. Frequency agile modems are used with analog broadband or carrierband type systems. They take digital signals from the workstation and modulate it onto the LAN media. The term frequency agile implies that the modem has the ability to vary its output frequency to conform to channel assignments for the network. Because broadband-based LANs have access to many channels on the coaxial media, this type of modem must be setup and tuned to the channel allocated for the LAN.

Hub. Hubs are typically used with CSMA (802.3 10BaseT) type networks and they provide connectivity for multiple CSMA workstations. Hubs can be either passive or active. Passive hubs simply provide connectivity for eight to 120 workstations. Active hubs not only provide connectivity, but also provide active monitoring and network management on the LAN. Multiple hubs may be linked together using repeaters.

Ethernet Repeaters

The maximum distance for a standard Ethernet is 500 meters. This is based on the IEEE 802.3 specifications. This will also vary depending on the type of media used (twisted-pair, thin or thick coaxial, etc.). Often, users will find themselves in locations outside of normal distances supported by a CSMA network. Several manufacturers make repeaters available for extending CSMA/Ethernet networks. They allow the LAN to be extended up to a maximum distance of four kilometers. Three repeaters are usually the maximum number allowed on one LAN.

Token Ring Repeaters

The maximum distance for a token-ring network using MSAU devices is 375 meters. This is based on the IEEE 802.5 specifications. Token-ring repeaters are available to extend the ring typically to 750 meters, however, some repeaters will go further.

LAN SOFTWARE

Various software products have been developed for use in a local area network environment and a wide variety exists. This software can be categorized as: systems software, communications software, and applications software.

Because the end user generally assumes the responsibility for purchasing, installation, and maintenance of the local area network, care must be taken with the selection of software components. This is because software is designed to provide specific activities for specific environments and some may not be suitable for certain applications.

NETWORK INTERFACE SOFTWARE

Network interface software is responsible for ensuring that communications can take place between workstations on the network as well as between multiple networks and/or host computer systems. Depending upon the type of LAN installed, different network software components will be used.

NetBIOS. Originally developed by IBM and Sytek, network basic input output system (NetBIOS) can be considered a layer of software designed to link a network operating system with specific hardware. NetBIOS allow programs written using the NetBIOS interface to operate on token-ring and Ethernet networks. NetBIOS has been adopted as a de-facto standard and now its common to refer to NetBIOS-compatible LANs. It offers LAN applications, a variety of hooks to carry out inter-application communications, and data transfer. Essentially, NetBIOS is a way for application programs to talk to the LAN network. Other applications interfaces also exist such as IBM's APPC and Novells ODI. To run an application that works with NetBIOS, a non-NetBIOS LAN operating system or network interface card must offer a NetBIOS emulator. Today, many vendors provide a version of NetBIOS to interface with their hardware or emulate its transport-layer communications services in their network products. For medialess workstations, the NetBIOS is contained in a special chip along with code for establishing and maintaining a connection to the server when power is applied to the workstation.

There is no inherent routing or internetworking capability in NetBIOS. There are two NetBIOS modes. The datagram mode is the fastest, but does not guarantee delivery. It uses a self-contained packet with send and receive name, usually limited to 512 bytes. If the recipient device is not listening for messages, the datagram is lost. The session mode establishes a connection until broken. It guarantees delivery of messages up to 64 KB in length. NetBIOS applications can be run in any network that supports the NetBIOS protocol.

NETBEUI. NetBIOS Extended User Interface (NetBEUI), is another interface driver similar to NetBIOS, however, it provides an interface to the IEEE 802.2 LLC2 protocol for network-to-network communications. Implementation of the NetBIOS transport protocol within LAN Manager, LAN Server, and Windows NT communicates to the network adapter via NDIS which is described below.

ODI. Open Data-link Interface (ODI) is a device driver standard from Novell. ODI provides the ability to run multiple protocols on the same network adapter card. It is an interconnectivity strategy that adds functionality to Novell's NetWare and network computing environments by supporting multiple protocols and drivers. ODI's benefits allow you to: expand your network by using multiple protocols without adding network boards to the workstation; ODI creates a logical network board to send different packet types over one board and wire; it has the ability to communicate with a variety of workstations, file servers, and mainframe computers via different protocols without rebooting your workstation. For example, a PC can connect to a NetWare network running SPX/IPX and a UNIX network running TCP/IP; it also allows one transport protocol to run over different network adapters; for example, SPX/IPX over Ethernet and token-ring.

NDIS. Network Driver Interface Specification. A device driver specification co-developed by Microsoft and 3Com and supported by many network card vendors. Besides providing hardware and protocol independence for network drivers, NDIS supports Windows 95, MS-DOS, and OS/2. It offers protocol multiplexing so that multiple protocol stacks can coexist in the same host. NDIS is conceptually similar to ODI.

NETWORK OPERATING SYSTEM SOFTWARE

Network operating system (NOS) software is responsible for managing many functions within the local area network environment. Installed on each node of the LAN, it controls access to shared resources, insures the integrity of the network, provides security, and supports many common application functions such as sending and receiving faxes and sharing of resources. Figure 17.19 illustrates the typical functions of a network operating system.

Systems acting as servers will have access to NOS functions not available to clients. This is because servers are responsible for providing client management activities. NOS software in the client workstation will provide access to the resources of the server, it also controls media access.

Network Operating System Functions

File Management. A major feature of all network operating systems is file management support. It is used for managing the shared disk resources the server and all workstations on the LAN need. Functions include: management and access control for both private as well as public files; file access security; read-only access; queuing of file I/O requests; and providing a repository for both systems, as well as applications software.

Print Management. Another activity that is typically part of the network operating system is printer management software. This software is used for providing access to the network printers and provides queuing of files to be printed (spooling), priority management, and printer management.

LAN Communications Services. Almost all network operating systems manufacturers provide communications software for interconnection to other local area networks as well host computers. This usually requires communications interfaces such as those provided by bridges, gateways, and routers. The communications software and services allow connectivity to other networks such as the IBM SNA environment, DECNet, and other local area networks. Sun Microsystems TOPS is an example of software that provides interconnection between networks in that it allows the interconnection of Apple Macintosh computers and IBM PCs.

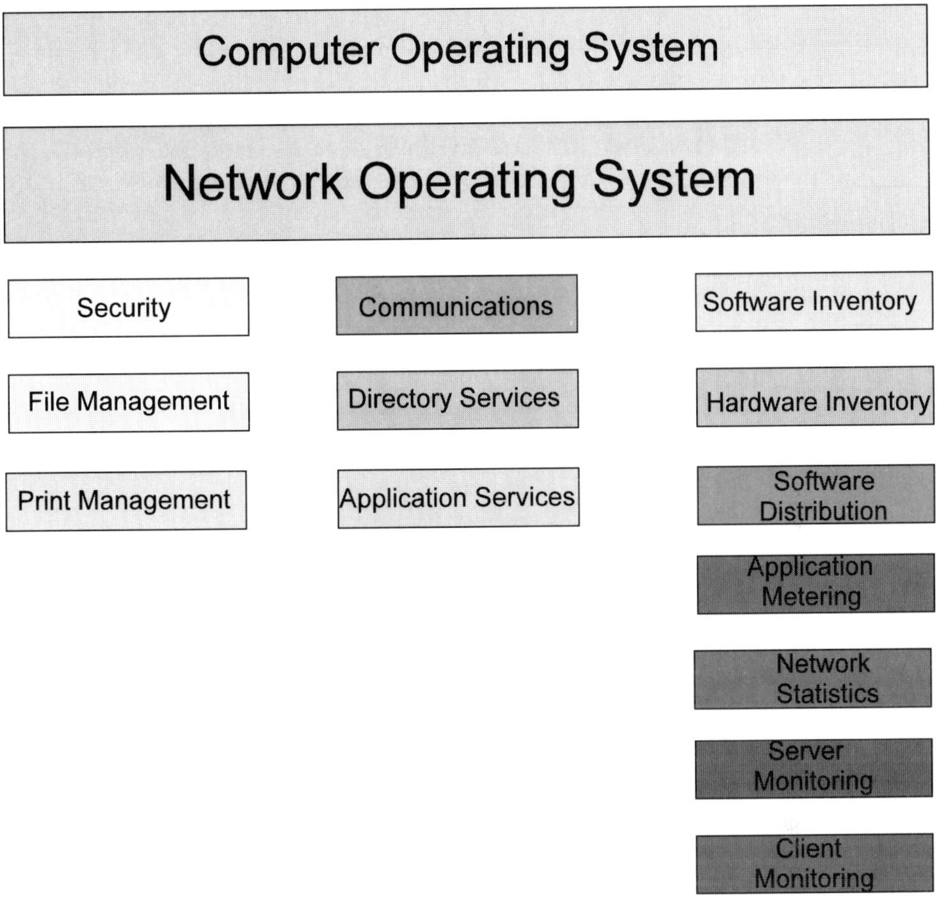

Figure 17.19 NOS Services

Security Services. Another feature of most LAN network operating systems is one of security. A typical LAN operating system will provide logon/logoff control, password protection at logon, and file control for both clients, as well as servers in the form of access protection for files and directories. It will allow users to establish private directories that require password control for access.

Network Management and Control

Network management and control varies from NOS-to-NOS; however, this would typically include the ability to monitor the performance of a network in terms of the number of good packets sent and received, the number of errors, the amount of delay, for Ethernet the number of collisions, etc.

Extensive network management capabilities above and beyond the basic features provided by the NOS are usually add-ons in the form of utility suites. A number of software provides offer these suites which include the capabilities to provide the following:

Hardware and Software Inventory. These tools provide network managers with the ability to keep track of what is installed on the network, and assists in managing and planning future upgrade needs. Hardware inventory information includes: types of hardware, amount of memory, BIOS configurations, hard disk sizes, and CPU speed. Software inventory keeps track of executable programs on each client, as well as version control.

Software Distribution. Software distribution allows the network administrator to perform remote upgrades to client machines. Typically, this is accomplished using scripts that manage the distribution and installation of packages prepared by the administrator.

Application Metering. Network applications are typically licensed for a given number of users. Application metering manages application usage to insure that no more than the maximum number of users for a license are active at any point in time.

Network Traffic. It is often desirable to have the ability to monitor specific clients as to their usage of resources. Network traffic monitoring capabilities above and beyond those that are provided by the NOS are desirable in order to pinpoint specific pockets of activity.

Virus Protection. This should be performed on both the server, as well as each client, on a regular basis. With the proliferation of viruses, this is a vital activity for any networking environment.

APPLICATIONS SOFTWARE

Hundreds of applications packages exist in support of local area networking. They include: network word processing systems, desktop publishing such as Corel's Ventura, computer aided design (CAD) packages (Autocad), accounting packages that include Lotus, as well as programs and systems that support virtually any area the user wishes to place in the LAN environment. Three general operating environments exist for client server-based applications.

- Standalone LAN-based Server with Micro Clients
  - No Connectivity to mainframe, all data resides on Server
  - No access to other LANs in the organization
- Standalone-based Server with Micro Clients with Download Access

    - Ability to download mainframe data, no cooperative processing
    - Requires redundancy
- Connected LANs in a Distributed Environment
    - Supports cooperative processing
    - Provides data sharing across the Enterprise

Database Server Software. Database management is usually a standalone activity because of the large amount of program code and I/O activity required to manage databases. A DBMS is used for organizing, storing, accessing, and retrieving data as requested by users on the network. Additional responsibilities include maintaining integrity of the data, providing security, and establishing/maintaining data relationships. Manufacturers of network DBMS software include Borland and ASK, as well as others.

Comm FAX Software. In a client-server environment, it is expensive and unnecessary to equip all workstations with communications capabilities. Placing network capable communications fax software on a server allows all workstations on the LAN to use these resources.

General Applications Software. General applications software is available in a variety of network ready configurations. They include: word processing, spreadsheet, graphics, desktop publishing, presentation, and multimedia software.

LAN PERFORMANCE

LAN performance is affected by system parameters such as transmission rate, cable length, data packet size, and software overhead. These factors determine the filtering rate or the packet per second (pps) rate of the LAN. Generally speaking, the bit rate specification of a particular LAN technology does not imply that the LAN can actually move data at that rate. The LAN's effective throughput rate is the only true indication of its actual performance. The effective throughput rate is a measurement of how many bytes of data can be transported to their destinations over a specific time interval. This is often a difficult figure to obtain due to the amount of overhead associated with a particular LAN product.

A LAN's throughput delay is important if optimal results are to be achieved. Throughput is of particular importance when LANs are interconnected using a gateway or bridge, because of the added overhead. Delay is a measure of the total transfer time of the data, defined as the time interval from the creation of the data at the origin to its reception at the destination node. This transfer time includes queuing and access delay at the sending node, the transmission time of the data, and the propagation delay created by the media length. The importance of proper design and planning of a LAN cannot be understated. Topology, access methods, media, growth, changes to the system, and costs are all important factors.

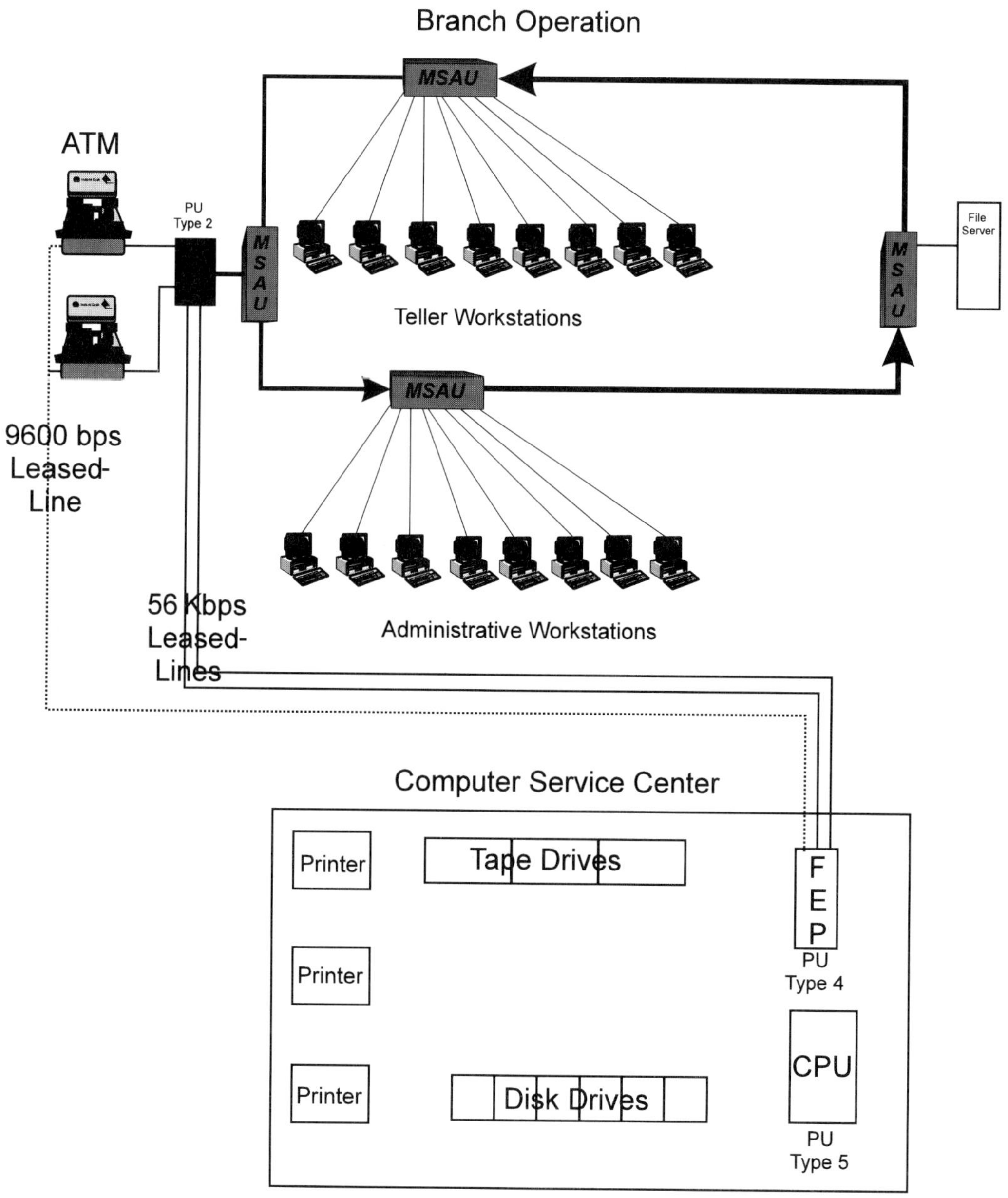

Figure 17.20 Bank of Scotts Valley Branch LAN

CASE STUDY (CONTINUED)

A major part of the Bank of Scotts Valley's branch data communications is planned around local area networking capabilities. Use of a LAN provides major benefits, including a reduction in the number of point-to-point links needed in each branch over what is required in a wide area network terminal environment. Another benefit is the reduction of costs associated with branch support. The incorporation of medialess workstations can reduce hardware expenses and still maintain improved response time because the workstations operate in a local environment.

A 16 Mbps token-ring LAN is selected for branch operations. Each teller window will have a medialess workstation connected to the ring. This workstation will be an Intel 80386-based PC with 4 Mb of memory. In addition to administrative workstations, which are fully configured PCs, each ring will have a LAN server that supports a laser printer and large disk storage. An additional administrative workstation will serve as a gateway for traffic between the service center and the branch, and will be used for routing activities. Operational software in support of the tellers will reside on the LAN server and provide inquiry, deposit, and withdrawal capabilities. The gateway node will function under control of the server. The server takes teller workstation requests and reformats them for transport by the gateway to the host computer.

Dual rings will be installed at each of the branches using unshielded twisted-pair wiring to take advantage of media already installed at each branch location. A dual-ring configuration was selected because it will provide a back-up capability if the primary ring fails. One administrative workstation within each branch will be set up to monitor and control the LAN. It will monitor the branch network, contain diagnostics, and maintain performance statistics that can be accessed by the service center if required.

The bank's information services department intends to continue to expand the use of local area networks and, at some point in the future, to have all users interconnected via token-ring LANs and communications processors located throughout the bank. Figure 17.20 illustrates a typical local area network installed at a branch location.

REVIEW QUESTIONS

1. What does the IEEE 802 standard cover?
2. What does CSMA stand for?
3. What standard is Ethernet an implementation of?
4. What is the principal transmission technique used by CSMA?
5. What does a token-ring network do?
6. What is the most popular transmission medium for LAN implementations?
7. Are LAN operating systems a required component?
8. Is spooling a function of the network operating system?
9. Does the NOS replace the basic operating system of the computer?

10. Do network operating systems provide interfaces between different LAN types?
11. Is security a network operating system capability?
12. Is e-mail a typical LAN application?
13. What is an MAU/MSAU device?
14. What function does the LAN interface card perform?
15. What distances are typically supported using twisted pair wiring?

CHAPTER 18

Network Troubleshooting

INTRODUCTION

Network failures happen! When a network goes down, the results can be serious. Today, companies are so highly automated that when communications or network failures occur, a potential loss of millions of dollars is not unrealistic.

Problems occurring within a data communications network are usually difficult to identify, isolate and correct in a timely manner. In addition, problems often exist that no one is aware of. Once a problem is detected, network components must be examined to determine exactly where the problem lies. A systematic approach to testing each component is needed to determine if the line or device is functioning correctly. During the analysis of network components, the following items should be examined and tested:

- Communications software parameters and activities
- Communications support hardware, such as multiplexers, routers, DSUs, and modems
- Front-end processors, terminals, and intelligent workstations
- Lines (both dedicated and switched)
- Connectors, cabling, repeaters, hubs, and media

After the problem or failed component is identified, correcting it may be a major task. This is because a wide variety of methodologies exist for problem identification and correction. Every organization should have a disaster recovery plan in place to address problems, regardless of the degree of failure encountered.

This chapter provides a template for error analysis and the development of a corrective action program to identify and resolve problems.

TYPES OF ERRORS

Specific categories that must be included in a problem-analysis methodology include the following:

- Software errors
- Computer and communications hardware errors
- Network link errors

Software Errors

Software problems can occur in any of the program components that make up the network environment. Application processes that are not properly tested prior to installation in the production environment can terminate abnormally when conditions occur that the program cannot handle. Often, these software errors are insidious in that a program may operate for an extended period of time without experiencing a failure. For example, consider a program that computes numerical values. It operates for an extended period of time without failure, but one day encounters a number larger than it can handle. When it tries to perform this arithmetic function, it may cause an arithmetic overflow error that abnormally terminates (crashes) the application process.

Communications system software is also subject to errors. Terminals or lines whose characteristics were incorrectly specified to the software certainly can create problems, including bit rates or operational values such as fall-back rates that are mismatched between devices. This type of error can take the device or line down. For example, a line time-out parameter value might be set lower than what is acceptable for occasional delays caused by congestion. When congestion on a link is encountered, a device attached to that line, when attempting to respond, will not be able to do so. This causes a time-out condition to occur in the control software. If the number of time-outs that take place is greater than what is specified, the link will be taken offline. Another example is an error threshold retry value not set high enough for a line that experiences frequent bit errors. If this limit is set lower than what is acceptable, the software will continually take the line down due to the excessive number of errors.

Hardware Errors

Communications hardware errors are often difficult to detect; hardware analyzers as well as software diagnostic programs are usually required to locate and correct them. One reason for this difficulty is the sheer number of hardware components that make up a typical network. Hardware also is difficult to diagnose because errors are often sporadic and the network requires extensive monitoring to locate and identify them. In many cases, hardware diagnosis requires that the network be taken down, which can compound the problem as well as upset users.

Network Link Errors

As with the other types of network errors, both software and hardware diagnostics are required to identify, locate, and correct line problems. Many times, link errors occur only once in a while, making this a difficult area to correct. An error may occur within the telephone company's network, requiring the telephone company to perform the diagnostics. However, the user must first discover the problem and determine that it does not lie in the organization's own software or hardware.

PROBLEM ANALYSIS PROCEDURES

To respond to problems on a network, you need a step-by-step procedure for error identification, resolution, documentation, and notification; in other words, it is important to have a systematic approach to identifying and resolving network errors.

Every organization should have a disaster recovery plan in place to address problems, regardless of the degree of failure encountered.

Documentation. The first component of a problem determination system is good documentation; the communications network must be mapped out in detail and kept up to date. The map must include transmission paths, routing, hardware identification, and software characteristics.

Transmission Paths. Each and every transmission path (both hard-wired and leased-line) must be documented. This includes the rated bit speed, the organization or vendor responsible for its maintenance, and a contact if a problem occurs. If it is an internal link, notes on accessibility must be included that describe access details.

Physical Routing. A routing map should show where the lines run and how to access them for repair. If a line traverses restricted areas, details on how to access it, including any legal restrictions, are necessary.

Hardware Identification. All network hardware must be identified as to the type, operational characteristics and line addresses used. Ownership and maintenance responsibilities also must be identified, as must contacts authorized to repair the components.

Hardware Monitoring Components. It is important to place network monitoring functions where they will provide complete visibility into the operation of the entire network. At minimum, major lines and hardware and software components must be monitored. In many cases, end points such as modems and terminals can be bypassed if cost is a concern. These points normally have self-contained diagnostics that can be run locally.

Software and Performance Characteristics. All software must be identified, and important configuration information must be accessible in order to determine operational parameters and performance characteristics. All staff members responsible for the care and maintenance of the software should be identified, along with information on how to contact them.

Problem Resolution Plan. A written plan that addresses problem resolution should be developed and kept updated. Step-by-step procedures must be defined for troubleshooting and resolution. Network schematics, as well as related hardware and software data, also must be part of the documentation.

Tools. You should have software in place for real-time performance monitoring of the overall software environment and the hardware components. Performance must be monitored because it is often difficult to detect errors when the network is experiencing problems but has not yet failed. Performance can be monitored with software monitors and hardware monitors attached to the network. Several software products have been developed to perform these functions. They include: IBM's NetView and Hewlett-Packard's OpenView products. Both of these have the ability to monitor hardware and performance activities across multiple networks and different hardware products.

Network Performance Parameters. To determine whether the network is experiencing a problem, operational values need to be developed for both normal and peak operational periods. Extensive testing and statistical analysis will be needed to establish benchmark figures for comparisons. This information should include activities such as average response time, I/O queue wait times, and channel/line busy activity. Transaction rates, numbers for errors encountered, and the number of retry on errors performed also should be tested.

Problem Determination

Once a problem has been detected, follow these steps:

1. Identify the problem area. Using the tools and procedures described earlier in this chapter, determine exactly where the problem resides.
2. Determine whether the problem is catastrophic or noncatastrophic. The nature of the problem normally determines how it is to be resolved. There should be procedures in place for catastrophic and noncatastrophic failures. Catastrophic failures require immediate attention to bring the network back online again. Noncatastrophic failures often can be bypassed and resolved during normal maintenance.
3. Determine whether an alternate path can be taken to bypass the error. In some cases, standby equipment is available and can be brought quickly into service. The availability of back-up links should be documented.
4. Determine whether the problem can be resolved without taking the network down. Regardless of the size of the network, bringing the network down should be the last

resort, unless it is planned well in advance and users have had ample notification. Downtime can be costly for a company.

5. Determine the mean time to repair. It is important to keep users informed of the status of the network. You should determine the mean time to repair the problem and notify users.

Problem Resolution

The following items are important for resolving network problems.

Trouble Tickets. When a problem is encountered, create (open) a trouble ticket to document the problem area and track the problem. The trouble ticket should specify the time the problem was first encountered, its location, and successful corrective actions taken.

Automatic Tracking System. An automatic tracking system is often developed to record the status of the troubleshooting activity. Categories of problems should be defined that govern the types of actions that must be taken (e.g., critical versus noncritical). After the problem is resolved, the trouble ticket must be closed, and its statistics should be included in a historical database for future reference.

Vendor Notification. If the problem occurs with leased or vendor-provided equipment, notify the vendor as soon as possible so that the vendor or equipment provider will participate in problem resolution activities and can take steps to prevent the problem from recurring.

Repair Scheduling. Often a network can be operationally maintained by bypassing the problem area. You should make sure that the component in question is scheduled for repair. This is an area that is easily overlooked if standby equipment is placed into service and the failed device is taken off the network.

User Notification. Keep users informed; an educated user understands that network failures are to be expected and will be assured that problems are being attended to if you provide status information. Such ongoing communication encourages user confidence.

SOFTWARE DIAGNOSTIC TOOLS

Software diagnostics are valuable for determining hardware, software, and network errors. They allow network providers to monitor and maintain an acceptable level of performance within the CPU and in the communications components of the network.

Software performance monitors, such as Hewlett-Packard's OpenView and IBM's NetView can take snapshots (view the operational status) at selected time intervals (usually one-second views) of all active network components and programs. The information gathered can be displayed in a real-time environment (monitored online), or can be collected and summarized for batch printouts. These are useful tools in determining real-time performance and for long-range planning purposes.

COMMUNICATIONS HARDWARE DIAGNOSTIC TOOLS

Hardware Monitors

To ensure that the communications network is performing at its best, it is important to be able to view hardware performance. Access to real-time hardware performance information gives you the ability to determine if the total environment is functioning properly. Hardware monitors provide a picture of the internal processing activity of the computer and communications network.

Modem Diagnostics

Modem diagnostics are generally included as part of the modem's circuitry. The diagnostics ensure the integrity of the modem and the attached lines. Diagnostic features are usually found in high-end modems, and depend on the intelligence and type of modem. Diagnostics that are usually internal to high-end modems include loopback testing, frequency bandwidth diagnostics, signal levels tests and system status indicators.

Loopback Testing. Loopback testing includes local and remote capabilities for both analog and digital transmission. It ensures that a signal can be propagated across the entire link. Loopback tests are usually provided in the form of a random bit generator within the modem that produces a special bit pattern that can be verified at both ends of the link.

Frequency Bandwidth Diagnostics. This type of testing is used to monitor and detect out-of-frequency conditions, because modems operating at specific frequencies may drift from their assigned frequency over time. Frequency bandwidth diagnostics make sure that frequency drift can be detected so that the operational spectrum is accurate.

Signal Level Tests. Modems must operate within a specific signal strength range as dictated by specifications set by international standards groups as well as federal regulatory agencies. An important test is the line signal level (DBM level) test, which checks the output signal strength of the modem. This provides information on whether the modem is operating within an acceptable signal strength range.

Signal Status Indicators. Many modems provide signal status indicators (front-panel LEDs) that validate the correct operation of the modem when it is active. Specific indicators are usually provided for major functions such as carrier detect, dataset ready, clear to send, request to send, and transmit/receive activities. These LEDs enable you to determine which of the modem's circuits are performing properly.

NETWORK HARDWARE DIAGNOSTICS

Hardware monitoring is desirable and may be required for many types of networks. This activity detects and allows correction of communications problems before they have the chance to substantially affect the network's performance. Network diagnostics cover a wide variety of activities; the major types of diagnostics include EIA-232/EIA-449 operation and interface performance, line monitoring for signal quality and noise, terminal monitoring using self-checking routines provided by the manufacturer and, on the network itself, the use of network controllers and monitors. Following are some of the hardware tools available to perform these functions.

Protocol Analyzers

Protocol analyzers are also referred to as line analyzers, data analyzers, and datascopes. These devices are used to monitor and evaluate line activities and data frame/packet formats. They can be used to determine the operational status of a line, device, or communications software process.

Generally speaking, a protocol analyzer is an intelligent device such as a personal computer with a CRT and keyboard. It usually provides a full-screen display, and is both programmable and menu driven. The manufacturer usually provides a simple programming language to set the analyzer to an operational status, and set alarms and triggers. Many analyzers include disk drive for offline storage, as well as a cartridge tape for backup. Most protocol analyzers also are able to support multiple host software protocols.

An analyzer can be set to capture and store information based on triggers programmed into it. A trigger is a predefined value such as a number that, when exceeded, indicates an unacceptable error level. The events that will be used to trigger the program can include specific line items or transaction types, specific packets or frames and certain types of error conditions. The frames that meet the set conditions either will be captured or counted for real-time or batch analysis. Batch capture and analysis lets you leave the analyzer unattended and still capture events and desirable information.

Many protocol analyzers include integrated timers that can be used to capture information such as line turnaround time, which is the number of milliseconds needed to reverse the direction of the line in half-duplex operations. Other information that can be captured includes:

- Response time—The amount of time it takes for a terminal to respond to a polling request
- Host response—The amount of time it takes the host to reply to a terminal activity

- Retransmit requests—The number of times a retransmit request is generated because of an error

The protocol analyzer setup is usually straightforward and must follow the manufacturer's format. Most manufacturers provide a simple, menu-driven set-up program that usually includes capabilities such as selection of a specific protocol, selection of the bit rate of the line to be analyzed, and selection of the events that need to be monitored, such as the number of retries on each transmission error.

Many companies produce protocol and data analyzers. When selecting analyzers, it is important to ensure that they are compatible with one another, because you might have to capture and integrate data for review from several remote analyzers. Companies providing this type of equipment are Digitech Industries, Atlantic Research, Hewlett-Packard, and Digilog.

Breakout Boxes

Breakout boxes allow you to view signals being transferred across the link between the terminal (DTE) and the modem (DCE). A breakout box typically has two rows of LEDs to monitor the electrical signals flowing between the two devices (DTE and DCE). These LEDs are sensitive only to electrical pulses present on the line and indicate signal activity.

The breakout box is powered by either a DC external power source or an internal battery. It usually includes two pairs of ribbon cables for attachment to the DTE and DCE. Some breakout boxes also have an audio monitor.

Many breakout boxes are supplied with bit error rate testers (BERTs), which can generate pseudo-random bit patterns that can be transferred across the link. Additional features include a three- or four-digit display for errors, error injection for proof of operation, and support for synchronous or asynchronous transmission.

Status Activity Monitors

Analog diagnostic status activity monitoring tools enable you to test and monitor the analog communications circuits for a variety of conditions. One of these conditions, signal loss, is the difference in the signal amplitude (power level) between the transmit and receive ends of the link. Loss is evaluated by measuring the amplitude at a reference frequency. Normal leased-line levels are −10 dBm at the transmit end and −26 dBm at the receiving end for an approximate 16 dBm loss.

Noise is another condition that needs to be analyzed. When line noise levels exceed the established signal-to-noise ratios, the devices at each end of the link cannot distinguish between the data signal and the noise present on the line.

Another type of problem is attenuated distortion, which causes different frequencies within an allocated bandwidth (channel) to shift their timing (speed up or slow down). This causes the information being carried to change its value.

Status activity monitors also can detect other analog problems such as delay distortion, phase jitter, nonlinearity, and transients (such as gain/phase hits and dropouts).

Testing analog circuits typically consists of level tests to ensure that all lines and components are functioning at the proper transmit and receive levels. Frequency response tests are implemented to ensure that all lines are functioning within the correct frequency bandwidth. Noise level tests determine whether the proper signal-to-noise levels are being maintained. Peak-to-average ratio tests are used to determine delay distortion, amplitude distortion, and noise.

Depending on the manufacturer, these devices are usually equipped with features such as an oscilloscope/CRT, automatic frequency sweep capabilities, a very wideband frequency range and intelligence to capture and summarize statistics. Many companies manufacture this equipment, including Hewlett-Packard and Halcyon.

Cable and Fiber Analyzers

Cable installation errors and manufacturing flaws can create continuity and bit rate problems if they remain undetected. Cable analyzers are designed to evaluate cable conditions to ensure that they comply with the manufacturer's operational specifications. Many types of analyzers are available for all types of media. They include simple devices such as those that evaluate interface cables as well as much more complex devices, such as time-domain fiber reflectometers, which analyze high-speed fiber-optic cable.

Network Control Monitors

Network control monitors, which can be either a part of a network control center operation or integrated into the front-end processor or CPU, provide visibility into the entire network for monitoring performance and detecting error conditions. For monitoring to be effective, it must be nonintrusive (i.e., it should not burden the system or affect performance of the network).

The network control monitor should provide a means for the rapid diagnosis and identification of the problem. It must enable operators to invoke tests that may or may not intrude on the network's operation. The network control monitor must be programmable to establish operational levels. When these levels are exceeded, the monitor will notify the operator. It also should provide information about error rates, network response times, and the status of all communications equipment on the network.

After the problem has been identified, the network control monitor must be permitted to isolate and take offline the portion of the network having problems. This means having the ability to shut down lines, modems, multiplexers, and even nodes. The network control monitor should enable the operator to reconfigure the network so that operations can resume.

Administrative responsibilities of the network control monitor include maintenance of a database for network configurations as well as the provision of online statistics. Port use, equipment reliability, and error rates also are important.

A second system or a shared backup facility that can be brought online via switches should be available if an emergency occurs. These systems are called hot standbys.

LEVELS OF MONITORING AND CONTROL

Several levels of monitoring and control can be implemented to achieve network visibility. One level is manual patching, which requires that modems and ports be wired through patch panels and manually switched in the event of a failure. Another level is electromechanical patching, which uses automated patch panels controlled by a local or remote terminal or computer. The use of a matrix switch is an alternate method that permits access to the communications lines. Matrix switches provide an electronic method for line control, and can be managed by a network monitor or network control center equipment.

For a higher level of network control, a network control center is required to integrate all monitoring and control functions. All points in the network are wired so they can be reached with network monitoring. Error conditions can be detected and corrected rapidly and the network can be reconfigured from a central point.

CASE STUDY (CONTINUED)

At the outset of its data processing and data communications planning and reorganization activity, the bank recognized that its number one priority is to keep the network up and running as much as possible. The planners realized that bank operations would come to a standstill if the network or service center came down. They also recognized that components do in fact fail, so they devised a plan to ensure that downtime would be kept to a minimum. The bank's approach to maintaining healthy network operation involves the following activities:

- The establishment of a help-desk staffed by data communications, network, and computer support personnel who can provide telephone assistance to any user who might need it. This might range from application support to troubleshooting activities that terminal, LAN administrators, and PC users can perform.
- The establishment of a troubleshooting system with several levels of response. The system defines actions that users and support technicians should follow if a problem occurs.
- The acquisition of local and remote software diagnostics to assist in problem determination. At each branch, an employee will be trained to run the diagnostics and interpret the results.
- The establishment of a regular schedule of preventive maintenance for branch and service center hardware.
- The complete testing of all software changes in parallel for an extended period of time before placing them into the production environment. This prevents changes from causing the network to crash due to unforeseen problems.
- The establishment of hardware reserves to ensure that critical components can be placed into immediate service if an operational unit fails. This necessitates addi-

tional funding; however, the bank believes that having standby components ready provides an additional level of safety.

It has also been decided to establish a test, development, and quality assurance lab to evaluate, test and certify any hardware and software product considered for use by the bank. This will ensure that all hardware and software considered for use goes through an extensive testing period prior to implementation.

REVIEW QUESTIONS

1. What are the three major types of errors encountered within a data communications network?
2. Name the basic procedures used to identify and resolve errors.
3. What devices perform analog and digital loopback testing?
4. What functions do protocol and data analyzers perform?
5. What function does a timer perform when used in a protocol analyzer?
6. What function does a breakout box perform on the network?
7. What is BERT an acronym for?
8. What functions do SAMs perform?
9. What activity is used for monitoring the entire communications network?
10. Why should network control centers be required within medium to large communications networks?

CHAPTER 19

Network Management and Control

INTRODUCTION

Network management has become a critical issue for the '90s. Global networks that use multivendor equipment, often in a complicated bridge and routing environment, are rapidly increasing in numbers. Controlling the resources of these networks is a vital aspect of data communications.

Managers responsible for network communications require network visibility and control to provide end users with the an acceptable quality of service. As discussed in Chapter 18, keeping the network up and providing acceptable levels of service are paramount to a successful communications environment.

Network management often is overlooked during the planning and implementation stages for one of the following reasons:

- The planned network is too small to need network management.
- The vendor will provide the necessary network management, along with the hardware and software.
- It is regarded as an unnecessary function.

But it is important to include provisions for network monitoring and management in any network plan, because the network will probably expand at some point and may require monitoring and control to maintain acceptable performance.

Proper planning is a major factor in successful network operation. Before a network is installed, consider how the network is to be maintained to achieve optimum performance. A network control center may be needed, depending on the size of the organization, the number of devices needing support and whether internetwork communications are required.

When deciding whether to implement a network control center, consider these questions:

- Can the size of the network justify the cost of implementing the center?
- Network control centers are expensive to maintain and staff. Will these costs be justified?
- What is the liability if the network goes down for a prolonged period of time?
- What will the costs be in terms of lost revenue if the network goes down and service cannot be restored quickly?
- If the network goes down, are alternate methods available for backup?
- Is network performance a key issue?

Small networks with light traffic loads may not require a network control center, however, having an ability to monitor the network is important. Even in the case of small LANs, you need to have some type of visability into its operation. Also, regardless of the environment, don't overlook the need for preventative maintenance and testing. One last point is that, often, network control centers may not be economical for smaller networks; in these cases, network control and monitoring can be contracted out to third-party vendors who will install remote monitoring capabilities on the network to be monitored.

NETWORK CONTROL CENTERS

Measurement and monitoring of networks that connect large quantities of dissimilar equipment are essential in controlling performance and keeping the network healthy. Specialized equipment and techniques are usually needed to gather and analyze performance data on fully distributed networks where there is no central control facility. The network control center should perform the following functions:

- Performance monitoring
- Configuration control
- Fault isolation
- Access security

Performance Monitoring

Most performance monitors provide programming capability to set parameters for monitoring a variety of conditions over specified periods of time. Factors that should be monitored to maintain peak performance of the network include message delay, channel throughput, message overhead, network stability, and traffic control.

Message Delay. The amount of packet delay (normal packet load versus abnormal packet load) that can occur during peak periods is a key factor in determining acceptable network performance. Delay is the result of congestion on the links caused by resource contention or equipment malfunction.

Channel Throughput. The total number of packets flowing over a specific period of time should be monitored and charted. This can be used to indicate performance degradation.

Message Overhead. The number of packet retransmissions required due to error conditions must be monitored continuously. This count is a major indicator that provides key information about any problems that might exist.

Network Stability. Statistics that relate to the number of network failures and restarts experienced over a specific period of time should be developed.

Network Response Degradation. Various traffic conditions must be identified and traffic activity monitored to indicate network or application degradation.

Configuration Control

Configuration control tools (hardware and software) aid in supporting and tracking user devices. The variety of equipment (e.g., modems, computers, terminals, storage devices, and printers) that can be connected to the communications network require a variety of interfaces.

Configuration software used with network management control tools defines and records values for node or device connections along with specific target information to be captured. This information is useful for discovering any changes to the network configuration by other than authorized personnel.

As an example, a user could change the bit rate (depending on the application) from 1,200 bps to dial out or 9,600 bps to access the host. The configuration software should be able to define parameters to allow or disallow this mode of operation, providing a level of control. At minimum, it should indicate the condition to the network manager.

Configuration tools can be used in two different ways: initializing the connection installation and using the update mode for day-to-day maintenance.

Fault Isolation

In addition to monitoring the network's configuration, network managers must be able to monitor the device connections to the cable and detect any device malfunction. Cable connectivity can be determined by simply polling the devices on the network. The network control center should be set up to initiate self-tests and to issue data loopback commands to devices on the network.

Automatic failure detection can be provided if the correct hardware and software is present. Automatic failure detection involves reporting a failure in the network to the network control center, which then executes testing functions to isolate and correct the problem. An inherent problem with any network is the failure of one or more components. However, with the proper test procedures, a fault can be isolated and corrected, ensuring the availability of the network.

Access Security

Data transmitted between computer resources may be sensitive or confidential. To ensure the protection of private information, access to devices on the network can be controlled through the software and the network control center.

Via program control in the network control center, a database can be set up to identify authorized users. This database file identifies access groups (who may talk to whom). All transmissions are routed to the network control center, which authorizes who is able to communicate on the system. A call scenario then might be as follows.

A specific user initiates the transmission. The network control center first verifies the validity of the calling node as a user of the system by examining an internal device address and a user password (ID). This provides a restriction that eliminates intruders from logging on to the system. If the sending device is invalid, the session is terminated immediately. If the check is valid, the sending device is acknowledged and queried for the destination address.

The destination address is then checked against the receiving device's user group. If common group membership is determined, a session is established. Protection of session data is ensured with encrypted data sessions; the node is equipped with an encryption chip and enhanced communications protocols.

NETWORK CONTROL CENTER ADVANTAGES

The network control center permits real-time analysis of network performance, allowing dynamic reconfiguration and modification of the network. It also allows quick response to problems, thereby minimizing downtime. Implementation of a network control center provides centralized control over maintenance decisions and assists in establishing performance criteria. Another advantage of the network control center is that it assists in reducing response time for repairs and helps to reduce maintenance expenditures. A network control center also maximizes network use before network growth creates costly modifications.

NETWORK CONTROL CENTER EQUIPMENT

Network Control Center Console Monitor

The network control equipment should be passive; it should not interfere with the operation of the network. Therefore, the manner in which the center is attached is important. The network control center normally uses a network monitor to gather statistical information. The monitor provides information to an intelligent device (usually a microcomputer) for data analysis. A storage device (hard disk) is needed during high network traffic conditions to store data for later report generation.

Artificial Traffic Generator

Artificial traffic generators consist of software or hardware designed to enable users to stress test the network to determine weak points. They can be used to present a known random distribution of traffic load to the network while data analysis software captures the information. The data gathered by the software generally includes:

- Time tagged frames—Test frames containing time and date stamps that can be used for performance tracking.
- Source and destination address—Both the sending and receiving node addresses should be included for tracking and routing activities.
- Retries—The number of retries required to move the packet from its source to its destination.
- Total transmission attempts with a valid CRC check—The total number of transmission attempts in which the packet is received correctly. The valid CRC check helps to track nondata packet errors that may result from time-outs, equipment problems, or congestion.

This data can provide valuable information for determining the quality of network performance. It also aids in long-term tracking to predict and plan for expansion. This stored statistical information serves two basic purposes: It provides information to the network manager on current and predicated network performance, and it is used to develop reports that provide information such as packet type, data size, interarrival time, channel delay, communications delay, collision count, and throughput.

The traffic generator (which is a processor in itself) is connected to communications nodes by links that generate experimental conditions. A console CRT and printer are used for network control center monitoring of real-time information and to produce hard-copy reports.

A variety of vendors provide network control center equipment. Software is offered by Proteon Inc. and IBM, among others. Hardware vendors include IBM, Codex, AT&T, and Network Equipment Technologies (NET). A variety of functions are provided by the equipment, including electronic software metering, audit trail and report generation, menu managers, and editors. Additional network control hardware functions include resource assignment matrix monitoring, universal network users, workstation inventories, resource databases, alarmed packet counts and reporting and automatic line test, and fault isolation.

CASE STUDY (CONTINUED)

The Bank of Scotts Valley cannot yet afford to establish a network control center for monitoring its communications network. This does not mean that the bank's managers think a monitoring function is unnecessary. Data communications operations and technical support personnel have implemented IBM's NetView software on the mainframe and the distributed network, enabling them to use the existing hardware for monitoring and control.

The bank's planners also have provided plans for a separate network control center to be located at a third site when it can be cost-justified. The remote network control cen-

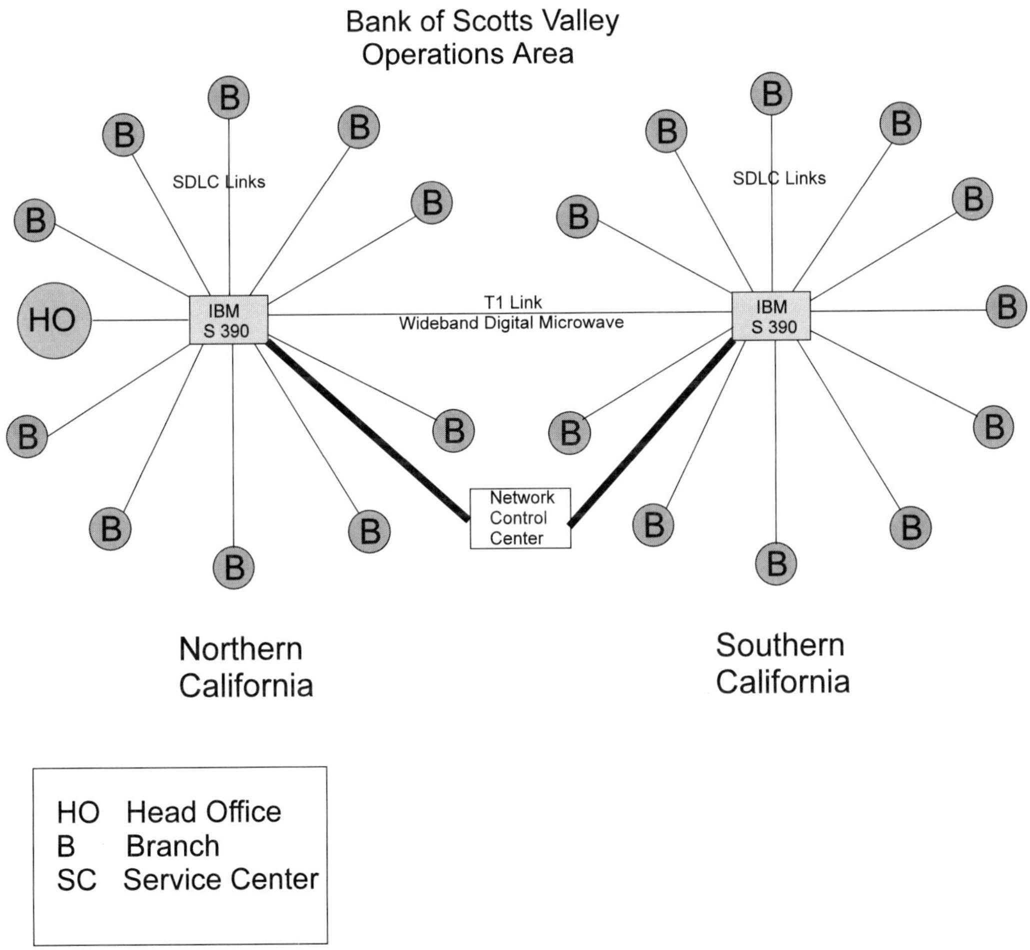

Figure 19.1 Bank of Scotts Valley Network Configuration

ter will provide network monitoring and dynamic network reconfiguration, and it will serve as a disaster recovery center should one of the service centers experience a failure.

Equipment to be installed will include a back-up IBM S/390 mainframe that will operate as a hot standby, an IBM 3745 communications processor, and redundant communications links to each of the service centers. A network controller will tie the entire network together, and all network links will pass through the network controller. (Figure 19.1 illustrates the bank's proposed network, which includes the remote network control center.)

REVIEW QUESTIONS

1. Is the network control monitor a required component in a communications network?
2. Are frame error counts important to performance?
3. Can dynamic configuration be accomplished using a network control monitor?
4. What are the functions of a network control center?
5. Are network control centers a required component for all communications networks?
6. Is a hot standby a required component in network management?

CHAPTER 20

Future Trends

THE NEXT GENERATION

Data communications continues to evolve at a rapid pace. Global heterogeneous networks are becoming the norm for many international organizations, and multiple networks that support different types of hardware and software are common today in large corporations. These companies are the driving force behind the development of new hardware, software, and standard link technologies. Users need new software and hardware to meet increasing demands for worldwide high-speed data communications.

New technical capabilities will be required to meet the ever-increasing demand for data. As the world becomes more digitized and embraces Internet and Intranet technology, bit rate requirements will grow exponentially, due in part to the demand for digital imaging and full motion digital video. Digital video and imaging require a large number of bits to represent each graphics frame; it is not uncommon to see a single video frame requiring millions of bits. Data processing and computer-developed information also will continue to grow and require communications as companies shift to remote distributed client server processing environments.

This chapter describes the hardware, software, and services we can expect in the future, based on current directions and demand. Some of these capabilities are peeking over the horizon, such as the SONET communications network. Others will not be operable for many years.

The next generation of communications hardware and software will be based on new high-speed link capabilities made available by public and private telephone service providers. Communications processors, multiplexers, and modems will be designed around the link service capabilities made available. High-speed fiber technology supporting speeds well into the gigabit-per-second rate is under development, and new hardware and software will be developed to use this high-speed capability.

Three of the basic limitations we have had to live with for some time are rapidly diminishing as hardware and software improve. These limitations include link capacity and the ability of equipment to manage communications activities at very high bit rates. Specifically, hardware cannot be built until manufacturers know what type of link and bit speeds will be available in the future. Software manufacturers also must develop new protocol and management software that can support the high bit rates provided by the new hardware. Finally, costs must continue to decline as technologies improve in order for companies to shift to them.

The next generation of data communications equipment probably will use high-speed, single-mode multifrequency fiber optics operating at very high bit rates. The fiber will be directly coupled to computers capable of handling gigabit-rated links and having processing rates well into the billions of instructions per second. Future hardware will mandate new microprocessors and greatly increased memory capacities to handle the high-speed data communications rates.

Computers

Performance limitations relating to memory capacity and processor speed are based on the physical properties of the materials and the etching technology used. Only so many circuits can be etched onto a specific size of silicon wafer (semiconductor chip). The smaller the etched circuit, the more heat generated by friction of the electron movement. When the heat reaches a certain temperature, the chip fails.

Another limitation to the development of higher performance computers and processors is the circuit speed of the chip. Depending on the circuit material used by the semiconductor manufacturer (e.g., gold, silver, etc.), the circuit speed of the chip can vary greatly between different chips on the printed circuit or etched circuit board. Also, manufacturing costs of specific circuits vary, depending on the material used in chip construction.

A new generation of computer is emerging with major improvements in capacity and performance. New architectures, new materials, and improved semiconductor/printed circuit board etching technology is enabling semiconductor manufacturers to reduce size substantially and increase the durability of processor and memory chips. A benefit of that reduction is a gain in circuit performance and, ultimately, a faster computer. Improvement in memory capacity (more memory in the same physical space) will allow much faster memory access speeds, contributing to the computer's performance.

Originally, memory and processor chips had only four-bit wide paths that were used for data movement within the CPU. Now it is standard to find computers supporting 32, 64, or higher bit paths for connecting the computer's internal components (memory, logic, I/O control). In the near future, we will probably see microprocessors and computers capable of supporting 128-, 256-, and 512-bit bus interfaces. This, however, will require new architectures and bus technologies to be designed to support the new bus structures. Larger bus pathways and new computer architectures will provide a capability for much faster data movement within the computer or processor.

Another rapidly developing area is parallel processor architecture. This type of processor architecture will support hundreds or thousands of interconnected microproces-

OPEN SYSTEMS ARCHITECTURE

In the future, vendors will have to adopt an open systems approach to systems development to satisfy users who need interoperability to maximize the investments they make in hardware and software. The following paragraphs describe the benefits open systems will offer.

Application Portability. Application portability benefits the end user and the hardware/software manufacturer. Portability allows end users to run the same applications across a broad range of equipment, regardless of manufacturer. In an open systems environment, users will be able to select the most appropriate system for each part of their data processing/data communications activities. They also will have a wider choice of applications from which to choose.

Software developers will no longer have to develop and support different versions of their applications. Software systems suppliers will be able to have a greater number of applications up and running sooner on new hardware platforms. This will be an increasingly important objective as hardware product life cycles become shorter and shorter.

An important aspect of an open systems architecture is that it is based on standards. Portable applications depend on standard interfaces. Therefore, conformance interfaces within an open systems environment must provide a substantial degree of application portability.

Interoperability. An open systems architecture will encourage independent software vendors to create newly distributed applications. In addition, users will be able to interconnect all of their systems into networks and communicate effectively within and among networks.

Flexibility. The restrictions of many older vendor-specific systems meant that meeting the remote needs of the end user was often not economical or even possible. By using an open systems approach, users will have the business benefits associated with flexibility. Individual machines will be bought as required from the supplier that offers the best deal at the time. Systems will be reconfigured to meet changing organizational needs. A distributed open systems architecture will make it possible to use large numbers of small systems in a variety of combinations to create a large communications environment. This means that special needs can be met much more cost-effectively and efficiently. Users will be able to buy the most appropriate system from an alternative supplier if the principal supplier does not have a system of that size at that particular time.

Independence from Suppliers. For the end users, a major benefit of an open systems architecture will be independence from hardware and software suppliers. Products will become more price competitive, and support and service will become key factors when users are no longer tied to a single provider.

Added Functions. With products that conform to an open systems architecture, users will be able to protect their investments in applications and data. Data is the lifeblood of many companies, retaining its value for many years. Software may also have a long life cycle. When new hardware is installed in a vendor-specific environment, both the data and the software often must be altered to operate on the new equipment. With open systems, users will be able to move from one processing/communications platform to another without converting data or applications.

Maximized Return on Investment. If users demand it, software and hardware manufacturers will have to invest heavily in open systems technologies and products to remain competitive. At the present time, most start-up companies are basing their product development on open systems specifications. There is a great deal of competition, resulting in more innovation and the introduction of more features than has been the case for some years.

An open systems architecture greatly influences the way vendors develop products. In particular, there is a marked increase in the use of standard systems and off-the-shelf components. Buying standard parts will allow system vendors to concentrate development resources on high-value product differentiations. For example, many products use chip-level implementations that can be purchased from various semiconductor companies.

THE INTERNET

With the passage of the U.S. Telecommunications Act of 1996, the Internet is positioned as a major world-wide force in providing information, services, and commerce-based activities. Because the Internet is a global network, it provides a level playing field for any business, anywhere, to compete in the worldwide marketplace. A one person operation will have the same opportunities as a major corporation such as Microsoft in bringing their products and services to the buying public.

At the present time, the Internet is primarily used for information and the exchange of ideas and software. This is rapidly changing with the introduction of newer high-speed access methods, secure transaction processing, and a host of new capabilities to display video and multimedia presentations.

More and more, companies and organizations are discovering that the Internet provides a tremendous opportunity for establishing visibility into their products and services and for providing a platform for delivering goods and services. Future services on the Internet will include:

Advertising and Company Visibility. The Internet is positioned to be excellent platform for reaching out to millions of people who regularly access the World Wide Web. Establishing a Web site or home page will allow the very same people to view information about your company and services on a 24 hour 7 day a week basis. Information such as company profiles, product and service's descriptions, special offerings, press announcements, and annual reports can easily be made available to anyone at any time.

Selling Services and Products. Companies will be able to establish electronics stores for people wishing to purchase products and services. Credit card information will be taken in a secure manner, electronic funds can be transferred, warrantees can be filled out, and user satisfaction surveys can be completed, all online.

Information Retrieval. Information retrieval is probably the best example of Internet usage today. This will continue to be a major feature of the Internet in the future. What is free now, will most likely become a fee-based service in the future. Electronic books, newspapers, and publishing tailored to specific audiences will be available 24 hours a day.

Technical and Product Support. Have a question on a specific product, get the answer by accessing the companies technical support site. Have a problem, get instant feedback.

Training and Education. The Internet and the Web are positioned to become the virtual schools of the future. Even today, you can take a course or get a degree over the Internet. Many colleges and universities have established online training programs geared toward users that work full time and cannot maintain a schedule that will allow them to physically attend school.

The Future of the Internet

The future of the Internet will be determined by several factors. They include: Software hardware, and communications advances.

Hardware. Computer hardware advances will play an important role in the future of the Internet. In order to accommodate the goals of Internet content providers, computer hardware will have to be much faster and provide features that are just beginning to appear they include:

1) Full motion video.
   While we have video capabilities at the present time, it sometimes leaves a lot to be desired. The computers of tomorrow will need to provide full screen, full-motion high-resolution video, and audio capabilities. Many companies are addressing this need at the present time by developing capabilities to provide this requirement. They include faster internal speeds, faster microprocessors, and faster video processing subsystems.
2) Greater disk storage.
   Hard disk storage capabilities will have to exponentially increase in order to provide the space needed for caching and storing content.

Software. The Web site of the future will have to incorporate audio, video, and multimedia to attract visitors. As well, clients will have to have the necessary software to be able to view those Web sites. Browsers will have to be able to support full motion video, audio, and multimedia but they also will have to be intelligent enough to recognize the capabilities of the computer system they are installed on and configured themselves to take advantage of its configuration.

Communications. This is perhaps the key ingredient for insuring the future of the Internet. Delivery bandwidth will play a major role in the expansion of the Internet. Much higher bandwidth is an absolute requirement for multimedia, video, and audio delivery. Telephone companies, cable television companies, and ISPs are all working on new delivery methodologies, because this will determine the success of the Internet of the future. We will need multi-megabit per second capabilities that can only be provided by new switch and wiring methodologies.

CASE STUDY CONCLUSION

At this point, the Bank of Scotts Valley's network is functioning well and provides a foundation for increased business. In real life, the establishment of the data centers, the creation of branch operations, and the implementation of local area networking (all of which are requirements for a bank's automated operations) would have taken years. While IBM equipment has been used at the Bank of Scotts Valley, an automated processing environment and its supporting network can be created using many vendors' products.

An open systems approach to networking will provide a foundation on which an extensive array of banking services can be offered in the future. The bank of the future will be in a position to offer customers virtually all services that currently require a visit to a branch office. In the area of commercial business activities, future bank services will include interbanking debit cards used in place of checking accounts, as well as international customer accounts linked to partner banks. This will enable customers to access funds from other banks as if they were at a branch of the bank they normally do business with.

The key to a successful network implementation is the thorough preplanning phase, in which the required software and hardware are identified. The development of operational procedures and training to support the new systems also must be considered essential to the planning process.

APPENDIX A

Terms and Definitions

The following represents definitions commonly used within the data communications/ data processing industry.

100BaseT: The specification proposed by the Fast Ethernet Alliance to the IEEE that defines a 100 Mbps implementation of Ethernet over two pairs of Category 5 twisted-pair cabling at segment distances of up to 100 meters.

100BaseVG: A proposal put forth by Hewlett Packard and AT&T for a star configured 100 Mbps network using four pairs of unshielded twisted-pair wiring (voice grade).

100BaseX: This specification covers the 100 Mbps operation of the MAC layer and CSMA/CD protocol and a media independent interface. Specifications like 100BaseT use this set of definitions and add media-dependent specifications (like Category 5 cabling) to define a network.

100VGAnyLAN: HP's marketing name for 100BaseVO.

10BaseF: Defined in IEEE 802.3 and implemented over two fiber-optic links (transmit and receive). 10BaseF segments may be up to 2 kilometers in length. Two implementations are contained within 100BaseF: 10BaseFL, which is interoperable with FOWL devices and is the most common implementation of 10BaseF; and 10BaseFS, which defines a synchronous fiber-optic signaling scheme.

10BaseT: The implementation of the IEEE 802.3 standard designed to operate in a star configuration over two pairs of UTP wiring at 10 Mbps. A hub or concentrator acts as the center of the star, and provides retiming and repeating for each of the segments (links). The 10BaseT standard limits segments to 100 meters.

10Base2: The implementation of the IEEE 802.3 standard also known as thinnet. 10Base2 networks use a bus configuration and operate over thin coaxial cable at 10 Mbps.

Devices are connected to the coaxial cable with BNC "T" connectors. The standard limits cable segments to 200 meters.

10Base5: The implementation of the IEEE 802.3 known as "standard" Ethernet. 10Base5 networks are a bus configuration, and run on a continuous length of thick coaxial cable at 10 Mbps. Devices are connected to the coaxial cable with transceivers using a "vampire tap" to make connection to the cable's conductors. The standard limits coaxial cable segments to 500 meters.

802.1: 802.1 is the IEEE standard for hardware-level network management. It includes spanning-tree algorithm for Ethernet MAC-layer bridges and the heterogeneous LAN management specification for managing Ethernet and token-ring hubs.

802.2: This IEEE standard specifies logical link control, which defines services for the transmission of data between two stations at the data link layer of the OSI model.

802.3: The IEEE standard that governs Carrier Sense Multiple Access/Collision Detection (CSMA/CD) networks, more commonly called Ethernet.

802.4: The IEEE standard that defines the use of the token-bus network access method. A token-bus network is typically used in manufacturing networks.

802.5: The IEEE specification standard that defines a logical ring network that uses a token-passing access method. It is commonly called token-ring. It comes in 4 Mbps and 16 Mbps speeds. Token-ring runs over shielded twisted-pair, unshielded twisted-pair, and fiber-optic cabling.

802.6: This IEEE specification standard defines metropolitan area networks (MAN). The MAN standard implements a distributed-queue, dual-bus access method over a fiber-optic cable plant. Switched Multimegabit Data Services (SMDS), an emerging high-speed WAN service, can run over a MAN physical network.

802.7: Defined by the IEEE, the 802.7 standard defines broadband LANs. They can carry video, data, and voice traffic. Broadband LANs are constructed of cable television-like components, and use radio frequency to transmit information in separate channels in a single cable. They are built using a tree topology. Broadband LANs are rarely used anymore.

802.9: The IEEE 802.9 standard defines integrated digital and video networking.

802.11: When finalized by the IEEE, the 802.11 standard will define wireless networking.

ABEND: The abnormal termination of a computer program due to an error in the logic or data associated with that program.

ABORT SEQUENCE: A sequence which prematurely terminates the transmission of a frame.

ACCESS METHOD: An access method is the set of rules by which the network arbitrates access among the nodes. Collision Sense Multiple Access/Collision Detection (CSMA/CD) and token passing are two access methods used in LANs.

ACOUSTIC COUPLER: A device that converts electrical signals into audio signals, enabling data to be transmitted over telephone lines using a conventional telephone handset.

ADDRESS: An address is a unique identification code assigned to a network device so it may independently send and receive messages.

ADDRESS FILTERING: A bridge function for allowing addresses of other networks to pass through unchanged to the next bridge while identifying the addresses of that bridges subnetworks.

ADDRESS RESOLUTION PROTOCOL (ARP): Within TCP/IP, ARP is the protocol that determines whether a packet's source and destination addresses are in the data link control or Internet Protocol format.

ADPCM: Adaptive digital pulse coded modulation, a modulation technique designed to carry digital voice and video signals.

ADVANCED PEER-TO-PEER NETWORKING (APPN): APPN is the network architecture within the IBM Systems Application Architecture that provides for peer-to-peer access among computers. Under APPN, a mainframe host is not required. It also implements concepts such as dynamic network directories and dynamic routing in a Systems Network Architecture network.

ADVANCED PROGRAM-TO-PROGRAM COMMUNICATIONS (APPC): APPC is the protocol suite in the IBM Systems Application Architecture that provides peer-to-peer access, enabling PCs and midrange hosts to communicate directly with mainframes. APPC is key for distributed computing in an IBM environment. APPC can be used over a Systems Network Architecture, token-ring, Ethernet, or X.25 network.

ALOHA: The multiple access broadcast technique that permits users to transmit packets whenever they desire.

AMPLIFIER: Similar to a bit repeater, but is used for analog transmission. Amplifies both the signal and the noise.

AMPLITUDE MODULATION (AM): A method of modifying the amplitude of a carrier signal to transmit data signals.

ANALOG: A representation of real world events. The term analog refers to real-world conditions such as sound, temperature, and movement having many different conditions.

ANALOG TRANSMISSION: A transmission method that uses continuously variable signals that are modulated and demodulated to represent on and off electrical signals.

AMERICAN NATIONAL STANDARDS INSTITUTE (ANSI): ANSI is the principal group in the United States for defining standards. ANSI represents the United States in ISO, the international standards-making body. Fiber Distributed Data Interface, a 100 Mbps network, is one network standard developed by ANSI.

API: Application programming interface, is a set of subroutines or functions that a program, or application, can call to tell the operating system to perform some task. The Windows API consists of more than 1,000 functions that programs written in C, C++, Pascal, and other languages can call to create windows, open files, and perform other essential tasks. An application that wants to display an on screen message can call Windows' Mes sageBox API function, for example.

APPLETALK: AppleTalk is the name of Apple Computer's networking specification. AppleTalk includes specifications for the physical layer as LocalTalk, EtherTalk, and

Token-Talk network and transport functions as Datagram Delivery Protocol and AppleTalk Session Protocol; addressing as Name Binding Protocol; file sharing as AppleShare; and remote access as AppleTalk Remote Access.

APPLICATION LAYER: The seventh and uppermost layer of the OSI model, the application layer allows users to transfer files, send mail, and perform other functions where they interact with the network components and services. It is the only layer users can communicate with directly.

ARB (Adaptive Rate-based Congestion Control): Part of IBM's High Performance Routing (HPR) enhancements to APPN. ARB is a rate-control and feedback mechanism in the RTP protocol.

ARCnet: An older token-passing LAN standard developed by Datapoint Corp. Uses RG-62 coaxial or twisted-pair cable. ARCnet LANs are usually star-wired with central hubs. They can also use a bus topology.

Arcuet (Short for Attached Resource Computer Network): Developed by Datapoint, Arcnet is one of the earliest LANs ever invented. A 2.5 Mbps baseband, modified token passing, media-access protocol, Arcnet runs on coaxial cable, UTP, and fiber-optic cable.

ARP (Address Resolution Protocol): The Internet protocol used to dynamically map Internet addresses to physical addresses on LANs. ARP is limited to networks that support hardware broadcast capability.

ARPA (Advanced Research Projects Agency): Now called DARPA, Defense Advanced Research Projects Agency. U.S. Government agency that funded ARPANET.

ARPANET: A packet-switched network developed in the early 1970s that was the predecessor of the Internet. ARPANET was decommissioned in June 1990.

ASCII (American Standard Code for Information Interchange): A seven- or eight-bit code set containing 128 characters and used extensively by most of the computer manufacturers in the world.

ASIC (Application-Specific Integrated Circuit): An integrated-circuit chip designed for a particular use rather than general use. Many video boards and modems use ASICs.

ASSEMBLER: A translator program that takes low level assembly language code and translates it into machine language code that can be understood by the computer.

ASYNCHRONOUS COMMUNICATION SERVER (ACS): An ACS is some combination of a computer motherboard, asynchronous modems, and software that enables multiple people to dial out of a LAN. ACSs also provide dial-in service, where remote users can use modems to call up their network services in the office. ACSs are also called dial-in/dial-out servers or modem servers.

ASYNCHRONOUS TRANSFER MODE (ATM): ATM is a method of data transmission. ATM transmission is defined as 53-octet, fixed-length packets over a cell-switched network. Speeds up to 2.2 gigabits per second are possible, and it is capable of carrying voice, video, and data. It is also called cell relay.

ASYNCHRONOUS TRANSMISSION: The transmission of data or bits without regard to time. The data or bits to be transmitted are wrapped (surrounded by start and stop bits) to indicate beginning and ending of the stream to be sent or received.

ATTENUATION: A measure, in decibels, of the loss in strength (amplitude) that an electrical signal experiences as the signal travels through a wire.

AUI (Attachment Unit Interface): The connector used to attach a device to an Ethernet transceiver.

BANDWIDTH: The range of frequencies assigned to a channel or system. The difference expressed in hertz (cycles per second) between the highest and lowest frequencies of the transmission band.

BASEBAND: The base or starting frequency for the transmission of channels of information (data and bits). This is also referred to as a category of communications network. Baseband networks are all digital networks which have a transmission bandwidth of up to 10 megabits per second (see Broadband).

BASIC: A very high-level source language that allows programmers to develop application programs easily and quickly.

BASIC RATE INTERFACE (BRI): An ISDN-defined level of service that provides two 64 Kbps links for the transfer of data and one 16 Kbps link for the transfer of control information.

BATCH: A procedure for executing application programs not involving communications or terminal interaction.

BAUD: Term for an analog line's signaling rate or modulation rate. Without compression baud is equal to bits per second (bps). With compression, bps is usually greater than baud depending upon the compression rate.

BBS (Bulletin Board Service): Usually a dedicated computer with one or more modems and special software that people dial up and log into.

BCD (Binary Coded Decimal): Binary coding scheme using a six-bit (six level) code.

BCS (Binary Compatibility Standard): Developed by the 88open Consortium, BCS gets all vendors to compile a single version of an application that will run, without change or recompilation, on all certified Motorola 88000-based processors.

BEACON (Beacon Process/Beconing): Process used in a token-ring network to recover the network when any attached station has sensed that the ring is inoperable due to a hard error. Stations can withdraw themselves from the ring if necessary. A station detecting ring failure upstream transmits (beacons) a special MAC frame that is used to isolate the location of the error using beacon transmit and beacon repeat modes. Beacons typically occur when a station set for 4 Mbps operation attempts to connect to a ring operating 16 Mbps.

BECN (Backward Explicit Congestion Notification): A field in the header of a Frame Relay packet that is set by a switching node experiencing congestion and sent to the dc stream source switching node to notify it of congestion conditions.

BER (Bit Error Rate): The rate at which errors occur in a data stream. Basic Encoding Rules. Rules for encoding data units described in 1. An abstract syntax description language encoding technique.

BERT (Bit Error Rate Tester): A network diagnostic instrument used to troubleshoot LANs.

BGP (Border Gateway Protocol): A routing protocol designed to replace EGP. Unlike EGP, BGP weights each route. BGP is used to interconnect organizations.

BINDERY: A NetWare 2.x/3.x database that holds data about various network elements.

BIOS (Basic Input/Output System): A set of low-level routines in a computer's ROM that application programs (and operating systems) can use to read characters from the keyboard, output characters to printers, and interact with the hardware in other ways. Many plug-in adapters include their own BIOS modules that work in conjunction with the BIOS on the system board.

BISDN (Broadband ISDN): The proposed advanced version of ISDN based on ATM technology that will provide speeds of 155.52 Mbps and higher. Standards and switching technology that will work this fast are being developed. See IEEE 802.6, SONET and SMDS.

BIT: The smallest indicator of the presence or absence of information in a computer system. The foundation for which data and information is represented in a computer. A single digit of data expressed as a binary value in the computer.

BIT-ORIENTED PROTOCOLS: A bit-oriented protocol is one that handles bits instead of characters, and uses bits to indicate control activity.

BNC: A bayonet-locking connector used with 10Base2 (thin net) networks. Said to be short for Bayonet Neill-Concelman, after the developers of the connector. Also commonly referred to as Barrel Nut Connector.

BRIDGE: A bridge connects and passes packets between networks or network segments, based on the packet's destination address. Each bridged network segment is part of a single LAN, with a single network address. Bridge's are protocol-independent and media-dependent. Bridge's can forward multiple protocols, but cannot link an Ethernet segment with a Token Ring segment.

BROADBAND: Refers to the transmission facilities whose bandwidth (range of frequencies they will handle) is greater than that available on voice grade facilities. Also known as wideband.

BROUTER: A brouter is a hybrid of bridges and routers. It performs some of the functions of each. Like a bridge, the brouter operates at the data link layer of the OSI model. It does not look at protocols, only at the destination addresses of packets. Like a router, the brouter is media-independent. It routes packets between LANs, choosing the most efficient path between them.

BSC (Binary Synchronous Protocol or BSC): BSC is a handshaking technique used between the cluster controller or remote communications processor and the host computer or front end processor.

DUAL HOMING: In FDDI, dual homing is a method of cabling concentrators and stations in a tree configuration that permits an alternate path to the FDDI network in case the primary connection fails.

E1: In Europe, E1 is the basic telecommunications carrier, and it operates at 2.048 Mbps. In the United States, the basic carrier is T1, which operates at 1.544 Mbps.

EBCDIC (Extended Binary Coded Decimal Interchange Code): An eight bit code set containing 256 characters developed and used by IBM.

EDO RAM: A form of DRAM that speeds accesses to consecutive locations in memory by assuming that the next memory access will target an address in the same transistor row as the previous one and latching data at the output of the chip so it can be read even as the inputs are being changed for the next memory location. EDO RAM reduces memory access times by an average of about 10 percent compared with standard DRAM chips, and costs only a little more to manufacture. EDO RAM has already replaced DRAM in many computers, and this trend is expected to continue.

EDRAM: A form of DRAM that boosts performance by placing a small complement of static RAM (SRAM) in each DRAM chip and using the SRAM as a cache. Also known as cached DRAM or CDRAM.

EEPROM: A special type of read-only memory (ROM) that can be erased and written electrically. EEPROM is frequently used for system-board BIOSes so that a computer's BIOS can be updated just like a piece of software (see also ROM, EPROM, and BIOS).

EGA: See Enhanced Graphics Adapters.

EGP (External Gateway Protocol): A routing protocol that is used to exchange network reachability information among organizational networks. EGP indicates whether a network is reachable; it does not weight that decision.

EIDE: An enhanced version of the IDE drive interface that expands the maximum disk size from 504 MB to 8.4 GB, more than doubles the maximum data transfer rate, and supports up to four drives per PC (as opposed to two in IDE systems). Now that hard disks with capacities of 1 GB or more are commonplace in PCs, EIDE is an extremely popular interface. EIDE's primary competitor is SCSI-2, which also supports large hard disks and high transfer rates (see also IDE and SCSI).

EISA: Extended Industry Standard Architecture. A 32 bit adaptation of the 8- and16-bit buses originally developed by IBM and now standard in almost all PCs that use Intel's 8086 and 80X86 microprocessors. The EISA bus is a joint development with Compaq and other PC manufacturers.

ELECTROMAGNETIC INTERFERENCE (EMI): EMI is noise on data transmission lines that reduces data integrity. It is caused by motors, machines, and other generators of electromagnetic radiation.

ELECTRONIC DATA INTERCHANGE (EDI): EDI is a method of electronically exchanging business documents, such as purchase orders, bills of lading, and invoices. Customers and their suppliers can set up EDI networks. EDI can be accomplished through OSI standards or through proprietary products.

ELECTRONIC MAIL: E-mail is an application that enables users to send messages and files over their computer networks. E-mail can range from a simple text-based system to a messaging system that accommodates graphics, faxes, forms processing, workflow, and more.

EMS (Expanded Memory Specification): A bank-switched memory management scheme developed by Intel, Lotus, and Microsoft that allows MS-DOS applications (normally limited to 640 K of memory) to access vast quantities of memory. The first widely accepted version of the EMS specification, Version 3.2, supported up to 8 MB of memory, and Version 4.0 increased the limit to 32 MB. Memory that conforms to this standard is often referred to as expanded memory. EMS memory has been all but made obsolete by protected-mode operating systems such as Windows.

EMULATION: The ability to program one machine to perform the identical functions of another machine.

ENCAPSULATION: Encapsulation or tunneling is the process of encasing one protocol into another protocol's format.

END SYSTEM: In Internet terminology, an end system is a host computer.

END SYSTEM TO INTERMEDIATE SYSTEM (ES-IS): ES-IS is an OSI routing protocol that provides the capabilities for hosts (or end systems) and routers (or intermediate systems) to find each other. ES-IS does not handle the router-to-router protocols, the Intermediate System to Intermediate System protocol does.

ENHANCED GRAPHICS ADAPTOR (EGA): A board for the personal computer that provides enhanced color and resolution capabilities for the CRT.

ENTERPRISE MANAGEMENT ARCHITECTURE (EMA): EMA is Digital Equipment's umbrella architecture for managing enterprise networks. EMA is a distributed approach.

ENTERPRISE NETWORK: An enterprise network is one that connects every computer in every location of a company and runs the company's mission-critical applications.

EPP (Enhanced Parallel Port): A parallel port that conforms to the EPP standard developed by the IEEE (Institute of Electrical and Electronics Engineers) 1284 standards committee. The EPP specification transforms a parallel port into an expansion bus that can handle up to 64 disk drives, tape drives, CD-ROM drives, and other mass-storage devices. EPPs are rapidly gaining acceptance as inexpensive means to connect portable drives to notebook computers.

EPROM (Externally Programmable Read Only Memory): Special ROM chips that can be reconfigured from an external device or downloaded from a network.

ESCON (Enterprise Systems Connectivity): A fiber-optic based mainframe communications channel architecture announced by IBM in 1992. Escon channels provide an effective data rate of 17 Mbps (136 Mbps) at distances up to 12.5 miles (20 kilometers) without repeaters and 37.5 miles (60 kilometers) with repeaters. One enhancement offered by Escon is the Escon Director, a high-speed circuit switch with an aggregate capacity of about 6 Gbps, which allows a single Escon channel to communicate with many different devices.

ETHERNET: Ethernet, is a 10 Mbps CSMA/CD network that runs over thick coax, thin coax, twisted-pair, or fiber-optic cable. DIX or Blue Book Ethernet is the name of the Digital Equipment, Intel, and Xerox specification; 802.3 is the IEEE's specification; and 8802/3 is the ISO's specification.

ETHERTALK: EtherTalk is Apple Computer's Ethernet implementation.

Ethernet Switch: A new class of multiport intelligent hubs that actually switch Ethernet packets between two ports based upon source and destination MAC addresses, rather that repeat an incoming packet to all other ports as do repeaters or the classic intelligent hub or concentrator. The ultimate effect of this packet switching is that each port receives dedicated 10 Mbps Ethernet bandwidth and does not have to share it with other users, as in the case of repeaters. Collisions are not passed through Ethernet Switches. This is a recent advancement in hub technology designed to improve performance of traditional Ethernet networks.

EWOS (European Workshop on Open Systems): Works on behalf of the European countries to review ISO's set of OSI standards to pare down the standards to provide a better fit for the respective countries they represent.

EXECUTE: The process of starting a program within the computer system.

EXECUTABLE CODE: A program that has been converted from a source format understandable by programmers, into a machine readable format.

FAT (File Allocation Table): The file system used by DOS and 16-bit versions of Windows to manage files stored on hard disks, floppy disks, and other disk media. The file system takes its name from an on-disk data structure known as the file allocation table, which records where individual portions of each file are located on the disk (see also VFAT).

FAQ: A frequently-asked-questions list is a document that covers basic information from a particular Usenet newsgroup or mailing list.

Fast Packet Switching: A general term applied to a variety of packet-mode concepts that offer vastly improved performance over conventional X.25 packet-switching.

FC (Frame Control): Data that supplies frame type on token ring networks.

FCC (Federal Communications Commission): U.S. regulatory and approval agency.

FCS (Frame Check Sequence): A field in the header of a Frame Relay (or other packet) during transmission on the communication link.

FDDI (Fiber Data Distributed Interface): A standard for a 100 Mbps LAN based on fiber-optic media configured as dual counter-rotating token-rings. FDDI is defined for fiber-optic cable, but is also available over shielded and unshielded twisted-pair cable.

FILTER: The ability to selectively forward data based on criteria specified by the network manager.

FLOW CONTROL: A method for controlling the flow of data between two communicating devices.

FPU (Floating-Point Unit): A formal term for the math coprocessors (also called numeric data processors or NDPs) found in many personal computers. The Intel 80387 is one example of an FPU. FPUs perform certain calculations faster than CPUs because they specialize in floating-point math, whereas CPUs are geared for integer math. Today, most FPUs are integrated with the CPU rather than packaged and sold separately (see also CPU).

FRAME: A message being sent or received which consists not only of data but also transmission information such as terminal location, length of text message and routing information.

FRAME RELAY: A variation on the X.25 interface standard that derives its name from use of the Data Link Layer, or "frame layer" (Layer 2 of the OSI model) to route, or "relay," a packet directly to its destination instead of terminating the packet at each switching node. Error correction is eliminated at the switching nodes, eliminating processing overhead, and increasing speeds. Based upon the LAPD CCITT standard, Frame Relay uses variable-length packets with a three byte header, up to 4,510 bytes of user data, and three byte trailer (current network offerings recommend a maximum of 4,096 bytes of user data).

FREQUENCY: Vibrations or cycles per second. The frequency of a carrier is the number of electrical vibrations per second expressed in hertz.

FREQUENCY DIVISION MULTIPLEXING (FDM): A process that divides the available transmission frequency range into narrower bands, each of which is used for a separate channel.

FRONT END PROCESSOR (FEP): A communications computer associated with a host computer whose specialized function is to handle communications to/from terminals and nodes in the network.

FT1: Abbreviation for Fractional T1.

FTAM (File Transfer Access and Management): An OSI application protocol that defines a file service and specifies a file protocol available within the application layer of the OSI Reference Model to allow files to be transferred and accessed from remote systems.

FTP (File Transfer Protocol): Application layer protocol which uses TCP/IP to transfer files between networks.

FULL DUPLEX (FDX): Transmission mode in which the channel operates in both directions at the same time.

GATEWAY: A networking device used to connect two otherwise incompatible networks together. Gateways perform a protocol conversion; that is, one protocol is terminated and the data is translated to a different protocol. An example would be Ethernet to IBM's System Network Architecture (SNA).

Gbps (Gigabits per second): One billion bits per second.

Gbyte (Gigabyte): One billion bytes.

GDA (Global Directory Agent): The GDA is a naming gateway that connects the DCE domain to other administrative domains through the X.500 worldwide directory service and Domain Name Service (DNS).

GDI (Graphics Device Interface): The component of Windows that permits applications to draw on screens, printers, and other output devices. The GDI provides hundreds of convenient functions for drawing lines, circles and polygons, rendering fonts, querying devices for their output capabilities, and more.

GDS (Global Directory Service): Based upon the X.500 standard, GDS provides a high level directory hierarchy for 0SF DCE networks.

GFC (Generic Flow Control): A field in an ATM cell header reserved for UNI to control traffic flow and alleviate short term overload conditions.

GIGABYTE (Gb): One billion bytes.

GOSIP (Government OSI Profile): A U.S. Government procurement specification for OSI protocols. At one time the Government mandated GOSIP compliance for computer and hardware networking purchases. However, by the end of 1993 they have all but eliminated any need for GOSIP compliance; and instead are focusing on TCP/IP, Internet, and SNMP compliant devices

GUARD BANDS: Guard bands are used to separate transmission channels from interfering with each other.

HALF-DUPLEX (HDX): A transmission mode in which the channel operates in either direction but not at the same time.

HDLC (High Level Data Link Control): A communications protocol which is a subset of SDLC and used in X.25 and OSI packet switching networks.

HERTZ: Frequency rating. The number of electrical vibrations per second is rated in hertz. One hertz is equal to one cycle per second.

HOST: This term is used to refer to a computer that is usually central to a terminal network and serves as the main controller for all activity required by each terminal or node in the network.

HPFS (High Performance File System): OS/2's native file system. HPFS offers superior performance compared with the FAT file system, supports long filenames and can efficiently handle hard disks of virtually any size. (See also FAT and NTFS.)

HTML (Hypertext Markup Language): An ASCII text-based, scriptlike language for creating hypertext documents like those on the Internet's World Wide Web.

HTTP (Hypertext Transport Protocol): This makes hypertextual browsing through the World Wide Web possible; the user clicks on links that are established in a Web document and moves to that document, even though it may be located on a different computer.

HYPERMEDIA: The ability to display a range of different media, accessible through hyperlinks. A World Wide Web page, for example, may contain photographs or drawings, textual formatting, and links to audio and video sources. Hypertext Data which provides links between key elements, allowing you to move through information non-sequentially.

IAB (Internet Activities Board): The technical body that oversees the development of the Internet suite of protocols (commonly referred to as TCP/IP). The IAB has two task forces—the WIF and the IRTF.

IBM AT: Third IBM Intel 80286-based 6 MHz personal computer with a 16-bit processor and 16-bit bus (ISA). Updated to 8 MHz speed. Today, it is used to describe 80286/386/486 clone PC architectures.

IBM PC: The first IBM personal computer based on Intel's 8088 processor running at 4.77 MHz using an 8-bit processor and an 8-bit bus.

IBM XT: The second IBM personal computer based on Intel's 8088/86 processor running at 4.77 MHZ using a 16-bit processor and an 8-bit bus.

ICMP (Internet Control Message Protocol): Allows status messages to be generated by an internetwork gateway, indicating a variety of errors or unusual conditions.

IDE (Integrated Development Environment): A term for products such as Microsoft's Visual C++ and Borland's Delphi that combine a program editor, a compiler, a debugger, and other development tools into one integrated unit. The forerunner of all modern IDEs, Borland's Turbo Pascal changed the way programmers write code by allowing programs to be edited and compiled within the same application.

IDE (Integrated Device Electronics or Intelligent Drive Electronics): A drive-interface specification for small to medium-size hard disks (disks with capacities up to 504 MB) in which all the drive's control electronics are part of the drive itself, rather than on a separate adapter connecting the drive to the expansion bus. This high level of integration shortens the signal paths between drives and controllers, permitting higher data transfer rates and simplifying adapter cards. IDE drives have virtually replaced the Enhanced Small Device Interface (ESDI) drives that enjoyed widespread acceptance in the late 1980s (see also EIDE and SCSI).

IEC (InterExchange Carrier): Telephony term that refers to the company that provides communication interfaces between LATAs (Inter-LATA).

IEEE (Institute of Electrical and Electronics Engineers): A U.S.-based organization formed to establish standards for the data communications industry. Its best known effort is Project 802 which attempted to define LAN standards (see IEEE 802.1–802.6).

IETF (Internet Engineering Task Force): The organization that produces the standards used in TCP/IP and the Internet.

IMPEDANCE: The resistance of a wire to the flow of alternating current. Mismatches in impedance along a cable cause distortions and reflections.

INTERLEAVE: Interleave is a term used in data communications transmission whereby messages to be sent on the link are inserted in some order so that each terminal has an opportunity to use the link.

INTERNET: The worldwide matrix of connecting computers using the TCP/IP protocols.

INTERNETWORK PROTOCOL (IP): An ISO network layer protocol used for data transmission between two networks.

INTERNET SOCIETY: Promotes the growth of the Internet and works to assist those groups involved in its use and evolution.

INTEGRATED IS-to-IS: A routing protocol that combines routing for TCP/IP and OSI protocols. It is a superset of IS-IS, the OSI routing technology.

INTERFACE: A device or scheme that allows the connection and operation of various types of equipment. An interface can also be a software interface that allows programs to communicate with each other.

I/O (Input/Output): A general term for reading and writing data on a computer. The term "file I/O," for example, refers to the act of reading or writing information in a disk file.

IP (Internet Protocol): A communications protocol for networks using Microsoft LAN Manager, NetWare, or VINES. IP software tracks the addresses of nodes, routes outgoing messages and recognizes incoming messages.

IPX: Although IPX stands for a particular protocol Internet Packet Exchange, it is normally used to refer to the entire Novell protocol suite. IPX is the Novell equivalent of the IP protocol. It is used to route NetWare packets between LANs.

IRQ (Interrupt Request): A signal from a hardware device such as a keyboard or a drive controller indicating that it needs the CPU's attention. IRQ signals are transmitted along IRQ lines, which connect peripheral devices to a programmable interrupt controller (PIC). The PIC prioritizes the incoming interrupt requests from different devices and delivers them to the CPU one at a time via a dedicated IRQ line connecting the PIC to the CPU.

ISA (Industry Standard Architecture): The 8- and 16-bit bus design featured in the IBM PC/AT and still used today in one form or another in most PCs (see also EISA and MCA).

ISDN (Integrated Services Digital Network): A project within the CCITT to standardize the operating parameters and interfaces for a network that will accommodate a variety of mixed digital transmission services. Two access channels defined are BRI and PM.

(ISO) INTERNATIONAL STANDARDS ORGANIZATION: A United Nations Organization chartered to develop worldwide standards.

ISSI (Inter-Switching System-Interface): The ATM term referring to the interface between two ATM switches in a public ATM network. Intra-LATA ISSI refers to switch-to-switch interfaces within an single RBOC; Inter-LATA ISSI refers to interfaces between an RBOC's ATM switch and that of an inter-exchange carrier like AT&T, Sprint, or MCI. Inter-LATA ISSI is also called B-ICI (Broadband Inter-Carrier Interface).

ITU (International Telecommunications Union): An organization affiliated with the United Nations that sets standards for data communications.

JABBER: Term used to denote continuous transmission of corrupted or random data onto a network usually caused by network interface card failures. Jabber can potentially saturate a network.

JITTER: The slight movement of a transmission signal in time or phase that can introduce errors and loss of synchronization in high-speed digital communications.

JPEG (Joint Photographic Experts Group): A lossy image-compression algorithm that often reduces the size of bitmapped images by a factor of 10 or more with little or no discernible image degradation. JPEG compression works by filtering out an image's high-frequency information to reduce the volume of data and then compressing the resulting data with a lossless compression algorithm. Low-frequency information does more to define the characteristics of an image, so losing some high-frequency information doesn't necessarily affect the image quality.

Kbps (Kilobits per second): One thousand bits per second.

Kbyte (Kilobyte): One thousand bytes.

KERBEROS: Kerberos is an encryption-based system designed to authenticate users and network connections. Kerberos was developed at MIT's Project Athena in the 1980s, and is named after the three-headed dog of Greek mythology that guards the entrance to Hades. Kerberos uses electronic tickets to authenticate a user to a server. The ticket, which is an encrypted message containing the name of the user and server, network addresses, time stamp and session key, is good only for a single server and a single user during a certain period of time. Once users receive a ticket, they can use it to access the server as many times as desired until the ticket expires. The user cannot decrypt the ticket, nor can anyone on the system read or modify the ticket without detection an invalidation.

KERNEL: A collection of system routines that interact with a computer's hardware. This is the essential core of UNIX.

KILOBYTE (Kb): One thousand bytes.

LAN Manager: LAN Manager is Microsoft's network operating system based on OS/2. It uses NetBEUI or TCP/IP network protocols. LAN Manager supports DOS, Windows, OS/2, and Macintosh clients.

LAN Server: LAN Server is IBM's network operating system based on the OS/2 operating system and the NetBIOS network protocol. LAN Server supports DOS, Windows, OS/2, and Macintosh clients.

LANtastic: LANtastic is Artisoft's peer-to-peer, NetBIOS-based network operating system. It supports Windows, DOS, Macintosh, and UNIX clients.

LASER PRINTER: A printer that uses a laser beam to impress toner material upon paper or film.

LCD (Liquid Crystal Display): A display technology that relies on polarizing filters and liquid-crystal cells, rather than phosphors, illuminated by electron beams to produce an on-screen image. To control the intensity of the red, green, and blue dots that comprise pixels, an LCD's control circuitry applies varying charges to the liquid-crystal cells through which polarized light passes on its way to the screen. The amount of light that makes it through to the screen depends on the amount of charge applied to the corresponding cell and thus the degree to which the light is "twisted" before it passes through a second polarizing filter and a red, green, or blue color mask. Laptops, notebooks, PDAs, and other types of portable computers commonly use LCDs (see also CRT).

LEASED LINE: A leased line is a transmission line reserved by a communications carrier for the private use of a customer. Examples of leased-line services are 56 Kbps and T1 lines.

LETTER QUALITY PRINTER: A printer that provides letter quality output using a daisy wheel, laser, or other similar technique.

LIM-EMS: A software technique that allows MS-DOS to access memory beyond 1 Mb by mapping the memory into a window in an area that MS-DOS can access. LIM stands for Lotus/Intel/Microsoft, the founding organizations that developed the Expanded Memory Specification.

LINE DRIVER: A converter and transmitter receiver used for digital transmission across limited distances.

LINE OF SIGHT: Laser, microwave, and infrared transmission systems require that no obstructions exist in the path between the transmitter and receiver. This path is called the line of sight.

LISTSERV: A program that manages mailing lists by responding automatically to e-mail requests and distributing new messages.

LLC: Logical Link Control. IEEE standard protocol for a subdivision of the OSI Data Link Layer for LANs (see IEEE 802.2).

LOBE: Token Ring term for the "link" between a Token Ring node and a CAU/LAM (hub).

LOCALTALK: Apple Computer's 230 Kbps baseband network used primarily with its Macintosh computers and LaserWriter printers. LocalTalk uses the CSMA/CA media access method. The older Phase I is limited to 254 nodes, while the current Phase II allows much larger (theoretically unlimited) networks (also see AppleTalk).

Local Area Network (LAN): A LAN is a group of computers, each equipped with the appropriate network adapter card and software and connected by cable, that share applications, data, and peripherals. Connections are made via cable or wireless media. A LAN does not use telephone services.

LU6.2 (Logical Unit 6.2): An SNA protocol that is a device-independent, process-to-process protocol that also supports asynchronous (store-and-forward) networking. LU6.2 provides peer-to-peer communications between two programs and provides mechanisms to manage resources and recover from network failures. Programs and small or mid-range computer systems use LU6.2 (along with PU2. 1) to establish and control sessions with each other without host mainframe participation.

LU (Logical Unit): Network addressable units (NAUs) in an SNA network that support port(s) for communication via the network. LUs perform transmission control, data control, and presentation services. An LU is the logical window or port through which end users access the SNA network. Specified LUs are: LU1—host program to keyboard, printer, disk drive; LU2—host program to 3270 display; LU3—host program to 3270 printer; LU4—host program to text processor; LU5—undefined at this time; LU6—general purpose program-to-program; and LU7—host program to 5250 display.

Local Area Transport (LAT): LAT is Digital Equipment's protocol suite for connecting terminals to an Ethernet network. Because LAT lacks a network layer, it must be bridged, not routed, in an enterprise network.

Logical Link Control (LLC): OSI Layer 2, the data link layer, is divided into the Logical Link Control and the Medium-Access Control sublayers. LLC, which is the upper portion, handles error control, flow control, and framing of the transmission between two stations. The most widely implemented LLC protocol is the IEEE 802.2 standard.

LOOP BACK DEVICE: A part of the cable test unit that takes a signal transmitted through a cable by the cable tester and returns it to the tester for analysis.

MAC (Media Access Control): The lower half of the data link layer of the OSI model that defines access to the transmission media (i.e., coaxial cable or twisted-pairs).

MAPI (Messaging Application Programming Interface): An API developed by Microsoft and other computer vendors that provides Windows applications with an implementation-independent interface to various messaging systems such as Microsoft Mail, Novell's MHS, and IBM's PROFS. A subset of MAPI known as Simple MAPI lets developers easily create "mail-aware" applications capable of exchanging messages and data files with other network clients.

MCI (Media Control Interface): The Windows component that allows multimedia devices such as CD-ROM drives and videodisk players to be programmed using high-level function calls that insulate the software from the nuances of the hardware. You can use the MCI play command, for example, to play a song on an audio CD or a video clip on a videodisc player. MCI drivers provided with Windows translate the play command into low-level commands specific to the intended output device.

MIDI (Musical Instrument Digital Interface): A specification that standardizes the interface between computers and digital devices that simulate musical instruments. Rather than transmit bulky digitized sound samples, a computer generates music on a MIDI synthesizer by sending commands just a few bytes in length characterizing the pitch and duration of sounds (and the instruments that produce them) through a unidirectional serial cable. Each channel of a MIDI synthesizer corresponds to a different instrument, or "voice," and you can program several channels simultaneously to produce symphonic sound.

MPEG (Motion Picture Experts Group): A multimedia video playback standard that allows digital video to be compressed using a combination of JPEG image compression and a sophisticated form of differencing—encoding a video sequence by recording differences between frames rather than entire images of each frame. There are two MPEG standards: MPEG-1, which supports a playback quality roughly equal to that of a VCR, and MPEG-2, which supports high-quality digital video. MPEG-1 is the form normally used with personal computers (see also JPEG).

MPP (Massively Parallel Processor): A computer architecture that employs many processors (potentially thousands) and distributed memory in its design.

MACH: A UNIX look-alike developed at Carnegie-Mellon University. It is at the core of OSFII.

MAIL-ENABLED APPLICATIONS: Mail-enabled applications are a class of software that incorporates e-mail's functionality but provides additional services, such as workflow automation, intelligent mail handling, or contact management software.

MAILING LIST: A group discussion carried on through electronic mail.

MAIN DISTRIBUTION FACILITY: In a structured wiring system, the main distribution facility is the portion of the wiring located in the computer room. From the main distribution facility extends the campus wiring subsystem, which runs to each building.

Management Information Base (MIB): A MIB is a repository of the characteristics and parameters managed in a device. Simple Network Management Protocol (SNMP) and Common Management Information Protocol (CMIP) use MIBs to contain the attributes of their managed systems.

Manufacturing Automation Protocol (MAP): MAP is an ISO protocol for communicating among different pieces of manufacturing equipment.

MARK: The presence of a signal. In telegraph, the mark represents a closed or current flowing condition which is equivalent to a bit being present.

MATRIX PRINTER: A printer that uses 9 to 24 pins striking a ribbon to create dots for characters and graphics.

Multistation Access Unit (MAU): A MAU is a multiport wiring hub for Token Ring networks. IBM calls MAUs that can be managed remotely Controlled Access Units, or CAUs.

Mbps (Megabits per second): One million bits per second.

Mbyte (Megabyte): One million bytes.

MCA (MicroChannel Architecture): See Microchannel.

MEDIUM ACCESS CONTROL (MAC): The MAC is the lower sublayer of the data link layer (Logical Link Control is the upper sublayer), and it governs access to the transmission media.

MEGABYTE (Mb): One million bytes.

MESH TOPOLOGY: In a mesh topology, any site can communicate directly with any other site.

Message Handling Service (MHS): MHS is another name for ISO's X.400 protocols for store-and-forward messaging. The User Agent (UA) is part of the users' e-mail programs. Through the UA, users interact with the messaging system. The Message Transfer Agent (MTA) and Message Store (MS) are usually embedded into the mail engine, and they handle the transmission and storage of messages.

Message Transfer Agent (MTA): In ISO's X.400 electronic messaging protocols, the MTA is responsible for storing messages then forwarding them to their destinations. The MTA is commonly implemented as the mail server.

Metropolitan Area Network (MAN): A MAN covers a limited geographic region, such as a city. The IEEE specifies a MAN standard, 802.6, which uses the Dual Queue, Dual Bus access method and transmits data at high speeds over distances up to 80 kilometers.

MS-DOS: Personal computer single-user operating system developed by Microsoft. Used as basis for IBM's PC-DOS.

MSNF (Multiple Systems Networking Facility): Software that allows an SNA network to have more than one System Services Control Point (SSCP).

MIB, MIB 1, MIB 11 (Management Information Base): The database used by SNMP to monitor and manage networks. MIB 1 and MIB 11 are revisions of the database used on TCP/IP networks. (The original MIB was renamed MIB 1 when MIB 11 was defined.)

MICROSECOND: One millionth of a second.

MICROPROCESSOR: A semiconductor chip which contains all computer processes and is the heart of a central processing unit.

MILLISECOND: One thousandth of a second.

MIPS: Millions of instructions per second.

MISSION-CRITICAL APPLICATION: A mission-critical application is one that is crucial to a company's continued operation.

MODEM (Modulator/Demodulator): A device that converts digital signals into analog signals in order to transmit data across telephone lines. A modem is located at each end of the line and connects to a host or communications controller.

MODULATION: The process of varying the characteristics of a wave in accordance with another wave or signal, usually to make user equipment signals compatible with communications facilities.

MONOCHROME DISPLAY: A black and white CRT display device.

MPP (Massively Parallel Processor): A computer architecture that employs many processors (potentially thousands) and distributed memory in its design.

MULTIPLEXOR: A networking device that provides the concurrent transmission of multiple information signals on a single data channel by apportioning the time available on the composite channel to the individual signals (time division) or by assigning specific frequencies to each information signal (frequency division).

MULTIPLEXING: Multiplexing is putting multiple signals on a single channel.

MULTICS: A predecessor of UNIX designed by MIT and General Electric in 1965.

MULTICAST: Multicast packets are single packets that are copied to a specific subset of network addresses. In contrast, broadcast packets are sent to all stations in a network.

MULTIMEDIA: Multimedia is the incorporation of graphics, text, and sound into a single application.

MULTIMODE FIBER: Multimode fiber optic cable uses light-emitting diodes (LEDs) to generate the light to transmit signals. Multimode fiber is prevalent in data transmission.

MULTIPROCESSING: A computer that includes two or more processors sharing common access to memory.

MULTIPROCESSOR: A computer having multiple CPUs or microprocessors.

MULTIPROGRAMMING: The ability to concurrently run several application programs at the same time on a single processor.

Multipurpose Internet Mail Extension (MIME): MIME is an Internet specification for sending multiplepart and multimedia messages. With a MIME-enabled e-mail application, users can send PostScript images, binary files, audio messages, and digital video over the Internet.

MULTITASKING: The ability to concurrently run more than one program in the CPU at the same time.

MULTIPOINT LINE: A communications line capable of being shared by several devices.

MVS (Multiple Virtual System): An IBM operating system used in mid size and large computer systems to manage application programs and hardware resources for all programs that require them.

MVS-XA (Multiple Virtual System Extended Architecture): See MVS.

NAK (Negative Acknowledge): A control character or message sent back to the sender that signifies the information was received with an error(s) and to re-send the data. Opposite of ACK.

NAMED PIPES: A communications channel used by applications on Microsoft LAN Manager networks.

NAMESERVER: A computer that manages Internet names and numeric addresses.

NANOSECOND: One billionth of a second.

NAS (Network Applications Support): Digital Equipment's architecture designed to allow creation of distributed applications across its own operating systems as well as those from other vendors.

NAU (Network Addressable Unit): A generic SNA term representing a Logical Unit (LU), Physical Unit (PU), or System Service Control Point (SSCP).

NCP (Network Control Program): IBM software that runs on IBM or compatible front end processors (FEPs). NCP, along with VTAM, controls an IBM SNA network.

NCP (NetWare Core Protocol): NetWare's format for requesting and replying to requests for file and print services.

NCS (Network Computing System): This is a major PPC standard developed by Apollo.

NetBIOS: NetBIOS is a protocol developed by IBM that governs data exchange and network access. Because NetBIOS lacks a network layer, it cannot be routed in a network, which makes building large internetworks of NetBIOS-based networks difficult. Examples of NetBIOS-based NOSs include IBM LAN Server and Artisoft LANtastic.

NetBEUI: Microsoft's version of NetBIOS is called NetBEUI. It is a protocol that governs data exchange and network access. Because NetBEUI lacks a network layer, it cannot be routed in a network, which makes building large internetworks of NetBEUI-based networks difficult. Windows for Workgroups uses NetBEUI.

NETIQUETTE: The etiquette of using the network.

NETVIEW: IBM's name for its mainframe-based network management software that runs in an SNA environment. NetView has since been modified to manage PCs and SNA attached token ring networks.

Newsgroup: An Internet usenet discussion group about a particular topic. Some 25,000 newsgroups now exist.

NARROWBAND ISDN: Narrowband ISDN is another name for ISDN. Narrowband ISDN offers a smaller bandwidth than the Broadband ISDN services, such as Asynchronous Transfer Mode (ATM) and Switched Multimegabit Data Services (SMDS).

NDIS (Network Driver Interface Specification): An API developed by Microsoft and 3Com and used by LAN Manager and VINES for LAN adapter drivers. NDIS helps different operating systems and adapters to work together and provides a way to use multiple protocol stacks in the same host.

NETWARE: NetWare is Novell's network operating system. NetWare uses IPX/SPX, NetBIOS, or TCP/IP network protocols. It supports DOS, Windows, OS/2, Macintosh, and UNIX clients. Through NetWare for UNIX, users can gain access to various UNIX hosts. NetWare versions 4.x and 3.x are 32-bit operating systems; NetWare 2.2 is a 16-bit operating system.

NetWare Loadable Module (NLM): An NLM is an application that resides in the NetWare server and coexists with the core NetWare operating system. NLMs provide better performance than applications that run outside the core.

NETWORK: A network is a system of computers, hardware, and software that is connected and over which data, files, and messages can be transmitted and end users communicate.

NETWORK LAYER: The OSI model's third layer is the network layer. It governs data routing. Examples of network-layer protocols are IP and IPX.

NETWORK FILE SYSTEM (NFS): NFS is Sun Microsystems' file-sharing protocol that works over TCP/IP.

NETWORK INTERFACE CARD (NIC): A NIC is the adapter card that plugs into computers and includes the electronics and software so the station can communicate over the network.

NETWORK OPERATING SYSTEM (NOS): A NOS is the software that runs on a file server and governs multiple users' access to the files and resources of the network. Examples of NOSs include Banyan's VINES, Novell's NetWare, and IBM's LAN Server.

NETWORK-AWARE APPLICATION: A network-aware application knows that it is running on a network and has file- and record-locking features.

NETWORK CONTROL CENTER (NCC): A location and function designed to monitor and manage network communications activities.

NETWORK-IGNORANT APPLICATION: A network-ignorant application has no knowledge that it is running on a network. It lacks file and record locking and cannot guarantee data integrity in a multiuser environment.

NETWORK-INTRINSIC APPLICATION: A network-intrinsic application knows it is running on a network and takes advantage of a network's distributed intelligence. For example, a client-server database is a LAN-intrinsic application.

NEXT (Near-end Crosstalk): The interference that occurs adjacent to the connector at either end of the cable. NEXT is measured close to the source of the test signal.

NIST: National Institute of Standards and Technology.

NOISE: The electrical interference that garbles the data signal passing through a cable. The interference can come from external electrical devices, such as motors and fluorescent lights, or from other wires inside the cable.

NODE: A computer system or terminal that is used as a junction or connection point. A node is an interface between user terminals and computers in a communications network.

NONVOLATILE: This term applies to a type of memory that retains its ability to store and remember its contents after power has been turned off.

NOS (Network Operating System): An operating system such as Novell NetWare that provides basic file system services and supervisory functions to computers connected by a network.

NTFS (NT File System): The file system that is native to Microsoft Windows NT. NTFS is probably the most advanced file system available for personal computers, featuring superior performance, excellent security and crash protection, and the ability to handle large volumes of data (see also FAT and HPFS).

NVP (Nominal Velocity of Propagation): The speed at which a signal flows through a cable, expressed as a fraction or percentage of the speed of light in a vacuum. Cable testers use this speed, along with the amount of time it takes for a signal to reflect back to the testing device, to calculate the length of a cable, open A break in a cable circuit.

NULL MODEM: An interface that emulates a DCE to DTE connection and allows two DTE devices to communicate.

OCR (Optical Character Recognition): A technology that allows dots or pixels comprising characters in a bitmapped image to be converted into ASCII text. OCR is frequently combined with scanners to scan documents into a computer and convert the resulting information into textual data.

ODBC (Open Database Connectivity): A standard promulgated by Microsoft that allows databases created by various database management programs—such as dBASE, Microsoft Access, Microsoft FoxPro, and Oracle—to be accessed using a common interface independent of the database file format. By relying on ODBC, one can write an application that uses the same code to read records from a dBASE file or a FoxPro file. Internally, ODBC drivers use a form of SQL to carry out database operations (see also SQL and WOSA).

ODI (Open Datalink Interface): An API developed by Novell for LAN adapter drivers. ODI makes it easy to use a NOS with various LAN adapters, and lets you use multiple protocol stacks in the same host.

OLE (Object Linking and Embedding): A specification that describes the interfaces used for such tasks as embedding objects created by one application within documents created by another, performing drag-and-drop data transfers within or between applications, creating automation servers that expose their inner functionality to other programs, extending the Windows 95 shell with custom DLLs, and much more. Version 1.0 of the specification was originally created for placing objects such as Excel spreadsheets inside documents created by other applications such as Microsoft Word for Windows. OLE 2.0 greatly expanded the scope of OLE and made the original name obsolete, but the name had achieved widespread recognition and was retained.

OSF (Open Software Foundation): Founded in 1988 and sponsored by the world's largest computer and software companies to promote standards for interoperability of software on multivendor hardware platforms. OSF's major deliverables have been OSFIl, OSF/Motif, and OSF DCE. Major development continues on the Distributed Computing Environment (DCE) and Distributed Management Environment (DME) standards.

OSFIl: Based on the Mach operating system, OSFIl provides a fully symmetric multiprocessing operating capability and uses OSF/Motif as the GUl.

OSF DCE (Open Software Foundation Distributed Computing Environment): See DCE.

OSI (Open Systems Interconnect): An international standardization program to facilitate communications among computers from different manufacturers.

OSI/CS (OSI Communications Subsystem): Software that provides connectivity between SNA and OSI.

OSI Reference Model: A logical structure (model) for network operations standardized within the ISO that contains a seven layer network architecture to define the network protocol standards to enable any OSI-compliant computer or device to communicate with any other 051-compliant computer or device. The seven layers, starting with the lowest, are the Physical Layer, Data Link Layer, Network Layer, Transport Layer, Session Layer, Presentation Layer and Applications Layer.

OSPF (Open Shortest Path First): TCP/IP routing protocol that employs user-defined parameters to determine the best or shortest path between routers on an internetwork.

Out-Of-Band: Term used to denote communication with a networked device using a protocol other than its LAN protocol. An example would be an EIA232 connection to a Concentrator to provide network management of a 10BaseT LAN.

OS-2 (Operating System—2): A multitasking operating system developed by Microsoft and IBM for use on the personal computer.

PACKET: A collection of data bits and associated control information bits that is transmitted as a whole unit. In LAN terminology, the basic unit of information transfer.

PACKET SWITCHING: The process of routing and relaying packet data across data communications networks.

PAD (Packet Assembly/Disassembly): Term used to describe the electronic function of creating the proper packets of information, adding headers and trailers to data, and then recovering received data by stripping off headers and trailers. Associated with packet-switched networks like X.25

PARALLEL TRANSMISSION: A method of data transfer in which all bits of a character or byte are transmitted simultaneously, either over separate wires or communication lines, or by different frequencies on the same communications line.

PARITY: A method used by the computer to aid in error detection by the addition of a bit to make the total number of bits turned on an even or odd number.

PCI (Peripheral Component Interface): A 32- or 64-bit local bus architecture developed by Digital, IBM, Intel, and others that is widely used in Pentium-based PCs. A PCI bus provides a high-bandwidth data channel between system-board components such as the CPU and devices such as hard disks and video adapters. PCI is one of two widely adopted local-bus standards. The other, the VL-Bus, is primarily used in 486 PCs (see also VLB).

PCL (Printer Control Language): The printer language that drives HP DeskJet, LaserJet, and other Hewlett-Packard printers. PCL defines a standard set of commands enabling applications to communicate with HP or HP-compatible printers. PCL has become a de facto standard for laser and ink jet printers and is supported by virtually all printer manufacturers. HP compatible or LaserJet compatible means that a printer supports the PCL command set.

PCMCIA (Personal Computer Memory Card International Association): A consortium of computer manufacturers that devised the standard for the credit card-size adapter cards used in many notebook computers. PCMCIA defines three card types: Type I cards can be up to 3.3 mm thick and are generally used for RAM and ROM expansion cards; Type II cards can be as thick as 5.5 mm and typically house modems and fax modems; Type III cards are the largest (up to 10.5 mm thick) and are mostly used for miniature hard disks. PCMCIA support is a major component of Windows 95's Plug and Play architecture, which automatically recognizes when PCMCIA devices are added and removed. (The simpler term PC Card has largely replaced this acronym to refer to these cards.)

PCM (Plug Compatible Manufacturer): A company that makes equipment which can operate with other manufacturers equipment by simply plugging it in.

PDA (Personal Digital Assistant): A hand-held computer, such as the Apple Newton. PDAs typically use pens for input, rather than keyboards, to conserve space.

PHASE MODULATION: An analog modulation technique used by modems to transmit multiple bits per cycle of frequency across a communications network.

PHYSICAL LAYER: A defined layer within the ISO architecture that specifies how devices and network nodes are connected (wired or interfaced).

PIC (Programmable Interrupt Controller): A chip or device that prioritizes interrupt requests generated by keyboards, serial ports, and other devices, and passes them on to the CPU in order of highest priority.

PIF (Program Information File): A binary file in which Windows stores configuration information about a DOS program. A PIF file includes such information as the path to the executable file, the amount of memory the program requires, and whether the window in which the program is run closes automatically when the program terminates.

PIM (Personal Information Manager): A software application designed to store and manage "personal" information such as schedules, phone lists, and contact notes.

PING: A test in which an ICMP echo request packet is sent to another router, the sending program waits for an indication that the packet has reached its destination.

PLCP (Physical Layer Convergence Protocol): The part of the physical layer that adapts the transmission facility to handle DQDB functions as defined in IEEE 802.6.

PLU (Primary Logic Unit): In an SNA LU-LU session, the LU that is responsible for setting the session parameters and for performing error recovery.

PM (Presentation Manager): The graphical user interface in IBM's OS/2 operating systems which includes windowing, graphics standards for the keyboard interface, and editors for icons and font.

PMD (Physical Medium Dependent): One of two parts of the Physical Layer in the ATM protocol reference model which is specific to a particular physical medium and includes definitions for proper cabling as well as bit timing.

PnP (Plug and Play): The technology that lets Windows 95 automatically detect and configure most of the adapters and peripherals connected to a PC. A fully Plug and Play-enabled PC requires three PnP components: a PnP BIOS, PnP adapters and peripherals, and a PnP operating system. Adding a PnP-compliant CD-ROM drive, hard disk, monitor, printer, or other device to a PnP PC requires little more than making the physical connection. The operating system, in conjunction with PnP logic present in the BIOS and in the device itself, handles the IRQ settings, I/O addresses and other technical aspects of the installation to ensure that the device doesn't conflict with other installed devices.

PORT: The entrance or physical access point to a computer, multiplexer, device, or network at which signals may be supplied, extracted or observed.

PORTING: The work involved in moving a software product from one proprietary computer operating system or hardware architecture to another.

PORTMAPPER: A protocol that defines a network service to provide a standard way for a client to look up the port number of any remote program supported by a server. A client sends an PPC call message to the portmapper, and it in turn returns the corresponding port number in an PPC reply message.

POINT-TO-POINT: A connection between two pieces of equipment.

POLLING (Poll): The process of checking each terminal on the network to see if it has traffic to send.

POSIX (Portable Operating System Interface for Computer Environment): The name of the IEEE's standard operating system interface specifications that provide a means to ensure source-level software portability. POSIX provides a broader definition than a typical UNIX kernel interface and includes shell (user interface) definitions as part of the specification. POSIX has been adopted by both ISO and MST as a standard.

POSIX/003.3 defines general rules for developing test methods to measure compliance to the standard.

PPP (POINT-TO-POINT PROTOCOL): The successor to SLIP. PPP provides router-to-router and host-to-network connections over both asynchronous and synchronous serial links.

PRESENTATION LAYER: The sixth layer of the OSI model that determines how application information is represented (i.e., encoded) while in transit between two end systems (see ISO Reference Model).

PRIMARY RATE INTERFACE (PRI): An ISDN defined level of service that provides 24 or 30 64 Kbps links for the transfer of data, and one 64 Kbps link for the transfer of control information.

PROTOCOL: A formal set of conventions governing the format and exchange of data and bits between two communication processes. A specification that describes the rules and procedures that products should follow to perform activities on a network. Protocols enable products from different vendors to communicate on the same network.

PROTOCOL DATA UNIT (PDU): Data and control information flowing through the OSI model layers are referred to as PDUs.

PUBLIC SWITCHED TELEPHONE NETWORK (PSTN): A public dialup telephone network used by the general public for voice and data communications.

QBE (Query By Example): A database query method in which the user forms a query by filling in a table with examples of the requested information. IBM created QBE in the 1970s to simplify the process of retrieving information from mainframe databases; it was later implemented on the PC platform in such products as dBASE and Paradox (see also SQL).

RAM (Random Access Memory): Also known as read-write memory. The memory used to execute application programs in a computer, may be read from and written to.

RARP (Reverse Address Resolution Protocol): A TCP/IP protocol that governs the translation of Data Link Control (DLC) addresses into Internet Protocol (IP) addresses.

REPEATER: A repeater regenerates digital signals to extend the distance of the transmission. Repeaters can be used to overcome the maximum distance barrier of a medium. All concentrators, bridges, routers, and gateways incorporate repeater functionality with their products.

RESISTANCE: The degree to which a wire opposes the flow of electricity in a cable, measured in ohms.

RESPONSE TIME: The amount of time required to return data to the transmitting device.

REVERSED PAIRS: A misfiring of a connector in a way that causes one of the wire pairs to carry a signal that the other wire pair should carry.

RFC (Request For Comment): The document series, begun in 1969, that describes the Internet suite of protocols and related experiments.

RFS (Remote File System): A UNIX System V network protocol developed by AT&T that is a distributed file system and allows users to share data and processing resources across a network of identical UNIX nodes. Like NFS, RFS uses TCP/IP, but differs in that PFS retains UNIX features like named pipes and direct device.

RF MODEM: A modulation/demodulation device used to propagate analog modulated signals onto a wired media between communicating devices.

RING: A topology used to interconnect communicating devices such as the token passing ring local area network onto a closed loop circular network.

RIP (Routing Information Protocol): A routing protocol used on TCP/IP networks that distributes the addresses of reachable networks and metrics reflecting the degree of difficulty involved in reaching particular networks from particular locations.

RG-58: 50-ohm coaxial cable for 10Base2 networks.

RG-59: 75-ohm coaxial cable for ARCnet or CATV LANs.

RG-62: 93-ohm coaxial cable for IBM terminals or ARCnet LANs.

RISC (Reduced Instruction-Set Computing): A microprocessor architecture that favors the speed at which individual instructions execute over the robustness of the instruction set. The PowerPC architecture is one example of a RISC microprocessor design.

RJ11: Popular name for the standard four-wire modular connector used for telephone connections.

RJ21: Popular name for a standard 50-pin data connector. Also called Telco connector.

RJ45: Popular name for the eight-pin modular connector used in the 10BaseT standard for UTP connections to workstations or smart wiring hubs. Actual connector is defined in ISO Standard 8877.

RMON or RLMON MIB (Remote Monitoring Management Information Base): A set of complex extensions to the first SNMP MIBs that allows network segments to be monitored for performance. In addition to the data provided by standard MIBs, RMON adds nine new functions: real-time statistics, historical statistics, host statistics, host threshold statistics, node-to-node communication statistics, filtering capabilities, packet capture and event notification and logging. RMON is defined in the IETF's RFC 1513.

ROM (Read Only Memory): Memory containing pre-loaded programs or data that cannot be rewritten or changed by a computer.

ROUTER: A router examines a packet's destination address and determines which of several paths network traffic will follow. A router is the link between separate LANs, each having its own network address. Routers segment network traffic based on network address. They are protocol-dependent and media-independent.

ROUTING: The process of determining appropriate directions for the transmission of packets.

ROUTING PROTOCOL: The messages exchanged by routers to indicate LAN reachability. Reachability can be defined in a variety of ways, depending on the routing protocol. In the case of RIP, reachability indicates the number of routers that sit between two nodes. This information is used to determine the optimum path to all of the LANs in an Internet.

SCSI (Small Computer Systems Interface): A computer-to-peripheral interface targeted for workstations developed at about the same time as the bus and tag interface. Maximum distances range from 20 feet (6 meters) for single-ended SCSI, to approximately 80 feet (25 meters) for differential SCSI. The specification now consists of several versions, each with different connectors and signaling schemes. SCSI-1 provides effective data rates of 2.1 Mbs to 5.0 Mbs, while SCSI-2 and fast wide SCSI-2 handles data rates of 4.2 Mbs to the theoretical maximum of 80 Mbs.

SDH (Synchronous Digital Hierarchy): Worldwide digital hierarchy standard for broadband transmission. In North America, SDH equates to SONET.

SDLC (Synchronous Data Link Control): An IBM bit oriented communications protocol used in SNA networks.

SDU (SMDS Data Unit): The user payload in an SMDS L3PDU packet. The SDU can contain up to 9,188 bytes.

SFF (System Fault Tolerant): A more complex version of Novell's NetWare that provides capabilities to store data in more that one drive for improved survivability.

SERIAL TRANSMISSION: A method of transmission in which each bit of information is sent one after another on the channel rather than simultaneously as in parallel transmission.

SERVER: A system that is requested to perform some activity by a client or a program that is accessed by a client to perform a specific task.

SESSION: In SNA, a temporary logical connection between two NAUs following agreed- upon rules. In OS/2, a group of processes that can occupy a screen which constitutes a single application.

SESSION LAYER: A defined layer of the ISO architecture that specifies methodologies for recovery, restart, and integrity in end-to-end communications across heterogeneous networks.

SHELL: The program by which a UNIX system implements its standard user interface.

SHINER: A condition in which a portion of the insulation that surrounds a cable is scraped off, thus exposing the copper wire, this usually leads to cable failure.

SHORT CIRCUIT: The condition of having a circuit completed at a point too close to the origin, often resulting in power surges and damage to the system. A short circuit can also be created when two wires are touching in a cable circuit.

SIMPLEX: Transmission mode in which a channel operates in one direction only.

SHIELDED TWISTED-PAIR (STP): A specially insulated cable whose wires have a minimum number of twists per foot. The greater the number of twists, the lower the crosstalk. STP wiring is commonly associated with Token ring networks.

SIDF (System Independent Data Format): A Novell standard to ensure that any SIDF-compliant storage media can be read by any SIDF-compliant product.

Signature: A short note, usually containing your name and address and a brief quotation, that appears at the end of e-mail or newsgroup messages you send.

SIP (SMDS Interface Protocol): The protocol defined in IEEE 802.6 for the interface between an SMDS network and the end-user (SNI). SIP specifies the SIP L2PDU packet at the SM connection.

SIP L2PDU: SMDS Interface Protocol Layer 2 Protocol Data Unit. The 53-byte (7-byte header, 44 bytes of user data, 2-byte trailer) unit of information processed by the second layer of the SIP.

SMB (Server Message Block): A formatted message used to request and reply to requests for file and print services in network systems.

SGML (Standard Generalized Markup Language): A text-based language for de scribing the content and structure of digital documents. HTML, which has gained fame as the language used to create World Wide Web pages on the Internet, is a descendant of SGML. SGML documents are viewed with transformers, which render SGML data the way Web browsers render HTML data (see also HTML and VRML).

SIMM (Single In-Line Memory Module): A form of chip packaging in which leads (pins) are arranged in a single row protruding from the chip. Connectors are attached to a stiff contact strip that permits a SIMM to be inserted into a slot like an expansion adapter. On PCs, SIMM-style RAM chips have virtually replaced the dual in-line package (DIP) chips, identifiable by two rows of protruding legs, that were popular in the 1980s.

SLIP (Serial Line Internet Protocol): Pronounced "slip." A protocol that allows a computer to connect to the Internet through a dial-in connection and enjoy most of the benefits of a direct connection, including the ability to run graphical front ends such as Mosaic and Netscape Navigator. SLIP is also used to run TCP/IP over phone lines (see also PPP and TCP/IP).

SNI (Subscriber Network Interface): The SIP (SMDS Interface Protocol) interface defined in IEEE 802.6 between an SMDS end-user's equipment and the network interface that directly serves the end-user. The SNI is characterized by fixed, 53 byte SIP L2PDU packets. SNI can support either a DS 1 or DS3 access arrangement. 2).

SNI (SNA Network Interconnection): An IBM software product that provides a gateway between two dissimilar SNA networks.

SNMP (Simple Network Management Protocol): The standard network management protocol defined by the IAB in RFC 1157 for TCP/IP based network management. Widely accepted as the defacto standard for LAN network management.

SNMPv2 (Simple Network Management Protocol version 2): The second version of SNMP proposed by the IAB in RFCs 1441-1450 that provides enhanced features like NMS station-to-agent security.

SNA (Systems Network Architecture): A seven level architecture which supports the IBM data processing data communications environment.

SOURCE ROUTING: Workstations maintain a table of network addresses which are inserted into the packet at creation time.

SOCKETS: The basic building blocks for inter-process communications in Berkeley's UNIX. Defined sockets are: Stream sockets—bi-directional, reliable, sequenced, unduplicated flow of data without record boundaries; Datagram sockets are bi-directional, unreliable, unsequenced, possibly duplicated flow of data with record boundaries; Raw sockets—direct access to communication protocols that support socket abstraction used mainly for development; Packet sockets are identical to stream sockets but with record boundaries (not implemented with TCP/IP); and Reliable delivered message sockets. Reliable datagram socket (not implemented with TCP/IP).

SONET (Synchronous Optical NETwork): An emerging broadband fiber network formed by a family of network elements that conform to the SONET interface requirements. SONET is a transport network of synchronously multiplexed tributary signals. The basic electrical signal of SONET runs at a rate of 55,840 Mbps (STS-l). SONET grows in multiples of the basic signal into the multi-gigabit range.

SPOOF: The term used to refer to a networking technique used by WAN devices (typically routers) to provide a local response to a computer's poll of a remote terminal (controller) rather than pass the poll over the network to save bandwidth and prevent timeouts. At the remote end the other WAN device provides a local poll to the remote terminal.

SPOOLING: A technique used to temporally store output destined for a printer from programs allowing the programs to continue processing.

SPACE: A telegraph term denoting the absence of a signal as opposed to a mark which denotes the presence of a signal.

SPX (Sequential Packet Exchange): Protocol used in Novell's NetWare network operating system.

SQL (Structured Query Language): A database programming language that promotes the portability of application programs among different installations. SQL is suited for databases employing the relational data model which require flexibility in the data structures and access paths.

SunOS (Sun Operating System): A SunOS UNIX variant developed and marketed by Sun Microsystems.

SRAM (Static Random Access Memory): A form of RAM that retains its data without the constant refreshing that DRAM requires. SRAM is generally preferable to DRAM, because it offers faster memory access times (a critical element in a PC's performance), but it is also more expensive to manufacture because it contains more electrical components. The most common use for SRAM is to cache data traveling between the CPU and a RAM subsystem populated with DRAM; this boosts performance by reducing the number of DRAM accesses required (see also DRAM).

SRB (Source-Route Bridge): A procedure in which a node sending a packet places a path in the packet header to define the path that packet should take through the network.

SRT (Source-Route Transparent): A proposed integration of source-route and transparent bridge functions.

STAR: A topology used to interconnect communications devices onto a network.

STATISTICAL TIME DIVISION MULTIPLEXING (STDM): A method of multiplexing whereby time slots are allocated to each sending device. If a device is not active, the time slots are reassigned to other active devices.

STP (Shielded Twisted-Pair Wire): Two or more twisted-pair wires wrapped in a braided or solid metal shield used to eliminate interference and noise.

SUBNET MASK: A configuration feature of a router that is used to select the portion of a 32-bit TCP/IP node address that refers to the LAN and the portion that refers to the node on the LAN.

SVC (Switched Virtual Circuit): The term that refers to a switched, or temporary, connection between two endpoints in an packet switched network. SVCs can provide "bandwidth on demand" as multiple 56 Kbps or 64 Kbps can be established to handle bursts of data. SVCs can be setup and torn down by the network on demand, without manual intervention.

SVGA (Super Video Graphics Array): An extension of the VGA video standard. SVGA enables video adapters to support resolutions of 1,024 by 768 pixels and higher with up to 16.7 million simultaneous colors (known as true color) (see also VGA).

SVID (System V Interface Definition): The specification for UNIX Systems Laboratories (USL) UNIX System V operating system. The specification provides an operating environment that distinguishes between base systems (system and library calls) and kernel extensions, and support source code portability across hardware platforms.

SYNCHRONOUS: A data exchange format in which data characters and bits are transmitted at a fixed rate with the transmitter and receiver synchronized by sharing a single clocking signal. This eliminates the need for individual start and stop bits surrounding each byte (character), and provides greater efficiency.

SYNCHRONOUS TRANSMISSION: Transmission of data characters or bits at a fixed rate based upon a clock synchronization between the transmitter and the receiver.

T1: A U.S. and Japanese standard for high-speed data transmission at 1.544 Mbps. Twenty-four 64 Kbps (user) channels plus 8 Kbps of control information are provided. Also referred to as DS1.

T3: A U.S. standard for high-speed data transmission at 44.736 Mbps. Provides the equivalent bandwidth of 28 T1 circuits. Also referred to as DS3.

T4: A U.S. standard for high-speed data transmission at 274.176 Mbps. Provides the equivalent bandwidth of 168 11 circuits. Also referred to as DS4.

TAPI (Telephony Application Programming Interface): A set of functions supported by Windows that permits Windows applications to program telephone line-based devices such as modems and fax machines in a device-independent manner. One of numerous high-level device interfaces that Windows offers as part of the Windows Open Services Architecture (WOSA), TAPI simplifies the process of writing a telephony application that works with a wide variety of modems and other devices supported by TAPI drivers.

TC (Transmission Convergence): One of two pans of the Physical Layer in the ATM protocol reference model which is mostly common to all physical medium, and includes cell delineation, cell rate decoupling, HEC generation, and checking and various operations and maintenance functions.

TRANSPORT LAYER: The fourth layer of the OSI Model that is responsible for reliable end-to-end data transfer between end systems (see OSI Reference Model).

TAP: A device used to connect workstations onto local area networks.

T-CONNECTOR: A T-shaped connector for coaxial cable. Each end of a T-connector attaches to an Ethernet network adapter. At either end of the network, the unused sides of the T-connectors are attached to terminators.

TCP/IP (Transmission Control Protocol/Internet Protocol): TCP and IP are two of the standard Internet protocols: the term TCP/IP is often used to refer to the entire Internet protocol suite. TCP/IP is used to exchange data between nodes that support the TCP/IP protocol suite. The IP portion works at the OSI network layer to direct data between nodes on LANs, and TCP provides a reliable transport service. TCP/IP is the Internet equivalent of IPX.

TDR (Time-Domain Reflectometer): A sonar or radar-like diagnostic technique that analyzes reflected electrical pulses to detect cable faults. A TDR measures the length of a cable by measuring the time it takes for the reflected pulse to return to the TDR, multiplied by the nominal velocity of propagation.

TERMINAL: A device which may be connected to a computer system that provides computational and data access services.

TERMINATOR: A resistor, usually 50 ohms, connected to either end of a thin Ethernet coaxial cable. The terminator matches the characteristic impedance of the cable and dissipates the electrical signal, preventing reflections.

TELNET: The TCP/IP remote terminal protocol.

TERMINAL EMULATION: Communication with a remote computer by which your PC acts as a terminal connected to it.

THICKNET: A standard coaxial Ethernet cable, about as thick as your thumb.

THINNET: Also referred to as cheaper net, a coaxial Ethernet cable about as thick as your little finger.

THYTE (Terabyte): One trillion bytes.

TIFF (Tagged Image File Format): A popular file format for bitmapped graphics that stores the information defining graphical images in discrete blocks called tags. Each tag describes a particular attribute of the image, such as its width or height, the compression method used (if any), a textual description of the image, or offsets from the start of the file to "strips" containing pixel data. The TIFF format is generic enough to describe virtually any type of bitmap generated on any computer, but it is also complex and difficult to write software for.

TIME DIVISION MULTIPLEXING (TDM): A method of multiplexing whereby each device is allocated a set amount of link time and shares the link with other devices.

TOKEN PASSING: A local area network access methodology employed in ring or bus architectures.

TOPOLOGY: A term used to describe the organization of links and nodes to form a communications network.

TRAILER: A term used to denote a collection of unique bits at the end of a packet. Trailers are required if variable length packets are used in a data communications so the network devices can determine the end of a packet. Trailers may be single or multiple byte fields and usually contain the Frame Check Sequence (CRC field) used to validate the accuracy of the packet.

TRANSCEIVER (Transmitter-Receiver): The device that connects a host interface to a LAN such as Ethernet. Ethernet transceivers contain electronics that apply signals to the cable and sense collisions.

TRANSPARENT BRIDGES: The bridging device appears transparent to the workstation and copies the network address from the incoming frame.

TREE: A topology scheme used to interconnect communications devices into a network.

TSR (Terminate and Stay Resident): Refers to programs that remain in memory so that they can be instantly popped up over some other application by pressing a hotkey. The program is displayed either as a small window on top of the existing text or image, or it takes up the full screen. When the program is exited, the previous screen contents are restored.

TWAIN: A memory-resident program used to interface scanners to application programs.

TUNNELING: The encapsulation of one protocol within another. Tunneling is often used to transport a LAN protocol across a backbone network that does not support the LAN protocol.

TWINAX: A type of coaxial cable having duel center conductors instead of a single conductor.

TWISTED PAIR: Small, insulated low grade wires used extensively in electronics and telephone connections.

UART (Universal Asynchronous Receiver/Transmitter): The chip that drives a serial port. IBM chose the National INS8250, better known simply as "the 8250," for the serial ports in its original PC. Enhanced versions of that chip—or custom logic that mimics the functions of the 8250—are still used in PCs today.

UDP (User Datagram Protocol): A transport protocol in the Internet suite of protocols. UDP, like TCP, uses II for delivery; however, unlike TCP, UDP provides for exchange of datagrams without acknowledgments or guaranteed delivery.

ULTRIX: Digital Equipment Corporation's (DEC) version of UNIX based on the Berkeley implementation and provides many system and library calls from UNIX System V. ULTRIX supports DCE and Ingres Relational DBMS. ULTRIX operates only on DEC hardware.

UI (UNIX International): A non-profit organization established to promote UNIX System V development. UI is the product management organization for USL and calls its development environment "Atlas". Atlas specifies 0SF DCE as one of its components for communication services.

UMB (Upper Memory Block): A block of memory (between 640K and 1 MB) created by a 386 memory manager. This is useful for loading TSRs and device drivers so they don't occupy the limited memory below 640K.

UNI (User to Network Interface): The ATM term referring to the interface between an ATM switch and an ATM endpoint (Private UNI); and to the interface between a customer's ATM equipment or switch and a public ATM network (public UNI).

UNICODE: A new standard method of encoding written characters for electronic storage or transmission. Unlike ASCII, which specifies coding rules for the Latin alphabet, Unicode includes all characters and mathematic/scientific symbols used for written languages throughout the world.

UNICODE CONSORTIUM: A non-profit organization established in 1991 to promote the use of Unicode as a character standard.

UNIFORUM: A non-profit organization founded in 1980 to serve as a forum for users, developers, and vendors to exchange information about open systems, UNIX, software, and applications.

UNIX: Operating system originally developed by Bell Laboratories for communicating multiuser, 32-bit minicomputers.

URL (Uniform Resource Locator): A logical address that identifies a resource on the Internet. For example, the URL http://www.zdnet.com/~pcmag/utils/utility.htm is the Internet address of a Web page that gives you access to PC Magazine's utilities.

USENET: A worldwide network of newsgroups on thousands of subjects.

UTP (Unshielded-twisted-pair): A popular twisted-pair cable that looks like telephone wire and does not have an insulating shield or sheath. It is commonly associated with Ethernet (10BaseT) and Token ring networks.

VBR (Variable Bit Rate): An ATM term used to define an ATM Quality of Service (QoS) level to support user information that is bursty in nature. Examples are client-server file transfers and LAN-to-LAN interconnection.

VESA (Video Electronics Standards Association): The consortium of computer manufacturers responsible for the SVGA video standard and the VL-Bus, a local-bus architecture (see also SVGA and VLB).

VFAT (Virtual File Allocation Table): The 32-bit file system that Windows 95 uses to manage information stored on disks. An extension of the FAT file system, VFAT supports long filenames while retaining compatibility with (and many of the limitations of) FAT volumes (see also FAT).

VGA (Video Graphics Array): Also referred to as Video Graphics Adapter. The standard for the graphics adapter built into most IBM PS/2 systems, VGA quickly replaced earlier standards such as CGA (Color Graphics Adapter), and EGA (Enhanced Graphics Adapter) and made the 640-by-480 display showing 16 colors the norm. Other manufacturers have since extended the VGA standard to support more pixels and colors (see also SVGA).

VLB (VESA Local Bus or VL-Bus): The local-bus standard created by the Video Electronics Standards Association (VESA) to provide a fast data connection between CPUs and local-bus devices. The VL-Bus was widely used in 486 PCs, but most Pentium PCs use PCI local buses instead (see also PCI).

VOICE BAND (VOICE GRADE CHANNEL): A channel used for transmission of speech. Usually with an audio frequency range of 300 to 3400 hertz (cycles per second). It is also used for transmission of analog and digital data with data rates up to 14,400 bits per second.

VIRTUAL CIRCUIT: A described service within the OSI architecture model that defines end-to end sequenced guaranteed delivery of packets.

VM (Virtual Machine): An IBM time sharing operating system found on many IBM computers.

VR (Virtual Reality): The technology that uses computer-generated sights and sounds to transport users to new worlds that exist only inside the computer. A key component of virtual reality is immersion blocking out the real world so as to make virtual worlds seem more real. A VR headset that combines computer-generated images displayed inches from the eyes with an audio component is a commonly used immersion tool.

VRML (Virtual Reality Modeling Language): An open standard for 3D imaging on the World Wide Web that paves the way for virtual reality on the Internet. The way VRML code describes a 3D scene is analogous to four points describing a square, or a center point and radius describing a sphere. VRML viewers, similar to HTML Web browsers, interpret VRML data downloaded from the Web and render it on your computer. This allows the bulk of the processing to be performed locally and drastically reduces the volume of information that must be transmitted from the Web a key consideration if rendering is to be performed in real time.

VRAM (Video Random Access Memory): A form of DRAM specially suited for video adapters. VRAM differs from common DRAM in that it features a "dual-ported" design allowing two devices to access it at once. Thus the CRT controller, which converts bits and bytes in video memory to pixels on the screen, and the CPU, which manipulates the contents of video memory, can access VRAM simultaneously. In video boards fitted with the less expensive DRAM, performance suffers somewhat because the CRT controller and the CPU must takes turns accessing the video buffer.

WAIS (Wide-Area Information Server): A system for searching huge distributed database servers across a network, usually the Internet. WAIS allows you to perform a keyword search; it's analogous to an index, whereas Gopher, which is sometimes used as a complement to WAIS, is analogous to a table of contents.

WAN (Wide Area Network): A geographically dispersed network formed by linking several computers or local area networks (LANs) together over long distances, usually using leased long-distance lines. WANs can connect systems across town, in different cities, or in different regions of the world.

WAVE DIVISION MULTIPLEXING: A method used with fiber optic transmission for multiplexing multiple data signals onto the fiber cable.

WINDOW (WINDOWING): A concept that allows multiple applications to be displayed on a terminal at the same time.

WORD: A term used to denote the number of bits that can be transferred internally with in the computer from one component to another.

WORKSTATION: Another term used to describe user terminals. Some work stations may be personal computers with terminal access capabilities.

APPENDIX B

Exercises

PROBLEM ONE

An organization wishes to implement a hybrid network which needs to provide data communications capability to link the following:

1. **London:**
   This is the corporate computer center and the host in a distributed network. It must be online 24 hours to Paris, Brussels, and Madrid. The equipment resident in this location consists of a high-end IBM system (390) and has the capabilities of linking to 25 other centers in a pure SNA environment. The company also wishes to implement an Intranet environment as well as centralized backup capabilities.
2. **Paris:**
   Remote site with a distributed network. Uses midsize IBM hardware (4300) and has a terminal network consisting of 20 PCs and 50 synchronous terminals. All terminals are located within 5,000 feet of the data center. A primary link, as well as backup links, are required to link to the corporate center. Transaction rates are approximately 50 per minute with each transaction consisting of approximately 500 bytes.
3. **Brussels:**
   Regional center with distributed nodes around the EEC. These nodes require 10 hours of service per day and can be low speed links. Equipment consists of a DEC minicomputer in the Amsterdam site and microcomputers at other locations. Volume between the microcomputers and the DEC system is approximately 26,000 bytes per minute (both directions) and 44,000 bytes per second back to London.
4. **Madrid:**
   This is a backup center to the corporate center and has the same hardware configuration as the corporate center. There are 100 terminals attached to this computer using remote access and they require access 20 hours per day.

5. **Other Locations:**
All other locations are networked PCs in sales offices. Some offices have up to 8 PCs with two remote printers. Workstations need to be online during working hours, and are networked using token-rings.

Based upon the information just presented, provide the following:

1. Draw a diagram of the network and describe the topologies that need to be employed.
2. Define the specific communications equipment needed at each of the nodes including the host node.
3. Define the methods needed for connectivity. (The number and speed of the links.)
5. Calculate the anticipated bit per second requirements for an 18 month period. Growth is expected to be around 15 percent for the corporation.
6. Cost justify the above.

PROBLEM TWO

A start-up company wishes to create and offer public data network services which will link the major cities of Europe together. It desires to offer line capabilities ranging from 300 bps up to T1 speeds to its subscribers. Additionally, it wishes to offer gateway services for codeset and protocol conversions. It also wishes to provide PAD capabilities for access to X.25 networks.

Most of the equipment will be IBM hardware, however in five of the smaller cities, DEC, and Tandem computer equipment will be used. It also wishes to offer connection service to PC users who wish to access information providers and bulletin boards.

1. Select 10 cities and develop a design showing the topologies to be used to create and backup the network.
2. Define the types of hardware required to support a minimum of 1,000 connections, and to communicate across the network to all other nodes.
3. Describe the various carrier and service levels required to provide low- and high-speed communications.

PROBLEM THREE

The XYZ company wishes to create and implement a network control center to monitor and control its data communications network. The various data centers are located in: London, Paris, Berlin, Toronto, and New York. The same equipment is used in all locations, and all links are computer-to-computer links on a point-to-point configuration. Each center has a terminal network that spans a 20 mile radius and consists of 130 terminals in groups of 20 per site. The network must be up 24 hours a day. Each center performs regional distributed processing in a batch mode as well as having an online network.

1. Describe the hardware that would be required at each location.
2. Define the various communications links required to support the network.
3. Define and diagram the topologies that would be needed to support this network.
4. Describe the functions performed by the network control center.

PROBLEM FOUR

A new international airline is being started and it initially intends to offer limited services from five major cities. The corporate offices will be located in Dallas Texas, as well as the maintenance base. Other locations will include: London, Paris, Singapore, Rio, and Washington, D.C.

1. Define the topology required including the backup links necessary to maintain 24-hour service.
2. Describe the equipment required at the corporate location to provide reservation support and business processing.
3. Describe the equipment needed at each of the remote locations to support the reservation system.
4. Describe the communications links required to support the remote sites.
5. Describe the primary functions performed at the corporate data center.

PROBLEM FIVE

A new local bank is expanding its business and needs a full line of data processing services. It plans to offer ATM services at approximately 130 locations. Additionally, it requires teller terminals at 30 branch offices. (Each branch requires 10 to 15 terminals.) The bank also wishes to offer home banking services on a dialup basis. The equipment selected will be IBM.

1. Describe the communications software required at the corporate computer center.
2. Describe the hardware required to support the communications network.
3. Describe the types of services required at both the branch level as well as the ATM level.
4. Define the method and facilities required to support home banking.
5. Describe future services that might be implemented.

APPENDIX C

Standards

This appendix lists various standards that are referenced throughout this book. To obtain the actual standards definition contact the appropriate agency.

1. CCITT ANALOG STANDARDS (PARTIAL LIST)

The following V. recommendations are some of the more important standards that apply to modems and data transmission; note: The letters "bis" following the V designation indicate the second version of a particular standard; "ter" indicates the third.

V.1	Defines binary 0/1 bits as space/mark line conditions.
V.2	Specifications of power levels for data transmission over telephone lines.
V.4	Sequence of bits within a character as transmitted.
V.5	Specification of synchronous data signaling rates over the switched network.
V.6	Specification of synchronous data signaling rates over leased lines.
V.7	Definition of other key terms used in the V series recommendations.
V.10	Description of an unbalanced physical level interchange circuit.
V.11	Description of a balanced physical level interchange circuit.
V.13	Simulated carrier control; FDX modem used as half duplex.
V.14	Asynchronous to synchronous conversion.
V.15	Acoustic couplers.
V.16	Electrocardiogram transmission on phone lines.
V.17	Application-specific modulation scheme for Group 3 fax.
V.19	CCITT parallel data transmission modems.
V.20	Parallel data transmission modems.
V.21	CCITT physical layer digital interface standard.

V.21bis	CCITT RS-232 compatible interface.
V.22	1,200/600 bps FDX modems.
V.22bis	2,400 bps modems.
V.23	1,200/75 bps, also supports 600 bps in the high channel speed.
V.24	Definition of the interchange circuit pins between DTE and DCE.
V.25	Automatic answering equipment and parallel automatic dialing (defines the 2,100 hz "answer tone" that modems send).
V.25bis	Serial automatic calling and answering. CCITT equivalent of Hayes AT command set used in the USA.
V.26	CCITT 2,400 bps four wire leased circuits.
V.26bis	2,400/1,200 bps HDX modems.
V.26ter	2,400/1,200 bps FDX modems.
V.27	CCITT 4,800 bps for switched/leased circuits.
V.27bis	4,800/2,400 bps 4W modems.
V.27ter	4,800/2,400 bps FDX modems. Also used in half-duplex two-wire mode to implement the 2,400 and 4,800 bps transmission schemes in Group 3 fax.
V.28	Description of unbalanced interchange circuits operating below 20 Kbps.
V.29	CCITT 9,600 bps for leased circuits. (Standard also used for Group III fax.)
V.31	Older electrical characteristics, rarely used.
V.31bis	V.31 using opto-couplers.
V.32	9,600/4,800 bps FDX modems.
V.32bis	4,800, 7,200, 9,600, 12,000, and 14,400 bps modems and rapid rate re-negotiation.
V.33	14,400 bps (and 12,000 bps for four-wire leased lines).
V.34	28,800/33,300 bps standard for two-wire dial-up communications.
V.35	48 Kbps four-wire modems. The CCITT no longer recommends the use of this standard. It was made obsolete by V.36.
V.36	48 Kbps 4W modems.
V.37	72 Kbps four-wire; V.36 and V.37 are not really "four-wire modems," they are group band modems, which means they combine several telephone channels.
V.40	Specifies how teletypes indicate parity errors.
V.41	An older, obsolete error control scheme.
V.42	Error correcting procedures for modems using asynch-to-synch conversion (V.22, V.22bis, V.26ter, V.32, V.32bis); defines LAPM protocol and provides fallback to MNP4.
V.42bis	Lempel-Ziv-based data compression scheme for use with the V.42 LAPM.
V.50	Standard limits for transmission quality for modems.
V.51	Maintenance of international data circuits.
V.52	Apparatus for measuring distortion and error rates for data transmission.
V.53	Impairment limits for data circuits.
V.54	Loop test devices for modems.
V.55	Impulse noise measuring equipment.
V.56	Comparative testing of modems.

V.57	Comprehensive test set for high speed data transmission.
V.100	Interconnection between PDNS (Public Data Networks) and PSTNs (Public Switched Telephone Networks).
V.110	Specifies ISDN terminal adaption.
V.120	ISDN terminal adaption with statistical multiplexing.
V.230	General data communications interface, layer 1.

2. CCITT DIGITAL STANDARDS (PARTIAL LIST)

X.3	CCITT Packet Assembly/Disassembly functions.
X.20	CCITT Interface between DCE and DTE for asynchronous data transmission on public data networks.
X.21	CCITT Interface between DCE and DTE for synchronous data transmission on public data networks.
X.25	Interface between data terminal equipment (DTE) and data terminating equipment (DCE) for terminals operating in a packet mode connected to a public data network.
X.28	CCITT terminal-PAD communications formats.
X.29	CCITT host-to-PAD communications formats.
X.75	Terminal and transit call control procedures for data transfer on international circuits between packet switched networks.
X.95	CCITT network parameters in a public data network.
X.121	CCITT international addressing standard.
X.200	Reference model of open systems interconnection for CCITT applications.
X.400	CCITT Interoffice electronic mail standards.

3. CCITT RECOMMENDED ISDN STANDARDS (PARTIAL LIST)

I.431	Layer 1 Physical specifications for primary rate DS1 connections.
I.441/451	ISDN primary rate interface.
I.515	Parameters for ISDN internetworking.
Q.700	Signaling System Number 7. (CCS#7) Specifications.
Q.921	Layer 2 (Data Link Protocol) for D channel (Link Access Procedure-D).
Q.931	Layer 3 (Network) ISDN user network interface specifications.
Q.932	Generic procedures for the control of ISDN supplementary procedures.
V.110	B channel procedures.
V.120	B channel procedures.

4. EIA INTERFACE STANDARDS (PARTIAL LIST ONLY)

EIA-232	25-pin interface between DTE and DCE.
EIA-269	Synchronous signaling rates for data transmission.

EIA-422	Electrical characteristics of balanced voltage digital interface circuits.
EIA-423	Electrical characteristics of unbalanced voltage digital interface circuits.
EIA-442	Electrical characteristics for balanced pairs.
EIA-449	37-pin general interface for DCE to DTE.
EIA-496	Interface between DTE and the public switched telephone network.
EIA-530	High-speed interface for DTE to DCE equipment.

5. ANSI TELECOMMUNICATIONS STANDARDS (PARTIAL LIST)

X3.1	Synchronous signaling rates for data transmission.
X3.4	American National Standard for Information Interchange.
X3.15	ASCII code bit structure.
X3.16	Character structure and character parity sense for serial by bit data communications in ASCII.
X3.24	Signal quality for synchronous data transmission.
X3.25	Character structure and parity sense for ASCII parallel transmission.
X3.28	Procedures for using ASCII communications control characters.
X3.36	Synchronous data signaling rates between DTE and DCE.
X3.41	Codeset extension techniques.
X3.57	Message header formats using communications control characters.
X3.66	Standard for bit-oriented link control protocol, ADCCP.

6. ISO STANDARDS (PARTIAL LIST)

ISO3309	Data communications HDLC control procedures and frame structure.
ISO7498	The seven layer ISO/OSI reference model.
ISO8072	Transport service definition.
ISO8208	X.25 packet level protocol for data terminating equipment.

7. INSTITUTE OF ELECTRONICS AND ELECTRICAL ENGINEERS STANDARDS (PARTIAL LIST)

802.1	Architecture and Internetworking Standard. This standard defines the layers of the architecture and the rules for interconnection of different LAN protocols.
802.2	Logical Link Control Standard. This standard defines the equivalent of data link control as specified in the OSI reference model and is also the protocol used for data transfer.
802.3	CSMA/CD Access Method and Physical Layer Specifications. This standard defines the carrier sense multiple access method with collision detection used for data transfer.

802.4	Token Passing Bus Access Method and Physical Layer Specifications. This standard defines the characteristics of the token passing bus network access method used for data transfer in token passing LANs.
802.5	Token Passing Ring Access Method and Physical Layer Specifications. This standard applies to those systems using token passing ring LANs and defines the access methods employed in data transfer.
802.6	Metropolitan Network Access Method and Physical Layer Specifications.
802.7	Sets standards for the definition of a broadband cable plant design. It also establishes guidelines for LAN construction within a building or plant.
802.8	(Working Group). This group was established to assess the impact of fiber optics within the LAN environment and recommend standards.
802.9	IVD Integrated Voice and Data Working Group.
802.10	IEEE Standards for Local and Metropolitan Area Networks: Interoperable LAN/MAN Security (SILS).
802.11	Standard for Wireless LAN Medium Access Control (MAC) and Physical Layer (PHY).
802.12	Demand Priority Access Method Physical Layer and Repeater Specification for 100 Mbps operation.

APPENDIX D

Bibliography

Cronin, Mary J. (1994). *Doing Business on the Internet.* New York, NY: Van Nostrand Reinhold Press.

Data Communications Magazine. A monthly publication of McGraw-Hill: New York.

Feit, S. (1994). "TCP/IP Architecture, Protocols, and Implementation." New York, NY: McGraw-Hill Inc.

Gelber, S. (Dec. 1989). "Local Area Network Technology Concepts." Falkner Technical Publications.

Gelber, S., and Morley, J. (1996). "The Emerging Digital Future." Boston, Mass: CTI Inc. A division of Thompson Publishing.

Harley H., and Stout R. (1995). "The Internet Yellow Pages." Berkeley, CA: Osborne McGraw-Hill Press.

Harley H., and Stout R. (1995). "The Internet Complete Reference." Berkeley, CA: Osborne McGraw-Hill Press.

Huitema, C. (1995). "Routing on the Internet." Englewood Cliffs, NJ: Prentice Hall.

IEEE Standards for Local Area Networks. (1985). *Logical Link Control IEEE.* New York.

Joshi, S., and Lyer, V. (July 1984). "New Standards for Local Area Networks Push Upper Limits for Lightwave Data." *Data Communications.*

Lan Magazine. *A Monthly Publication of Telecom Library.* New York.

Miller, M. (1991). "Internetworking, A Guide to Network Communications." New York, NY: M&T Books.

Naugle, M. G. (1994). "Network Protocol Handbook." New York, NY: McGraw-Hill, Inc.

Nagule, M. G. (1994). " The Illustrated Network Book." New York, NY: Thompson Publishing.

Network World. Framingham Ma: A Weekly Publication of CW Publishing Inc.

Pike, M. A. (1995). "Using the Internet." Indianapolis, IN: QUE Corporation.

Sherman, K. *Data Communications, A Users Guide* (3rd Edition). Englewood Cliffs, NJ: Prentice-Hall Inc.

Stallings, W. "Handbook of Computer Communications Standards." New York, NY: Macmillan Publishing Company.

Stamper, D. (1986). "Business Data Communications." Menlo Park, CA: The Benijaman Cummings Publishing Company.

Strole, N. (1991). "The IBM Token Ring Network—A Functional Overview." IEEE Network.

APPENDIX E

Review Question Answers

CHAPTER ONE

1. Why has the data communications industry grown at such an accelerated rate?
   - Due to the tremendous growth and demand of the data processing industry, the entertainment industry, and the Internet.
2. What does the term "data communications hardware" refer to?
   - The components and equipment required to perform data communications activities. This includes the lines, the modems, and the computer—communications hardware.
3. What are the advantages of implementing a data communications network?
   - Immediate access to current information.
   - Reduction in the amount of redundant data required to process data into information.
   - Business decisions can be based on a single version of the truth.
4. What are the disadvantages associated with data communications?
   - Costs.
   - Additional personal required.
   - Additional maintenance.
5. Does backup need to be considered in establishing communications between different locations?
   - Backup is an important part of any network environment. With proper backup, outages, and network down time will be minimized.
6. What types of software are typically used for data communications?
   - Network communications software.
   - On-line applications' software.
   - Network operating system (NOS).

7. What are the two transmission methods used today for data communications?
   - Analog.
   - Digital.
8. Is cost the only factor that needs to be considered in the implementation of a communications network?
   - No, technical support, systems programming, training, and hardware.
   - Software functionality.
9. What is the basic responsibility of a remote computer center?
   - It manages the business activities of the remote location and provides backup for the primary site.
10. What considerations should be evaluated when planning a data communications network?
    - Cost.
    - Links.
    - Speed.
    - Hardware.
    - Software.
    - Personal.

CHAPTER TWO

1. What did Samuel Morse invent?
   - Samuel Morse invented the telegraph system in 1856.
2. With what is Baudot credited?
   - Developing the Baudot Code for use with a automated telegraph system.
3. What two code sets are currently used in data communications?
   - ASCII.
   - EBCDIC.
4. What was the prime catalyst for the data communications industry?
   - The data processing industry.
5. What factors must be evaluated when selecting data communications hardware?
   - Costs.
   - Speed.
   - Capacity.
   - Maintenance.
   - Upgrade capability.
6. What are the three major components required for data communications?
   - Hardware.
   - Software.
   - Communications services.

7. What software needs to be considered for data communications?
   - Systems software.
   - Application software.
   - Communications control software.
   - Network control software.
8. What function does application control software perform?
   - Application control software manages all on-line applications and ensures each has the ability to communicate and process data.
9. What do communications service providers offer?
   - End-to-end links.
   - Network management.
10. What role does the telephone company play in providing communications services?
    - The telephone company provides value added services, low- and high-speed communications lines, and data communications services.

CHAPTER THREE

1. What functions does communications software perform?
   - Communications software controls, monitors, and manages the links, application programs, and communications pathways.
2. What functions does system software perform?
   - Systems software manages the hardware resources for all programs and processes.
3. What activity is application software responsible for?
   - Applications software is responsible for manipulating data into meaningful information.
4. Name four major functions performed by the communications control software.
   - Polling.
   - Selection.
   - Error detection.
   - Application management.
5. What does the term "conversational mode processing" describe?
   - Establishing and maintaining a communications session until the user terminates the connection.
6. What functions do BTAM and VTAM perform in the IBM environment?
   - BTAM and VTAM are IBM access methods that provide the pathways for terminal connections.
7. What functions do network control programs perform?
   - Network control programs provide network management and internetwork communications.
8. True or false: Some overlap of functions exists between communications software and system software.
   - True.

9. What is the difference between a terminal and a LAN workstation to the host application software?
   - From the application standpoint, no difference exists unless it is a Client-Server aware application.
10. Communications and system software are usually purchased from the hardware manufacturer or a software vendor as opposed to the facility developing the software itself. Why?
   - The software is very complex and requires years to develop.
   - It must be updated and maintained which can be very costly when new hardware is implemented.

CHAPTER FOUR

1. What are the differences between transaction based communications and time-sharing based communications?
   - Communications control software manage terminal application processes in a conversational or transaction based environment.
   - Time-sharing software allocate slices of time to each user allowing the user to control application program activity.
2. Can time-sharing software be used in place of transaction based communications software, and why would one want to do so?
   - Yes, however, it is a waste of resources to use time sharing services for transaction or conversational processing because of slow response time.
3. What special considerations do application programmers need to take into account when writing programs that perform online functions?
   - On-line programs must not modify themselves during execution.
   - On-line programs should not perform long record searches.
4. What is the major difference between a conversational online system and a transaction-based online system?
   - Conversational systems are used by programmers for session based activities.
   - Transaction-based networks are designed to process single transactions in a random order.
5. What is a time-sharing system primarily designed to do?
   - Time-sharing systems provide CPU access to users who need the resources of a large mainframe to perform processing activity.

CHAPTER FIVE

1. Define the term "channel."
   - A pathway between the CPU and device controllers such as a communications processor.

2. What is baseband transmission?
   - Baseband transmission is usually a single transmission channel on the media.
3. True or false: Simplex transmission is transmission in both directions but not simultaneously.
   - True.
4. True or false: Full-duplex transmission is the most common transmission mode.
   - True.
5. True or false: Half-duplex transmission allows two-way communication at the same time.
   - False.
6. What are the two grades of transmission service available for data transmission?
   - Voice grade.
   - Wideband service.
7. What type of transmission, analog or digital, are current telephone transmission networks designed to support?
   - From residence or small business users, analog transmission is used.
   - From large organizations and telephone facilities, digital transmission is used.
8. What transmission mode are most types of data communications equipment designed to support?
   - Half-duplex.
9. What are the two transmission methods available for data communications?
   - Synchronous.
   - Asynchronous.
10. Which transmission method requires clocking information?
    - Synchronous.

CHAPTER SIX

1. Name the three basic modulation techniques used in data communications.
   - Amplitude Modulation.
   - Frequency Modulation.
   - Phase Modulation.
2. Describe frequency modulation.
   - Frequency modulation varies the timing of the data signal but keeps the amplitude the same.
3. Describe amplitude modulation.
   - In amplitude modulation, the frequency remains the same however the amplitude of the signal changes in relation to the value of the bits.
4. What is the difference between baud and bps?
   - The term baud and bps represent different things—baud means signal changes per second and bps, bits per second.

5. What are dibits, tribits, and quadbits?
   - Dibits are 2-bits, tribits are 3-bits, and quadbits are 4-bits.
6. What is phase modulation?
   - Phase modulation describes a transmission technique where both the amplitude and frequency change according to the information bits to be transmitted.
7. Does TCM include a coding technique?
   - Yes, TCM is both a modulation technique as well as an encoding technique.
8. A pure digital signal can be defined as?
   - Voltage changes.
9. What is PCM?
   - PCM or pulse coded modulation is used for digital voice applications.
10. Name two digital transmission techniques.
    - Return to Zero.
    - Non-return to Zero.

CHAPTER SEVEN

1. Name some of the basic transmission media used in data communications.
   - Twisted-pairs wire.
   - Coaxial cable.
   - Optical-fiber.
2. What is the main advantage of coaxial cable over twisted-pair wire?
   - More manageable signal attenuation.
   - Better noise control.
   - Better quality cable.
3. Why is fiber-optics considered the wave of the future?
   - Signals travel at the speed of light and speed is only limited by the equipment located at each end of the link.
4. Are microwave and satellite transmission considered line-of-sight transmission methods?
   - Yes.
5. What transmission medium does "cladding" refer?
   - Optical-fiber.
6. What function does a transponder perform?
   - It is a receiver transmitter component found on satellites and microwave units.
7. For what function are twisted pairs primarily used?
   - Local area networks and terminal environments.
8. To what does line-of-sight transmission apply?
   - The transmitter must be able to see the receiver.
   - Used for satellite and microwave transmission activities.

9. In what manner is distortion a problem in data communications?
   - Distortion can alter the values of analog or digital signals (changing ones into zeros, etc.)
10. True or false: Distortion is really a general term for many types of errors.
    - True.

CHAPTER EIGHT

1. What is a datagram?
   - A data packet that has no relationship to any other packet being transported across a network.
2. How many types of datagrams exists?
   - Reliable.
   - Unreliable.
3. What is a connectionless service?
   - A connection does not need to be established between the sender and the receiver prior to data transmission.
   - The packet carries the sending and receiving address.
4. What is a virtual circuit?
   - A temporary end-to-end connection between the sender and the receiver for the duration of the transmission.
5. Name the two types of switching techniques discussed in this chapter?
   - Packet switching.
   - Circuit switching.
6. What is the difference between a datagram and a packet?
   - They are basically the same thing.
7. What is a Cell-Relay packet?
   - A cell-relay packet is a fixed length (53 byte) packet used for high speed communications.
8. What is STM?
   - Synchronous transmission mode. This is cell relay on a virtual circuit.
9. What is Frame Relay?
   - Frame relay is a fast packet service designed to replace slower packet services such as those associated with X.25 networks.
10. What is the difference between Frame Relay mode 1 and mode 2?
    - One is a public network provided by telephone companies and the other, a private network.

CHAPTER NINE

1. What three companies provide 95 percent of all carrier services in North America?
   - AT&T, MCI, Sprint.

2. What services do international carriers provide?
    - They provide international links between different countries.
3. What services do public data network vendors typically provide?
    - They provide end-to-end links for companies that do not want to implement their own networks.
    - They also provide network management, redundant links, and detailed billing.
4. What types of packets are available for use in packet-switching networks?
    - Reliable and unreliable datagrams.
5. What does PSTN refer to?
    - Public switched telephone network.
6. What is a type 3002 line?
    - A type 3002 line is a GTE term for a leased line.
7. Name the two types of conditioning available with leased lines.
    - C and D conditioning.
8. What transmission service do most common carriers support?
    - Leased line voice grade services.
9. The design criteria for ISDN were based on what?
    - All digital network.
    - International compatibility.
    - Bandwidth management.
10. What is 2B+D considered to be?
    - Basic rate interface, two 64 Kbps bearer channels, and one 16 Kbps data channel.
11. For what activity is the D channel used?
    - This is used for out-of-band signaling.
12. What is the bit rate for the ISDN 3H0+D channel?
    - 384 Kbps, used for video transmission.

CHAPTER TEN

1. What activities of the host computer are relieved by the communications processor?
    - Managing the physical links.
    - Managing the logical links.
    - Freeing memory in the host processor for application processing.
    - Reducing host CPU cycles needed for communications functions.
2. Name three major functions of the FEP.
    - Polling and selection.
    - Error control.
    - Network management.
3. True or false: FDM stands for frequency-directing modulation.
    - False.
4. True or false: Multiplexers provide four different techniques for the transfer of data.
    - False.

5. True or false: STDM is the most current multiplexing scheme.
   - False.
6. Wave-division multiplexing is used with what medium?
   - Optical-fiber.
7. True or false: Network controllers provide a focal point for monitoring the communications network.
   - True.
8. True or false: The terminal controller requires software to function properly.
   - True.
9. True or false: Smart and dumb are the two classifications for terminal types.
   - True.
10. True or false: Remote job entry systems are normally used for batch data transfer activities.
    - True.

CHAPTER ELEVEN

1. What are the basic functions of a modem?
   - To modulate digital signals into analog signals on the sending side and demodulate analog signals into digital signals on the receiving side.
2. True or false: The two classes of modems found in use today are asynchronous and synchronous.
   - True.
3. Why does synchronous data transmission require a clock?
   - To synchronize the computer or DTE device with the modem or DCE device.
4. Describe auto-dial and auto-answer.
   - A modem that supports automatic dialing and auto-answer features.
5. What CCITT type modem would be used in high-speed data communications over full-duplex lines?
   - V.35 modems.
6. True or false: Standard modems do not have a distance limitation.
   - True.
7. True or false: Some modems are limited to half duplex only.
   - True.
8. True or false: Modem eliminators are not really modems at all.
   - True.
9. True or false: Security is a function of the computer, not the modem.
   - True.
10. True or false: Auto-answer will automatically make a call for the user.
    - False.

11. What are the four loopback tests available for testing lines and modems?
    - Local and remote analog loopback.
    - Local and remote digital loopback.
12. Name two diagnostics available with most modems.
    - Bit error rate test (BERT).
    - Frequency drift monitoring.

CHAPTER TWELVE

1. What CCITT interface is considered to be the industry standard for DCE-to-DTE connections?
    - V.24.
2. What is the bit rate for the RS-449 interface?
    - 2 Mbps.
3. How many connector pins does the EIA-232-D interface have?
    - This is a 25-pin interface.
4. What interface uses balanced pairs?
    - EIA-449.
5. To what is the V.24 interface equivalent?
    - EIA-232.
6. What are the distance and speed limitations for the EIA-232-D interface?
    - 50 feet and 19 Kbps.
7. Name the three phases of activities associated with EIA-232-D?
    - Link activation.
    - Data transfer.
    - Link deactivation.
8. Why is the RS-449 interface better suited to high-speed data communications than the EIA-232-D interface?
    - Higher speed.
    - Balanced pairs reducing errors.
    - Support for 200 foot cable lengths.
9. Name the four signal types that are used for the transfer of information between DTE and DCE equipment.
    - Grounds.
    - Data transfer.
    - Timing.
    - Control.
10. What is CCITT V.35?
    - This is a leased-line high speed modem that supports full-duplex transmission.

CHAPTER THIRTEEN

1. Name the two ITU-T standard interfaces most commonly used today.
   - V. Series (Analog).
   - X. Series (Digital).
2. What is the ITU-T responsible for?
   - World-wide analog and digital communications standards for telephones, modems, and digital service units.
3. Why was the International Telecommunications Union created?
   - To provide a global focal point for establishing telephone and data communications standards.
4. What major contribution has the IEEE made to data communications?
   - Wire bus standards.
   - 802 LAN standards.
5. To what activity do the X.21, X.25, and X.29 standards apply?
   - Packet-switched networking.
6. What are the two main standards organizations in the world today?
   - International standards organization (ISO).
   - International telecommunications union for telecommunications (ITU-T).
7. Name two organizations responsible for de facto standards.
   - IBM.
   - Bell.
8. Name a standard that ANSI developed.
   - ASCII codeset standard.
9. What does the EIA do?
   - This is an industry association responsible for the EIA interface standards.
10. Why are standards necessary?
    - They allow equipment from different manufacturers to connect to each other.
    - They provide users with the ability to purchase equipment from different manufacturers that will work together.

CHAPTER FOURTEEN

1. What are the two basic types of line connections used in network communications?
   - Point-to-point.
   - Multipoint.
2. True or false: Multipoint topologies can only use dedicated links.
   - True.
3. Name the six basic network topologies.
   - Star.
   - Ring.

- Bus.
- Tree.
- Mesh.
- Mesh of trees.

4. What are the two main modes of polling in a multipoint line configuration?
   - List polling.
   - Hub polling.
5. Define the differences between polling and selection.
   - Polling is a request from a host to a secondary device.
   - Selection is a transmission from the host to a secondary device.
6. True or false: The mesh topology offers one of the better methods for backup and redundancy.
   - True.
7. Are economic considerations important in the design of a communications network?
   - Economic considerations play a major role in the design of a data communications network.
8. Name some of the factors that must be considered when designing a communications network.
   - Topology.
   - Cost.
   - Maintenance.
   - Equipment availability.
   - Usage.
   - Access.
9. Would a local network require different design criteria than a remote network? Why?
   - Used for servicing different environments.
   - Requires different wiring and link methodologies.
   - Requires different support and routing methodologies.
10. Define distributed processing.
    - Distributed processing can be defined as an environment where several computers in different locations share processing responsibilities so that processing occurs at the location most able to handle the traffic.

CHAPTER FIFTEEN

1. What does the ISO OSI model define?
   - A seven layer architecture that describes how connections can be established between dissimilar computing environments.
2. What is SNA?
   - This is IBM's systems network architecture. It defines how networking is established between local and remote IBM networking environments.

3. Are SNA and OSI architectures compatible?
   - No, while they appear to perform similar functions, they are not compatible.
4. SDLC is a protocol for what architecture?
   - SDLC is the data link layer for SNA.
5. What is layer two also known as?
   - Data link layer.
6. True or false: The OSI model is designed to encompass both data processing and data communications.
   - True.
7. For what is X.25 used?
   - International low-speed packet networking.
8. What do the first three layers of X.25 define?
   - Physical.
   - Data link.
   - Network.
9. What protocol is used with the X.25 architecture?
   - Link access protocol (LAP).
10. What is MAP?
    - Manufacturing automation protocol.

CHAPTER SIXTEEN

1. Define the term "protocol."
   - A protocol is considered a handshake between two computers, to networks, or two endpoints.
2. What functions does data link control accomplish?
   - Data transfer.
   - Timing control.
   - Direction control.
   - Error control.
   - Transparency.
3. What layer of the OSI or SNA protocol stack is the DLC protocol?
   - Layer two.
4. What older transmission method sends one byte at a time?
   - Asynchronous transmission.
5. What are the functions of control bits in bit-oriented protocols?
   - Requests and responses.
   - Error control.
   - Flow control.
6. What control codes in byte-oriented protocols accomplish line turnaround?
   - NAK.
   - Ack0, Ack1, EOT, RVI.

7. What does a NAK do?
   - Indicates a negative acknowledgment.
8. Name the three types of error detection in use.
   - VRC, LRC, and CRC.
9. What is the major difference between asynchronous and synchronous protocols?
   - Overhead for timing.
   - Bit V's byte configuration.
   - Synchronous protocols are more efficient.
10. What are IBM's current synchronous protocols?
    - SDLC.
    - BSC.
11. What BSC response does a secondary station send when it is not ready to receive the next frame or packet?
    - WACK (wait acknowledgment).
12. For which architecture are TP and IP defined?
    - OSI.
13. What is SNMP?
    - Simple network management protocol. This is a protocol sub-system used for network management and control.
14. What is RIP?
    - Routing information protocol. This is used for routing across TCP/IP based networks.

CHAPTER SEVENTEEN

1. What does the IEEE 802 standard cover?
   - Local area and metropolitan area networks.
2. What does CSMA stand for?
   - Carrier sense multiple access (used in Ethernet type networks).
3. What standard is Ethernet an implementation of?
   - The IEEE 802.3 standard.
4. What is the principal transmission technique used by CSMA?
   - Listen before talk.
5. What does a token-ring network do?
   - Uses a three byte packet (token) to indicate to a workstation network availability so that workstations can use the ring for data transfer.
6. What is the most popular transmission medium for LAN implementations?
   - Twisted-pairs wiring.
7. Are LAN operating systems a required component?
   - Yes.
8. Is spooling a function of the network operating system?
   - Yes.

9. Does the NOS replace the basic operating system of the computer?
   - Depending upon the operating system, the NOS may co-exist with the operating system, or in cases such as UNIX and Windows NT is part of the operating system.
10. Do network operating systems provide interfaces between different LAN types?
    - Most NOS provide compatibility between different operating systems.
11. Is security a network operating system capability?
    - Yes.
12. Is e-mail a typical LAN application?
    - Yes.
13. What is an MAU/MSAU device?
    - Multistation access unit (used to interface the workstation to the media).
14. What function does the LAN interface card perform?
    - provides an interface between the workstation and the hub or MSAU.
15. What distances are typically supported using twisted pair wiring?
    - 375 meters.

CHAPTER EIGHTEEN

1. What are the three major types of errors encountered within a data communications network?
   - Hardware.
   - Software.
   - Communication link.
2. Name the basic procedures used to identify and resolve errors.
   - Identify the problem area.
   - Determine the nature of the problem.
   - Determine if an alternate path can be taken.
   - Determine time to repair.
3. What devices perform analog and digital loopback testing?
   - Modems.
   - Hardware diagnostic equipment.
4. What functions do protocol and data analyzers perform?
   - Allow the user to examine data packets in real-time.
5. What function does a timer perform when used in a protocol analyzer?
   - Allows the user to set threshold values based on time parameters.
6. What function does a breakout box perform on the network?
   - Helps to identify signaling problems on interface cables.
7. What is BERT an acronym for?
   - Bit error rate tester.
8. What functions do SAMs perform?
   - Evaluates analog signals traveling over communications links.

9. What activity is used for monitoring the entire communications network?
   - Network control center.
10. Why should network control centers be required within medium to large communications networks?
    - They provide a focal point for monitoring the entire network from a single point. This allows quick response to problems.

CHAPTER NINETEEN

1. Is the network control monitor a required component in a communications network?
   - No, but is important for rapid diagnose and repair.
2. Are frame error counts important to performance?
   - Yes, they help to evaluate the health of the network.
3. Can dynamic configuration be accomplished using a network control monitor?
   - Yes, this allows the network control center to bypass trouble spots.
4. What are the functions of a network control center?
   - Monitor the local and remote hardware for errors.
   - Monitor local and remote software for error conditions.
   - Identify potential link problem areas.
5. Are network control centers a required component for all communications networks?
   - No.
6. Is a hot standby a required component in network management?
   - No.

Index